(*continued on back*)

Quantitative
Structure–Chromatographic Retention
Relationships

CHEMICAL ANALYSIS

A SERIES OF MONOGRAPHS ON
ANALYTICAL CHEMISTRY AND ITS APPLICATIONS

Editor

J. D. WINEFORDNER

Editor Emeritus: **I. M. KOLTHOFF**

VOLUME 93

A WILEY-INTERSCIENCE PUBLICATION

JOHN WILEY & SONS

New York / Chichester / Brisbane / Toronto / Singapore

Quantitative Structure–Chromatographic Retention Relationships

ROMAN KALISZAN

Medical Academy
Gdańsk, Poland

A WILEY-INTERSCIENCE PUBLICATION

JOHN WILEY & SONS

New York / Chichester / Brisbane / Toronto / Singapore

Library of Congress Cataloging in Publication Data:

Kaliszan, Roman.
    Quantitative structure–chromatographic retention
relationships.

    (Chemical analysis; v. 93)
    "A Wiley-Interscience publication."
    Includes bibliographies and index.
    1. Chromatographic analysis. 2. Solution
(Chemistry) I. Title. II. Series.

QD117.C5K35 1987   543'.089   87–14802
ISBN 0–471–85983–4

Printed in the United States of America

10 9 8 7 6 5 4 3 2 1

In memory of my father

PREFACE

The quantitative relationships between the structure of solutes and their chromatographic retention data have been extensively studied recently for three main reasons: explanation of the mechanism of chromatographic separations; prediction of retention coefficients; and characterization of the solute physicochemical properties of importance for reactivity and especially for bioactivity. Thus, this research area is truly an interdisciplinary one, and publications may be found in chromatographic, analytical, physicochemical, biochemical, pharmacological, and pharmaceutical journals.

At constant temperature, three main variables determine the distribution of a solute between mobile and stationary chromatographic phases: the chemical structure of the solute, the physicochemical properties of the mobile phase, and the physicochemical properties of the stationary phase. The solute distribution in various chromatographic techniques and modes is easily quantified by means of several free-energy-related retention parameters.

For any given solute the relationships between retention data and the composition of mobile and/or stationary phase have early been observed and met a more or less rigorous theoretical treatment based on chemical thermodynamics. Except for simple homologous series chromatographed under identical conditions, however, attempts to relate quantitatively the structure of an individual solute to its retention parameter were unsuccessful.

There has been an unquestionable trend in chemistry for the past few decades toward quantitation of chemical, physicochemical, and biological activities of various compounds. When computers became commonly available in the 1960s, the studies flourished because of pioneering works by Hansch and others on quantitative structure–biological activity relationships (QSARs). The QSAR methodology, that is, the means of characterizing solute molecular structure numerically, and the statistical procedures applied or developed for QSAR purposes have been successfully employed for quantitative structure–retention relationship (QSRR) studies. Since the late 1970s hundreds of papers have been published that may be categorized under the term QSRR. Certainly now, after about 10 years of intensive development, QSRR studies deserve a thorough review and critical discussion. Such an attempt is undertaken here.

Among the numerous QSRR equations published, there are many with little or no information value. Some may even be misleading if statistical requirements were not fulfilled when the appropriate equations were derived.

vii

Nonetheless, the vast majority of the QSRR studies yielded results of importance for physical, analytical, and medicinal chemistry. From these points of view the collected data are discussed here.

Chromatography is a unique system for studying structure–activity relationships involving intermolecular interactions. In a chromatographic process all the conditions may be kept constant or controlled, and thus the solute structure is the single independent variable in the system. Contrary to biological determinations, chromatography is able to yield readily a great amount of unequivocal, precise, and reproducible data. It may be anticipated that through QSRR studies the still more precise methods of solute structure parameterization will be worked out, which will next be applied to derive reliable QSAR equations allowing rational design of new drugs.

A knowledge of the physicochemical principles of solute–stationary/mobile phase interactions is required for proper understanding of the basis of chromatographic separations. Therefore, a description of the intermolecular interactions known in chemistry is presented first. After this, a chapter is devoted to the most commonly known general theories of the chromatographic distribution process. Then follows a brief characterization of the factors influencing the retention data for an individual solute. Next, the mathematical models employed in QSRR studies are reviewed and their formal requirements are discussed. In the following chapters the reported QSRR are reviewed successively in terms of individual molecular structure descriptors of the solutes. In another chapter the relationships between liquid chromatographic retention and partition coefficients are critically discussed. Finally, application of QSRRs in medicinal chemistry is reviewed in detail.

To help the reader use the literature cited, complete titles of the articles quoted are given. The literature up until March 1986 has been considered.

I am very grateful to Professor Richard A. Hartwick, Rutgers University, Piscataway, New Jersey, who encouraged the writing of this monograph. Thanks are also due to the authors and publishers of copyrighted materials. Without the patience of my wife, Anna, this book would not have been written.

ROMAN KALISZAN

Gdańsk, Poland
1987

CONTENTS

Quantitative
Structure–Chromatographic Retention
Relationships

CHAPTER

1

INTRODUCTION

The beginning studies of quantitative relationships between the structure of solutes and their chromatographic retention may be dated back as early as 1949. At that time, Martin [1], in his fundamental paper, suggested that a substituent changes the partition coefficient of a solute by a factor that depends on the nature of the substituent and both the mobile and stationary phases employed but not on the remaining part of the molecule. Since practically the birth of chromatography, regularities have been observed of retention behavior among the more or less closely related solutes. Evident have been simple relationships between chromatographic parameters and, for example, carbon number for a series of homologues.

Following Green et al. [2], who found that substituent increments to the thin-layer chromatographic parameter $R_M = \log(1/R_f - 1)$ are additive for a number of benzenoid compounds, Iwasa et al. [3] suggested in 1965 that chromatographic data for studies of quantitative structure–biological activity relationships (QSAR) may be useful. Since that time chromatography has been extensively employed for quantitation of hydrophobicity of bioactive agents [4–7].

In 1977 publications [8–10] appeared in which the QSAR methodology was directly applied for analysis of chromatographic retention data of a series of solutes. Soon the number of structure–retention correlations in the literature started to increase exponentially with time. This has been the result of common accessibility of personal computers as well as of the appropriate programs of statistical calculations. By analogy to QSAR, the term quantitative structure–retention relationships (QSRR) has been proposed [11] to comprise the new expanding area of chromatographic science.

The chromatographic conditions can be modified in several ways. One can change the stationary and/or mobile phase, and the changes in the relative retentions of investigated compounds caused by these modifications can be treated as a source of information about the ability of the compounds to undergo different kinds of intermolecular interactions on a thermodynamic basis [12]. Another approach consists in selecting a suitable group of test compounds for which chromatographic data would be determined at constant conditions or the numerical retention parameters obtained would be normalized to some standard conditions. In that second approach the differences in chromatographic data reflect the differences in the solute structure. Relationships between chromatographic retention data and the quantities related to the solute structure cannot be solved in strict thermodynamic terms. Such relationships are of the so-called *extrathermodynamic* type.

As Prausnitz has remarked [13], "Classical thermodynamics is revered, honoured and admired, but in practice is inadequate." Prausnitz has suggested the use of molecular thermodynamics for solving real problems—with molecular thermodynamics being seen as a synthesis of classical approaches, statistical thermodynamics, molecular physics, and physical thermodynamics. One way in which such results can be achieved is to employ *extrathermodynamics*.

The term extrathermodynamics means that the science lies outside the formal structure of thermodynamics, although the approach resembles that of thermodynamics in that detailed microscopic mechanisms do not need to be explicitly identified during use [14]. Extrathermodynamic approaches are combinations of detailed models with the concepts of thermodynamics. Since it involves model building, this kind of approach lacks the rigor of thermodynamics, but it can provide information not otherwise accessible. The manifestations of extrathermodynamic relationships are the linear free-energy relationships (LFER). Although LFERs are not a necessary consequence of thermodynamics, their occurrence suggests the presence of a real connection between the correlated quantities, and the nature of this connection can be explored [15].

Generally, LFERs may be regarded as linear relationships between the logarithms of the rate or equilibria constants for one reaction series and those for a second reaction series subjected to the same variation in reactant structure or reaction conditions [16, 17]. Thus, plotting the logarithms of rate or equilibrium constant for one reaction series against the corresponding constants for a second related series frequently gives a straight line, which can be expressed by

$$\log k_i^B = m \log k_i^A + c \tag{1.1}$$

where k_i^A and k_i^B are rate or equilibrium constants of two reaction series A and B that are subject to the same changes in the structure of reactants or the surrounding medium. It is often convenient to express LFER in terms of ratios of constants by referring all members of a reaction series to a reference member of the series; thus, the correlation in Eq. (1.1) can also be expressed by

$$\log (k_i^B / k_0^B) = m \log (k_i^A / k_0^A) \tag{1.2}$$

where k_0^A and k_0^B are the constants for the reference substituent or the reference solvent.

The chromatographic retention parameters used in correlation studies are normally assumed to be proportional to the free-energy change associated with the chromatographic distribution process. Not all chromatographic data, however, are suitable for QSRR studies. As is well known, free-energy changes, ΔG, are related to enthalpy, ΔH, and entropy, ΔS, changes by the Gibbs equation

$$\Delta G = \Delta H - T\,\Delta S \tag{1.3}$$

where T is temperature. For LFERs to be found between real and model systems, changes either in entropy or enthalpy must be constant, or the enthalpy changes must be linearly related to entropy changes [18]:

$$\Delta H = \beta\,\Delta S + \Delta G_\beta \quad (\text{at } T = \beta) \tag{1.4}$$

When enthalpy–entropy compensation is observed with a family of compounds in a particular chemical transformation, the values of β and ΔG are invariant and β is called the compensation temperature.

Using the Gibbs relationships [Eq. (1.3)], one can rewrite Eq. (1.4) in order to express the free-energy change ΔG_T measured at a fixed temperature T for isoequilibrium process as

$$\Delta G_T = \Delta H \left(1 - \frac{T}{\beta} \right) + \frac{T\,\Delta G_\beta}{\beta} \tag{1.5}$$

In liquid chromatography the retention parameter, the so-called capacity factor k', is related to the thermodynamic equilibrium constant K for solute binding by $k' = \phi K$, where ϕ is the phase ratio of the column. The free-energy change for the chromatographic process is expressed by

$$\Delta G = -RT \ln K = -RT \ln(k'/\phi) \tag{1.6}$$

where R is a gas constant. As shown by Melander et al. [18], the substitution of Eq. (1.6) into Eq. (1.3) yields, for the capacity factor:

$$\ln k' = -\frac{\Delta H}{RT} + \frac{\Delta S}{R} + \ln \phi \tag{1.7}$$

If the mechanism of the process is invariant over the temperature range studied and the enthalpy is constant, a van't Hoff plot of $\ln k'$ against $1/T$ yields a straight line. From the slope of the line, the enthalpy change ΔH for a given solute can be assessed.

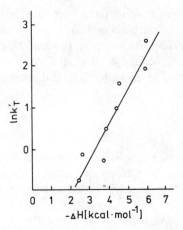

Figure 1.1. Enthalpy–entropy compensation plot for a group of aromatic acids chromatographed in different reversed-phase high-performance liquid chromatographic systems. (After
W. Melander, D. E. Campbell, and Cs. Horváth, *J. Chromatogr.*, **158**, 213, 1978. With permission.)

Equations (1.5) and (1.6) can be combined to give

$$\ln k'_T = -\frac{\Delta H}{R}\left(\frac{1}{T} - \frac{1}{\beta}\right) - \frac{\Delta G_\beta}{R\beta} + \ln\phi \qquad (1.8)$$

where k'_T is the capacity factor at temperature T. According to Eq. (1.8), plots
of $\ln k'_T$ of various solutes measured at a given temperature T under different
conditions against the corresponding enthalpy change are linear when the
enthalpy–entropy compensation occurs (Fig. 1.1). In such a situation, the
reversible binding of the solutes by a stationary phase involves essentially
the same mechanism. To avoid statistical artifacts, it is recommended that the
reference temperature T in Eq. (1.8) be near the harmonic mean of the
experimental temperatures used for the evaluation of the enthalpies by
Eq. (1.7) [18–20].

From the slope of compensation plot of $\ln k'_T$ versus ΔH [Eq. (1.8)], the
compensation temperature β may be obtained. If values of β for different
chromatographic systems approach their 95% confidence limits, the retention
mechanism is assumed to be the same [18, 21, 22].

It should be observed here that the prevailing majority of QSRRs reported
concern retention parameters as obtained in routine chromatographic
measurements. Enthalpy–entropy compensation is checked only occasionally.
Reservations are often justified as far as the chromatographic retention
parameters are defined and/or determined for individual solutes at specified

conditions. Unfortunately, for many authors the only criterion of value of the QSRR equations derived is the more or less exact conformability of the observed and calculated data. Even that conformability is sometimes treated quite mechanically; that is, the statistical significance of the statistically derived relationships is not analyzed. Nonetheless, certainly enough material has been collected to make a critical discussion worthwhile. It may be anticipated that studies on QSRR will further expand, and the aim of the following chapters is to help the interested reader clarify the present status of QSRR science and to encourage deeper, critical, and scientificaly productive analysis of chromatographic data.

References

1. A. J. P. Martin, Some theoretical aspects of partition chromatography, *Biochem. Soc. Symp.*, **3**, 4, 1950.
2. J. Green, S. Marcinkiewicz, and D. McHale, Paper chromatography and chemical structure. III. The correlation of complex and simple molecules. The calculation of R_M values for tocopherols, vitamins K, ubiquinones and ubichromenols from R_M (phenol). Effects of unsaturation and chain branching, *J. Chromatogr.*, **10**, 158, 1963.
3. J. Iwasa, T. Fujita, and C. Hansch, Substituent constants for aliphatic functions obtained from partition coefficients, *J. Med. Chem.*, **8**, 150, 1965.
4. E. Tomlinson, Chromatographic hydrophobic parameters in correlation analysis of structure–activity relationships, *J. Chromatogr.*, **113**, 1, 1975.
5. R. Kaliszan, Chromatography in studies of quantitative structure–activity relationships, *J. Chromatogr.*, **220**, 71, 1981.
6. R. Kaliszan, High performance liquid chromatography as a source of structural information for medicinal chemistry, *J. Chromatogr. Sci.*, **22**, 362, 1984.
7. R. Kaliszan, Quantitative relationships between molecular structure and chromatographic retention. Implications in physical, analytical, and medicinal chemistry, *Crit. Rev. Anal. Chem.*, **16**, 323, 1986.
8. R. Kaliszan and H. Foks, The relationship between the R_M values and the connectivity indices for pyrazine carbothioamide derivatives, *Chromatographia*, **10**, 346, 1977.
9. R. Kaliszan, Correlation between the retention indices and the connectivity indices of alcohols and methyl esters with complex cyclic structure, *Chromatographia*, **10**, 529, 1977.
10. I. Michotte and D. L. Massart, Molecular connectivity and retention indexes, *J. Pharm. Sci.*, **66**, 1630, 1977.
11. B.-K. Chen and Cs. Horváth, Evaluation of substituent contributions to chromatographic retention: Quantitative structure–retention relationships, *J. Chromatogr.*, **171**, 15, 1979.

12. P. J. Schoenmakers, H. A. H. Billiet, and L. De Galan, The solubility parameter as a tool in understanding liquid chromatography, *Chromatographia*, **15**, 205, 1982.

13. J. M. Prausnitz, Molecular thermodynamics for chemical process design, *Science*, **205**, 759, 1979.

14. E. Tomlinson, Boxes in boxes: Cases for extrathermodynamics, *British Pharmaceutical Conference Science Award Lecture*, Brighton, 1981.

15. C. Reichardt, *Solvent Effects in Organic Chemistry*, Verlag Chemie, Weinheim, New York, 1979, p. 227.

16. J. E. Leffler and E. Grunwald, *Rates and Equilibria of Organic Reactions*, Wiley, New York, 1963.

17. J. Shorter, *Correlation Analysis in Organic Chemistry: An Introduction to Linear Free-Energy Relationships*, Clarendon, Oxford, 1973, p. 64.

18. W. Melander, D. E. Campbell, and Cs. Horváth, Enthalpy–entropy compensation in reversed-phase chromatography, *J. Chromatogr.*, **158**, 213, 1978.

19. R. R. Krug, W. G. Hunter, and R. A. Grieger, Enthalpy–entropy compensation. I. Some fundamental statistical problems associated with the analysis of van't Hoff and Arrhenius data, *J. Phys. Chem.*, **80**, 2335, 1976.

20. R. R. Krug, W. G. Hunter, and R. A. Grieger, Enthalpy–entropy compensation. II. Separation of the chemical from the statistical effects, *J. Phys. Chem.*, **80**, 2341, 1976.

21. E. Tomlinson, H. Poppe, and J. C. Kraak, Thermodynamics of functional groups in reversed-phase high-performance liquid–solid chromatography, *Int. J. Pharm.*, **7**, 225, 1981.

22. Gy. Vigh and Z. Varga-Puchony, Influence of temperature on the retention behaviour of members of homologous series in reversed-phase high-performance liquid chromatography, *J. Chromatogr.*, **196**, 1, 1980.

CHAPTER

2

NATURE OF CHROMATOGRAPHIC INTERACTIONS

The distribution of a solute between a mobile phase and a stationary phase during the chromatographic separation process results from the forces that operate between solute molecules and the molecules of each phase. Thus, if the nature of the interaction between the solute molecule and the two phases can be determined, the behavior of a particular solute in a given chromatographic system could be predicted. The situation seems to be a bit more simple in gas chromatography (GC) as the interactions in the gas phase are relatively less important as compared to those in the stationary phase. It does not mean, however, that the carrier gas in GC is an absolutely inert medium.

What is the nature of intermolecular interactions governing chromatographic separation? Certainly, these are not the interactions leading to definite chemical alterations of the solute molecules through protonation, oxidation, reduction, complex formation, or other chemical processes. Specifically, in ion exchange chromatography, the separation-determining forces are ionic in nature (but may include the other intermolecular forces as well). In the other chromatographic techniques and modes the Coulomb interactions between ions are not involved, but the following generally known types of intermolecular interactions must be taken into consideration: directional interactions between dipoles, inductive, dispersion, hydrogen bonding, and electron pair donor–electron pair acceptor interactions, as well as solvophobic interactions.

Intermolecular forces, or forces that can occur between closed-shell molecules, are also called van der Waals forces, since van der Waals recognized them as the reason for the nonideal behavior of real gases. More often, the name van der Waals forces is reserved for the subgroup of intermolecular forces that are considered as nonspecific and that cannot be completely saturated. This category comprises the so-called directional, induction, and dispersion forces. One can say that these forces are "more physical" in nature, whereas the second group of forces, comprising hydrogen-bonding forces and the electron pair donor–acceptor forces, have a "more chemical" character. The latter group comprises specific, directional forces that can be saturated and lead to stoichiometric molecular compounds [1–10].

Most generally, the potential energy $V(r)$ of interacting molecules as a function of intermolecular distance r is described by the classical

7

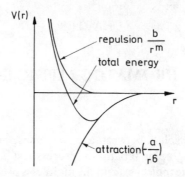

Figure 2.1. Potential energy of intermolecular interactions as function of intermolecular distance.

Lennard–Jones equation (Fig. 2.1)

$$V(r) = -\frac{a}{r^6} + \frac{b}{r^m} \tag{2.1}$$

Where a and b are constants, and the exponent m is significantly greater than 6; usually a value of 9–12 is observed for m experimentally.

As evident from Eq. (2.1), at closer distances repulsion forces predominate. The nature of these repulsive forces is rather obscure so far. Intermolecular attractive interactions are of low energy. Usually the energy of the van der Waals "bond" is lower than 1000–2000 J/mol. At room temperature the thermal energy is about 2500 J/mol; thus, van der Waals interactions between electrically neutral molecules are unable to form stable chemical compounds. Usually, weak interactions are intermolecular; that is, they are attractions between portions of different molecules. However, there can also be weak intramolecular attractive forces between different portions of the same molecule.

2.1. ION–DIPOLE INTERACTIONS

In fact, Coulomb forces between ions and electrically neutral molecules with permanent dipole moments do not belong to the intermolecular forces in the narrower sense [1]. The forces, however, may play some role in specific chromatographic processes and will be considered here for the sake of completeness.

The unequal sharing of electrons in a bond between two atoms of substantially different electronegativity will lead to a permanent displacement of the electron probability in the bond toward the more electronegative atom.

An unsymmetrical charge distribution in an electrically neutral molecule leads to formation of multipoles, like dipoles, quadrupoles, and octapoles. (Fig. 2.2). Multipoles are characterized by means of the multipole moment. If the magnitude of the two equal and opposite charges of a molecular dipole is denoted by q, and the distance of separation is d, the dipole moment is given by $\mu = qd$. The unit is the coulomb meter (C m), for example, for water $\mu = 6.07 \times 10^{-3}$ C m.

When placed in the electric field resulting from an ion, the dipole will orient itself so that the end with the charge opposite to that of the ion will be directed toward the ion, and the other repulsive end will be directed away. The potential energy of an ion–dipole interaction, E_{i-d}, is given by

$$E_{i-d} = -W^2 \frac{z\mu \cos \alpha}{\varepsilon r^2} \qquad (2.2)$$

where z is the charge on the ion, μ is dipole moment of the neutral molecule, r the distance from the ion to the center of the dipole, and α the dipole angle relative to the line r joining the ion and the center of the dipole. The higher the relative electric permittivity ε of the medium (solvent), the lower the energy of attractive ion–dipole interactions. Thus, decrease of ε by adding ethanol ($\varepsilon = 27$) to water ($\varepsilon = 81$) should increase electrostatic interactions. That dependence of intermolecular interactions on ε explains why, for example, the solubility of ionic crystals in organic solvents is low as compared to water. The coefficient W in Eq. (2.2) depends on the unit system applied; for the SI unit system $W = 1/4\pi\varepsilon_0$, where $\varepsilon_0 = 8.854 \times 10^{-12}$ F m^{-1} is the electric permittivity of vacuum.

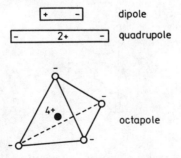

Figure 2.2. Unsymmetrical charge distribution in electrically neutral molecules.

2.2. DIPOLE–DIPOLE INTERACTIONS

The term *van der Waals interactions* is commonly understood to mean the long-distance forces resulting from the electric field generated by the molecules. The type of field depends on whether the molecules considered form stable dipoles (generally multipoles) or their charge distribution is spherically symmetrical. Thus, van der Waals interactions may be divided into three groups: the orientation interactions (Keesom effect), the inductive interactions (Debey effect), and the dispersive interactions (London effect).

For the very polar molecules the orientation or dipole–dipole interactions are characteristic. However, generally, the effect of orientation interactions is lower than the effect of additive dispersion interactions.

Orientation interactions concern molecules that possess a permanent dipole moment μ. When two dipolar molecules are optionally oriented to each other (Fig. 2.3), the force of attraction is proportional to $1/r^3$, where r is the distance between the dipoles. If all possible dipole orientations were equally probable, the attraction and repulsion would compensate each other. Not every orientation of one molecule with respect to the other is equally probable. The low energetic configurations related to the attraction are statistically favored as compared to high-energy repulsive configurations. Thus, the net attraction, which is strongly temperature dependent, may be calculated assuming the Boltzmann distribution of the relative dipole orientation. It should be noted

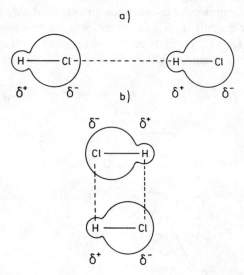

Figure 2.3. Dipole–dipole attraction of two HCl molecules; (*a*) "head-to-tail" arrangement; (*b*) antiparallel arrangement.

here that with temperature T the probability of energetically favored orientations decreases, and at very high temperatures all dipole orientations are equally populated and the potential energy is zero.

The potential energy of dipole–dipole interactions, E_{d-d} (taking no account of multipoles), between the molecules of dipole moments μ_1 and μ_2 is given by

$$E_{d-d} = -W^2 \frac{2}{3kT} \frac{\mu_1^2 \mu_2^2}{\varepsilon r^6} \qquad (2.3)$$

where k is the Boltzmann constant, and T absolute temperature; the other symbols are as explained for Eq. (2.2). The distance r between the interacting molecules is tacitly assumed to be much larger than the dimension of an individual dipole.

The temperature dependence of E_{d-d} requires special discussion from the chromatographic point of view. It may be expected that the dipole–dipole interactions in gas chromatography are the less pronounced the higher the column temperature. If the dipole–dipole interactions are of importance for separation of a given group of substances at a given temperature, then at some higher temperature their input to the separation forces may become negligible. In other words, different structure parameters may determine retention when measured at different temperatures. The respective QSRR equations may then depend on temperature.

2.3. DIPOLE–INDUCED DIPOLE INTERACTIONS

Attraction between a molecule possessing a permanent dipole moment and a nonpolar molecule was postulated by Debey in 1920. The electric dipole in one molecule can induce a dipole moment in a neighboring molecule (Fig. 2.4). Such an induced dipole moment may also be produced in polar molecules, that is, in molecules already possessing a permanent dipole moment, but then such an induced dipole moment is relatively small as compared to the permanent one. Similarly, a charged particle such as an ion introduced into the neighborhood of an uncharged, apolar molecule will distort the electron cloud of this molecule. The induced moment always lies in the direction of the inducing dipole. Thus, attraction always exists between the two entities, which is independent of temperature.

The induced dipole moment will be larger the higher the polarizability α of the apolar molecule. The polarizability α is defined as the dipole moment induced in the given molecule by an electric field of 1 V/cm. In fact, molecular polarizability is different in different directions, and thus all relative molecular orientations should be considered. For practical reasons the approximation

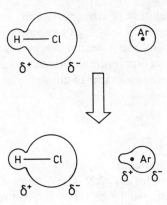

Figure 2.4. Dipole-induced dipole attraction. Polar HCl induces dipole in normally nonpolar argon atom.

assumed in Eq. (2.4) is satisfactory, as the inductive effect appears, at least to some extent, for every orientation of a molecule. Assuming the polarizability α as an isotropic (scalar) quantity one gets, for molecule 1 in which a dipole is induced when placed in an electric field E_2 generated by molecule 2, the energy of the induced dipole is $\frac{1}{2}\alpha_1 E_2^2$; and analogously for molecule 2 the energy is $\frac{1}{2}\alpha_2 E_1^2$. Thus, the potential energy of interactions, E_{d-id}, is described by

$$E_{d-id} = -W^2[\tfrac{1}{2}(\alpha_1 E_2^2 + \alpha_2 E_1^2)] \tag{2.4}$$

Specifically, if the electric field of a dipole is of concern, one gets

$$E_{d-id} = -W^2 \frac{\alpha_1 \mu_2^2 + \alpha_2 \mu_1^2}{\varepsilon r^6} \tag{2.5}$$

where α_1 and α_2 denote polarizabilities of the interacting molecules. The remaining symbols are as explained earlier.

The dipole–induced dipole interactions do not depend on temperature and are not additive.

2.4. INSTANTANEOUS DIPOLE–INDUCED DIPOLE INTERACTIONS

Classical electrodynamics does not explain the fact that two electrically neutral molecules possessing no permanent dipole moment attract one another at close distances. Yet, even noble gases may exist in the liquid state, which means that their atoms, characterized by spherically symmetrical charge distribution, interact. This type of universal interaction was explained in 1930 by London, basing his explanation on quantum mechanics. A specific feature of these so-called dispersion interactions or charge fluctuation interactions is their additivity and catholicity. Thus, these forces are associated with all ions and molecules, either possessing a stable dipole moment or not.

Even in atoms and molecules without permanent dipole moments, the continuous electron density fluctuations result, at any instant, in a small dipole moment that can fluctuatingly polarize the electron system of the neighboring atoms or molecules. This coupling causes the electronic movements to be synchronized in such a way that a mutual attraction results. To calculate precisely the energy of dispersion interactions, the wave functions of all excited states should be known, which generally is impossible. A detailed analysis leads to variously expressed results, but the potential energy of interactions is always proportional to a product of molecular polarizabilities. The dispersive forces are highly sensitive to distance, falling off with at least the sixth power of the distance and probably, due to retardation phenomena, at a greater power of the distance. An approximate expression for calculation of the potential energy of dispersion interactions, E_d, is

$$E_d = -W^2 \frac{3}{2} \frac{I_1 I_2}{I_1 + I_2} \frac{\alpha_1 \alpha_2}{\varepsilon r^6} \tag{2.6}$$

where I_1 and I_2 are respective ionization energies of the interacting molecules 1 and 2; the other symbols are as described previously.

Ionization potentials of different molecules are generally about 10 eV, and the expression $I_1 I_2 / (I_1 + I_2)$ is similar for various systems. Thus, the polarizabilities of the molecules involved are critical for dispersion interactions.

Because of the high polarizability of π-electrons, especially strong dispersion forces exist between molecules with conjugated π-electron systems (e.g., aromatic hydrocarbons).

In the London equation [Eq. (2.6)] molecular polarizabilities are given as scalar quantities. Thus, the dispersion interactions cannot be saturated, and their energy does not depend on temperature as the changes in molecular orientation have no effect on E_d. In fact, the π-electron systems form a special case as the anisotropy of electric properties in such systems is evident. At the same time, in π-electron systems polarizability is abnormally high. Thus, in π-electron systems the molecular orientations are especially favored, and the dispersion interactions are directional to some extent.

A special form of dispersion interaction is observed in large molecules. In such molecules the length of the fluctuating instantaneous dipole is so great that actually interacting are individual charges, changing their positions and not the instantaneous dipole moments.

Dispersion forces are often prevailing among the three kinds of van der Waals interactions. Generally, the potential energy of van der Waals interactions (neglecting multipoles) in the gas or liquid state may be expressed by combining Eqs. (2.3), (2.5), and (2.6):

$$E_{vdW} = E_{d-d} + E_{d-id} + E_d = -\frac{W^2}{\varepsilon r^6}\left(\frac{2}{3kT}\mu_1^2\mu_2^2 + \alpha_2\mu_1^2\right.$$

$$\left. + \alpha_1\mu_2^2 + \frac{3}{2}\frac{I_1 I_2}{I_1 + I_2}\alpha_1\alpha_2\right) \qquad (2.7)$$

The relative contribution of $E_{d-d} + E_{d-id}$ increases with the dipole moment, and usually E_{d-d} is larger than E_{d-id} (except in the cases when the dipole moment is very small, e.g., for carbon oxide). However, the dispersion energy contribution to E_{vdW} is normally much greater than the sum $E_{d-d} + E_{d-id}$. For example, the calculated cohesion energy of liquid 2-butanone at 40°C consists of 8% orientation energy, 14% induction energy, and 78% dispersion energy [11]. Two molecules with $\alpha = 3 \times 10^{-24}$ cm^3, $I = 20 \times 10^{-19}$ J, and $r = 3 \times 10^{-8}$ cm have dispersion interaction potential of -11.3 kJ/mol. These values of α, I, and r correspond to those for liquid HCl. For two HCl dipoles, both with dipole moments of 3.3×10^{-30} C m, separated by a distance of $r = 3 \times 10^{-8}$ cm and oriented according to "head-to-tail" arrangement (Fig. 2.3a), the interaction energy E_{d-d} is only -5.3 kJ/mol. Thus, for HCl and most other compounds, the dispersion forces are considerably stronger than dipole orientation forces of nearest-neighbor distances in the liquid state. However, at larger distances the dispersion energy falls off rapidly.

It should be noted here that the van der Waals interactions increase with pressure as the intermolecular distances decrease, which explains the increased boiling temperatures at higher pressure.

2.5. HYDROGEN BONDING

The van der Waals interactions are classified as nonspecific intermolecular interactions. There are also intermolecular interactions resulting from the constitutive features of the molecules involved. From the chromatographic point of view, hydrogen-bonding interactions and electron pair donor–acceptor interactions are important. These interactions are approximately 10 times stronger than the nonspecific intermolecular interaction forces but approximately 10 times weaker than covalent single bonds.

When a covalently bound hydrogen atom forms a second bond to another atom, the second bond is called a *hydrogen bond* [12]. This hydrogen bond, or hydrogen bridge, is characterized by a coordinative divalency of the hydrogen atom involved.

The concept of hydrogen bonding was formulated to explain abnormally high boiling points of liquids possessing hydroxy groups or other groups with a hydrogen atom bound to an electronegative atom. The forces responsible for a strong association of a proton donor and a hydrogen bond acceptor may be formed between two molecules or within an individual molecule. Both inter- and intramolecular hydrogen bonding are possible. When two or more identical molecules associate, the so-called homo-intermolecular hydrogen bonds are formed. The association of different molecules results in hetero-intermolecular hydrogen bonds.

The best known proton donor groups are –O–H, =N–H, –S–H, Cl–H, Br–H, F–H, I–H, and sometimes ≡C–H. The most important electron pair donors (or hydrogen bond acceptors) are the oxygen atoms in alcohols, ethers, carbonyl compounds, halogen atoms and ions, as well as nitrogen atoms in amines and the π-electron systems of aromatic compounds.

There are several interesting features of hydrogen bonding [7]. The length of a hydrogen bond, r, equals the distance between the proton donor and acceptor and is markedly smaller than the sum of van der Walls radii of the groups AH and B (Fig. 2.5). Thus, the contact between the directly interacting atoms is much closer than in ordinary, nonspecific interactions.

Formation of a hydrogen bond leads usually to a change of dipole moment of the interacting species. The direction of this change, $\Delta\mu$, is always from proton acceptor to the donor (Fig. 2.5).

An average energy of the O–H – – – O bond is 20–25 kJ/mol. For the strongest hydrogen bonds the energy may exceed 150 kJ/mol and thus attain a value comparable to that of covalent bonds, that is, 210–420 kJ/mol. In general, the geometry of the intermolecular hydrogen bridge is linear—the deviation from linearity is usually less than 15°. In the case of intermolecular hydrogen bonds six- or five-membered rings are formed for energetic reasons.

Most often, hydrogen bonding is nonsymmetrical; that is, the proton is closer to atom A or B.

Hydrogen bonds are partly electrostatic and partly valence in nature. As atom A is highly electronegative, the σ-bond between atoms A and H is

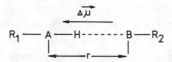

Figure 2.5. Hydrogen bonding between proton donor R_1A–H (Brønsted acid) and an electron pair donor BR_2 (Brønsted base). Bond length r and direction of dipole moment change $\Delta\mu$.

strongly polarized. Thus, the bonding electron pair is shifted toward atom A. According to the valence bond theory, there is a marked contribution of the ionic structure A^-H^+. Between the positive charge of the H atom and the negative charge of the free electron pair of atom B a strong electrostatic attraction occurs. However, there are reasons to believe that more is involved in hydrogen bonding than an electrostatic interaction. The approximate quantum-mechanical description may be found elsewhere [13].

The unique features of hydrogen bonds result from the unique properties of the hydrogen atom. Similar interactions are known for the hydrogen isotope deuterium.

2.6. ELECTRON PAIR DONOR–ELECTRON PAIR ACCEPTOR INTERACTIONS

According to Gutmann [14], the term *electron pair donor–electron pair acceptor* (EPD–EPA) complex comprises all complexes whose formation is due to an interaction between electron pair donors (Lewis bases) and electron pair acceptors (Lewis acids), irrespective of the stabilities of the complexes or the charges of the components. These chemical entities were originally termed *charge transfer complexes* by Mulliken [15], and this name is still used in the chromatographic literature. The other synonyms are electron donor acceptor complex or molecular complex.

A formation of an additional bonding interaction between two valency-saturated molecules is possible only if an occupied molecular orbital of sufficiently high energy is present in the electron pair donor molecule and a sufficiently low unoccupied orbital is available in the electron pair acceptor molecule. Whereas in a normal chemical bond each atom supplies one electron to the bond, in EPD–EPA bonding one molecule (i.e., the donor) supplies the pair of electrons, while the second molecule (i.e., the acceptor) provides the vacant molecular orbital.

Mulliken [15] produced a simple valence bond description of weak EPD–EPA complexes. According to Mulliken, the electronic ground state of the complex can be considered as a hybrid of two limiting structures: $D----A \leftrightarrow D^+----A^-$. The nonionic structure $D---A$ represents a state without any donor–acceptor interactions in which only nonspecific van der Waals forces hold D and A together. The mesomeric structure D^+----A^- characterizes a state in which an ionic bond was formed by transfer of an electron from D to A. Although the ionic structure contributes only slightly to the ground state, this small contribution is sufficient in establishing an extra bonding interaction in addition to the dipole–dipole, dipole–induced dipole, and instantaneous dipole–induced dipole interactions.

The long-wavelength absorption bands associated with electron transfer from the donor to the acceptor molecules are characteristic for EPD–EPA complexes. This electron transfer will be easier the lower the ionization potential of the donor and the higher the electron affinity of the acceptor.

Another characteristic quantity describing the properties of the EPD–EPA complexes is dipole moment. If the vector sum of the dipole moments of the donor and the acceptor is μ_0, and the induced dipole moment is μ_i, the dipole moment of the ground-state complex μ_N, is

$$\mu_N = \mu_0 + \mu_i + k\,\mu_{ion} \tag{2.8}$$

where k is a constant, and μ_{ion} is the dipole moment of the mesomeric structure $D^+ - - - - A^-$ formed by transfer of an electron from D to A.

EPD–EPA interactions are classified based on the type of orbitals involved in the formation of the complex. All electron pair donor molecules can be divided into three groups: n-, σ-, and π- donors. The energetically highest orbitals are, respectively, the lone pair of the n-electrons (e.g., amines, ethers, sulfoxides, phosphines); the electron pair of a σ-bond (e.g., cyclopropane, halogen derivatives); and the pair of π-electrons of unsaturated and aromatic compounds (e.g., polycyclic aromatic compound). Similarly, electron pair acceptor compounds are classified as v-, σ-, and π-acceptors. The lowest orbital in the first group is a vacant valency orbital of a metal atom (e.g., Al, B, Ag organometallic compounds), in the second a nonbonding σ-orbital (e.g., Br_2, I_2), and in the third group a system of π-bonds (e.g., aromatic polynitro compounds, tetracyanoethylene). Because any donor may form a complex with any acceptor, there exist nine different types of EPD–EPA complexes. Some examples are given in Fig. 2.6.

The reaction enthalpies for the formation of strong EPD–EPA complexes lie between -42 and -188 kJ/mol [16]. The bond energies for weak π-EPD–π-EPA complexes between neutral molecules are smaller, for example, the enthalpy in the case of a benzene–1,3,5-trinitrobenzene complex is -8 kJ/mol [1].

EPD–EPA interactions have been exploited to improve gas–liquid chromatographic separation of several solutes by adding complexing agents to the "inert" liquid stationary phases [8]. Unfortunately, most additives that are recognized as strong complexing agents exhibit only weak miscibility with solvents of vapor pressure sufficiently low to be useful as gas–liquid chromatographic stationary phases. An example may be addition of the EPA substance tetracyanoethylene to the stationary phase di-n-butyl adipate to improve separation of EPD solutes, that is, alkylbenzenes [17]. EPD–EPA interactions are also of importance for liquid chromatographic distribution of solutes. It may be observed that generally the distribution coefficient k

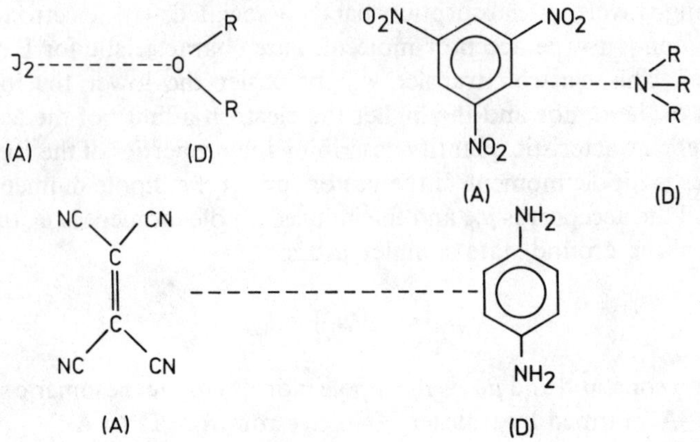

Figure 2.6. Examples of electron pair donor–electron pair acceptor complexes.

depends on the volume fraction of a complexing agent in a given chromato-graphic phase according to the relationship observed by Waksmundzki et al. [18]:

$$\log k = \phi_A \log k_A + \phi_S \log k_S \qquad (2.9)$$

where ϕ_A and ϕ_S are volume fractions of the complexing agent and the inert solvent, k_A and k_S are distribution coefficients determined with a pure complexing agent and a pure "inert" solvent, respectively. In view of the remarkable catholicity of Eq. (2.9), Laub and Purnell [19] suggested that solutions confirming to it be called *diachoric* (partitioned volume). Attempts to quantify the abilities of solutes to undergo chromatographic EPD–EPA interactions for QSRR purposes will be described in subsequent chapters.

2.7. SOLVOPHOBIC INTERACTIONS

Distribution of solutes between the mobile and the stationary zone in liquid chromatography, especially in the so-called reversed-phase systems, is often interpreted in terms of hydrophobic interactions. Because the phenomenon of hydrophobic (or, more generally, solvophobic [10]) interactions has been (since Kauzmann's pioneering works [20, 21]) the subject of interest and controversy [22], it deserves a brief description here.

According to Ben-Naim [10], the term *hydrophobic interaction* (or *hydrophobic bond*) is used in the literature to refer to two concepts. One is represented by the standard free energy of transfer of a solute S between water and some other nonaqueous solvent and is used to measure the relative tendency of the solute S to prefer one environment over the other. The second is used to measure the tendency of two (or more) solute molecules to aggregate in aqueous solutions. Both quantities have a common feature: they express a kind of "phobia" for the aqueous environment. However, they are quite different, and in general, one cannot infer anything about one quantity from the other. Certainly, the term hydrophobic interaction was applied by different authors to describe and interpret quite different phenomena occurring in aqueous solutions. Thus, the following questions are still justifiable: Are hydrophobic interactions a well-established experimental fact? Is there any relation between these interactions and the peculiar properties of water? Are hydrophobic interactions important in biological processes?

In the earlier literature it was commonplace to discuss hydrophobic interactions in terms of van der Waals forces between solute molecules. However, the van der Waals forces between, for example, two methane molecules in water are nearly the same as in any other solvent or in vacuum, and the peculiarities of the aqueous environment cannot be accounted for by using only van der Waals forces between the solute molecules.

The reader who is not familiar with the theory of liquids may find hydrophobic interactions a somewhat mysterious phenomenon. The reason is that one has been accustomed to the fact that a force between two particles is a property of the particles themselves. For example, Coulomb forces arise from the charges situated *in* the particles. In the case of hydrophobic interactions the forces are mainly dependent on the properties of the solvent and not on the solutes.

The structure of solvents (i.e., the structure of liquids) is the least known of all aggregation states [1]. Even for the most important solvent—water—the investigation of its inner fine structure is still the subject of current research. Numerous models were developed to describe the structure of water. However, these models fail to describe completely the physicochemical properties of water and to explain its anomalies. According to Horne [23], Fig. 2.7 should clarify the complexity of the inner structure of a solvent in the example of water.

Liquid water consists both of bound ordered regions of a regular lattice and regions where the water molecules are hydrogen bonded in a random array; it is permeated by monomeric water and interspersed with random holes, lattice vacancies, and cages. There are chains and small polymers as well as bound, free, and trapped water molecules. In this way a variety of models of water structure have been considered, and such concepts as "degree of icelikeness," "degree of crystallinity," or "amount of icebergs" appeared in many discussions and interpretations of phenomena in aqueous systems. In principle,

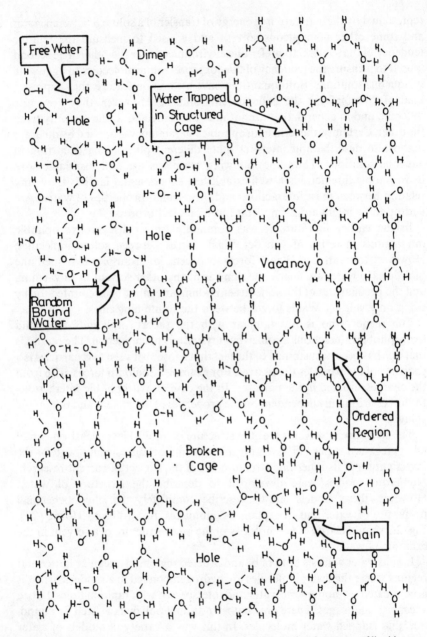

Figure 2.7. Two-dimensional schematic diagram of three-dimensional structure of liquid water. (After R. A. Horne, *Survey of Progress in Chemistry*, **4**, 1, 1968.)

organic solvents such as the protic ones should possess a similar complicated structure.

It has long been observed [24] that nonpolar molecules form in water the stable hydrates in which clathrate polyhedrons composed of water molecules comprise the solute molecules (Fig. 2.8). These polyhedrons (dodecahedra) can associate, yielding a kind of spatial lattice [6]. The space inside the dodecahedron is about 50 nm in diameter, enough to hold a nonpolar molecule. Characteristic for such hydrates is a large negative enthalpy of formation. The enthalpy of hydration is about 15 kcal/mol, independent on the type and size of the nonpolar molecule (e.g., Ar, Kr, CH_4, C_2H_2, C_2H_4, C_2H_6, and CH_3SH). Thus, the large contribution to enthalpy results from the liberation of energy due to an arrangement of water molecules into regular structure clathrates rather than due to immediate solute–solvent interactions.

The thermodynamics of the processes related to the interactions of a nonpolar solute with water was studied first by Kauzmann [21]. The thermodynamic parameters of the process of transfer of several hydrocarbons from nonpolar solvents to water are given in Table 2.1.

From the data in Table 2.1 it may be concluded that dissolution of hydrocarbons in water is energetically favored as compared to the nonpolar solvent ($\Delta H < 0$). However, a more negative entropy change accompanying the transfer to water results in a positive change of total free energy; thus, the process of transfer becomes unfavorable. For example, in the transfer of CH_4

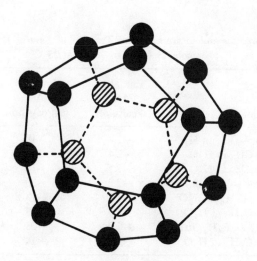

Figure 2.8. Dodecaedr formed by water molecules. (After M. V. Stackelberg and H. R. Müller, *Z. Elektrochem.*, **58**, 25, 1954.)

from benzene to water, 2.8 kcal/mol of heat is liberated, with an accompanying entropy change of -18 cal/deg mol, resulting is an increase in free energy of $+2.6$ kcal/mol.

Another thermodynamic analysis of the observed behavior of hydrocarbons in water may be as follows [1]. Hydrocarbons are extremely insoluble in water. Accordingly, the dissolution of a hydrocarbon in water is usually associated with an increase in the free energy of the system ($\Delta G > 0$). Since it is known experimentally that the dissolution of a hydrocarbon in water is exothermic ($\Delta H < 0$), it follows from $\Delta G = \Delta H - T \Delta S$ that the entropy of the system must decrease.

The thermodynamic data discussed above can be interpreted as a consequence of the highly ordered structure of the water molecules around the dissolved hydrocarbon molecules. In other words, the water molecules are more tightly packed around the dissolved hydrocarbon molecules than in pure water. This decreased entropy is avoided by the water and the nonpolar substances which remain associated with their own kind. The formation of a hydrophobic interaction between two molecules is illustrated in Fig. 2.9 [25].

Due to the contact between two nonpolar molecules, fewer water molecules are now in direct contact with the hydrophobic molecules. Thus, the ordering influence of the hydrophobic molecules will be diminished, and the entropy increases ($\Delta S > 0$). Although thermal energy is required for the destructuring of the hydration shells around the nonpolar molecules ($\Delta H > 0$), the free energy diminishes upon aggregation ($\Delta G < 0$). Therefore, it is energetically advan-

Table 2.1. Thermodynamic Parameters of Transfer of Several Hydrocarbons from Nonpolar Solvents to Water at 25° C[a]

Transfer Reaction	Enthalpy ΔH (kcal/mol)	Entropy ΔS (cal/deg mol)	Free Energy ΔG (kcal/mol)
CH_4 in benzene→CH_4 in H_2O	-2.8	-18	2.6
CH_4 in ether→CH_4 in H_2O	-2.4	-19	-3.3
CH_4 in CCl_4→CH_4 in H_2O	-2.5	-18	2.9
C_2H_6 in benzene→C_2H_6 in H_2O	-2.2	-20	3.8
C_2H_6 in CCl_4→C_2H_6 in H_2O	-1.7	-18	3.7
Liquid C_3H_8→C_3H_8 in H_2O	-1.8	-23	5.05
Liquid n-C_4H_{10}→n-C_4H_{10} in H_2O	-1.0	-23	5.85
Liquid C_6H_6→C_6H_6 in H_2O	0[b]	-14[b]	4.07[b]

[a] After Kauzmann [21].

[b] Measured at 18°C.

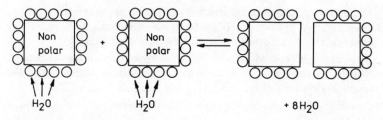

Figure 2.9. Formation of hydrophobic interaction between two nonpolar molecules.

tageous for apolar molecules or apolar groups in large molecules in water to aggregate with expulsion of water molecules from the hydration shells.

In principle, interactions similar to those discussed above should also apply to other solvents resembling water, and therefore, the more general term solvophobic interactions has been proposed [10]. In fact, analogous aggregations have been observed in solvents other than water (e.g., ethanol, glycerol, and D_2O). The relative phobia of hydrocarbons for H_2O is greater than for D_2O. Intramolecular hydrophobic interactions are probably of importance in biochemistry.

There is an explanation of hydrophobic interactions at the molecular level not involving the entropy considerations [26]. The forces binding water molecules are large, involving primarily electrostatic but including some dispersion forces as well. The forces of attraction between nonpolar groups, such as methyls or methylens, in a nonpolar chain can be considerable, especially if chains of some length are involved. Any interaction between a nonpolar group and water would involve a dispersion force. Thus, water bonding to water and nonpolar group bonding to nonpolar group is certainly favored from the comparative enthalpies. Thus, entropy considerations may not have to be involved to explain these phenomena.

References

1. C. Reichardt, *Solvent Effects in Organic Chemistry*, Verlag Chemie, Weinheim, New York, 1979, p. 9.

2. G. Kortüm, *Elektrochemia*, PWN, Warszawa, 1970, p. 120.

3. G. M. Barrow, *Chemia Fizyczna*, PWN, Warszawa, 1971, p. 460.

4. K. Pigoń and Z. Ruziewicz, *Chemia Fizyczna*, PWN, Warszawa, 1980, p. 638.

5. J. Dzięgielewski, *Chemia Nieorganiczna, Część I: Podstawy*, Silesian University, Katowice, 1985, p. 209.

6. C. R. Cantor and P. R. Schimmel, *Biophysical Chemistry*, Part I, (in Russian), Mir, Moskwa, 1984, p. 244.

7. L. Sobczyk, *Intermolecular Interactions* (in Polish), in *Chemia Fizyczna*, PWN, Warszawa, 1980, p. 429.

8. R. J. Laub and C. A. Wellington, Study of complexation phenomena by gas–liquid chromatography, in Foster, R. (Ed.), *Molecular Association Including Molecular Complexes*, Vol. 2, Academic, London, New York, San Francisco, 1979, p. 203.

9. E. M. Arnett and D. R. McKelvey, Solvent isotope effect on thermodynamics of nonreacting solutes, in J. F. Coetzee and C. D. Ritchie (Eds.), *Solute–Solvent Interactions*, Marcel Dekker, New York and London, 1969, p. 348.

10. A. Ben-Naim, *Hydrophobic Interactions*, Plenum, New York and London, 1980, p. 25.

11. E. F. Meyer and R. E. Wagner, Cohesive energies in polar organic liquids, *J. Phys. Chem.*, **70**, 3162, 1966.

12. M. D. Joesten and L. J. Schaad, *Hydrogen Bonding*, Marcel Dekker, New York, 1974.

13. H. Zimmermann, Wasserstoff brückenbindungen und ihre Bedeutung in der Biologie, *Chemie in unserer Zeit*, **4**, 69, 1970.

14. V. Gutmann, *The Donor–Acceptor Approach to Molecular Interactions*, Plenum, New York, 1978.

15. R. S. Mulliken and W. B. Person, *Molecular Complexes: A Lecture and Reprint Volume*, Wiley-Interscience, New York, 1969.

16. R. Paetzold, Schwache Elektronen-Donator-Akzeptor-Komplexe-Fortschritte und Probleme, *Z. Chem.*, **15**, 377, 1975.

17. D. E. Martire, Reconciliation of gas–liquid chromatographic and nuclear magnetic resonance measurements for association constants of organic complexes, *Anal. Chem.*, **48**, 398, 1976.

18. A. Waksmundzki, E. Soczewiński, and Z. Suprynowicz, On the relation between the composition of the mixed stationary phase and the retention time in gas–liquid and partition chromatography, *Collect. Czech. Chem. Commun.*, **27**, 2001, 1962.

19. R. J. Laub and J. H. Purnell, Solution and complexing studies. III. Further evidence for a microscopic partitioning theory of solution, *J. Am. Chem. Soc.*, **98**, 30, 1976.

20. W. Kauzmann, Denaturation of proteins and enzymes, in W. D. McElroy and B. Glass (Eds.), *A Symposium on the Mechanism of Enzyme Action*, John Hopkins University Press, Baltimore, MD, 1954.

21. W. Kauzmann, Some factors in the interpretation of protein denaturation, *Adv. Protein Chem.*, **14**, 1, 1959.

22. R. D. Cramer, "Hydrophobic interaction" and solvation energies: Discrepancies between theory and experimental data, *J. Am. Chem. Soc.*, **99**, 5408, 1977.

23. R. A. Horne, The structure of water and aqueous solutions, in A. F. Scott (Ed.), *Survey of Progress in Chemistry*, Vol. 4, Academic, London, New York, 1968.

24. M. V. Stackelberg and H. R. Müller, Feste Gashydrate II. Struktur und Raumchemie, *Z. Elektrochem.*, **58**, 25, 1954.

25. B. R. Baker, Interaction of enzymes and inhibitors, *J. Chem. Educ.*, **44**, 610, 1967.

26. L. B. Kier, *Molecular Orbital Theory in Drug Research*, Academic, New York and London, 1971, p. 8.

CHAPTER

3

MOLECULAR INTERPRETATION OF DISTRIBUTION PROCESSES IN CHROMATOGRAPHY

The number of publications that deal with the theory of solute distribution between two phases at the molecular level and provide an explicit equation relating distribution coefficient to phase composition is limited. The most common approach is to use the thermodynamic relationship that describes the distribution coefficient as a function of the exponent of the partial molar excess free energy associated with the removal of the solute from one phase to another. As pointed out by Scott [1], however, the thermodynamic approach has severe limitations from the practical point of view.

Thermodynamic properties are bulk properties and are a measure of the net interactive effects. Because there is usually more than one effect present, the individual interactions cannot easily be identified. Furthermore, it is difficult and time consuming to obtain reliable thermodynamic data.

As described in subsequent chapters, there are numerous schemes of retention prediction based on the more or less quantitative relationships between solute structure, properties of chromatographic phases, and retention parameters. These schemes for prediction and/or control of retention and selectivity can be quite useful in specific situations. However, a conceptually unified system for understanding and generalizing such schemes is required. Such a treatment would ideally be based on the fundamental intermolecular interactions common to various chromatographic systems. Furthermore, as noted by Karger et al. [2], there should be a means to relate the various classification parameters to more readily available (i.e., nonchromatographic) data, such as the properties of pure compounds. One of the first interpretations of chromatographic distribution in terms of intermolecular interactions has been promoted by Snyder [3, 4], who expanded the "solubility parameter" of Hildebrand and Scott's theory [5] to partial parameters related to the individual intermolecular interactions, like dispersion, induction, orientation, and so on. At the same time Kováts [6] analyzed chromatographic separation as a process involving intermolecular dispersion, dipole orientation, charge transfer, and hydrogen-bonding interactions.

The most complete studies on molecular mechanism of chromatographic distribution were published in 1976. In the present author's opinion these are the papers by Scott [1], Karger et al. [2], Tijssen et al. [7], and Horváth et al. [8]. In spite of some controversy the above publications are still valid and it should be interesting and instructive to summarize them here before analyzing the statistically derived quantitative structure–retention relationships.

3.1.   SCOTT'S APPROACH

The basic theory of distribution of a solute between two phases has been elaborated by Scott [1] in 1976. However, the theory appeared to fail with reversed-phase liquid chromatography. This was due to incorrectly assuming that the solvent mixture comprising water and methanol was a binary system, whereas, because of water–methanol association, the system was, in fact, ternary. Taking that into consideration, the theory has been modified to predict solute retention in reversed-phase chromatography. The extended version of the theory has recently been published by Scott and co-workers [9].

Scott [1] defined the distribution coefficient K of a solute between the two phases in a chromatographic system by

$$K = \frac{\text{magnitude of total forces acting on solute in stationary phase}}{\text{magnitude of total forces acting on solute in mobile phase}} \quad (3.1)$$

The forces in Eq. (3.1) are divided into two distinct groups: polar and dispersive. The polar forces are those arising from permanent or induced electric fields associated with both solute and solvent molecules; in ion exchange chromatography the forces acting on the solute molecules will be substantially ionic in nature but may include the so-called polar and dispersive forces as well. Dispersive (London) forces are operating between both polar and nonpolar molecules. The magnitude of the total forces acting on a solute in a given phase will be dependent on the following: the magnitude and nature of the force between each solute molecule and a given molecule of that phase, their probability of interaction, their probability of position of contact, and their thermal energy. Thus, for a series of three different types of interactions, namely, ionic, polar, and dispersive, the distribution coefficient K of a solute between the chromatographic phases is represented in the form

$$K = \frac{[\Phi_i F_i P_i f_i(T) + \Phi_p F_p P_p f_p(T) + \Phi_d F_d P_d f_d(T)]_s}{[\Phi_i F_i P_i f_i(T) + \Phi_p F_p P_p f_p(T) + \Phi_d F_d P_d f_d(T)]_m} \quad (3.2)$$

where Φ is a constant and incorporates the probability of position of contact of a solute molecule with another molecule of a specific type (Φ is decided by the size and geometry of the molecules concerned); F is the magnitude of the respective force between the solute molecule and the phase molecule; P is the probability of molecular interaction; and $f(T)$ is the thermal energy of the molecule at the time of contact and will contain another thermal probability factor but will be constant at constant temperature. The subscripts i, p, and d denote ionic, polar, and dispersion interactions, respectively, whereas the subscripts s and m denote the stationary and mobile phases, respectively.

Equation (3.2) is simplified if the separations are carried out at constant temperature as $f(T)$ becomes constant and can be incorporated in Φ'. There are no ionic forces effective in the majority of chromatographic separations, and the ionic term can be neglected for the sake of simplicity. In liquid chromatography the probability of interaction of a solute with one of the phases will be proportional to the concentration of the interacting moieties in each of the respective phases. Keeping the above in mind, one can rewrite Eq. (3.2) in the form

$$K = \frac{(\Phi'_p F_p c_p + \Phi'_d F_d c_d)_s}{(\Phi'_p F_p c_p + \Phi'_d F_d c_d)_m} \tag{3.3}$$

where c_p and c_d are the concentrations of polar moieties and dispersive moieties in the stationary and mobile phases.

Assuming that the dispersive forces are bulky (mass) properties, Scott [1] related directly the concentration of dispersive moieties, c_d, to the density of the dispersion medium,

$$c_d = Ad \tag{3.4}$$

where A is a constant and d is the density of the dispersion solvent in the mobile phase. With a new constant $\Phi''_d = \Phi'_d A$, Eq. (3.3) can be written in the form

$$K = \frac{(\Phi'_p F_p c_p + \Phi'_d F_d c_d)_s}{(\Phi'_p F_p c_p + \Phi''_d F_d d)_m} \tag{3.5}$$

The corrected retention volume V' is related to A_s, the volume of the stationary phase (or the total number of active moieties on the surface of a solid, depending on the chromatographic system and the manner in which the concentration of solute in or on the stationary phase is defined). This relation is

$$V' = KA_s \tag{3.6}$$

Thus, for the corrected retention volume of a solute, the following relation holds:

$$V' = KA_s = \frac{(\Phi'_p F_p c_p + \Phi'_d F_d c_d)_s A_s}{(\Phi'_p F_p c_p + \Phi''_d F_d d)_m} \tag{3.7}$$

or, in the inverse form,

$$\frac{1}{V'} = \frac{(\Phi_p' F_p c_p + \Phi_d'' F_d d)_m}{(\Phi_p' F_p c_p + \Phi_d' F_d c_d)_s A_s}$$ (3.8)

The dynamic equation for the distribution coefficient [Eq. (3.8)] has been verified experimentally and has been discussed in detail for liquid and gas–liquid chromatography [1].

To verify the reliability of Eq. (3.8) in describing the role of polar interactions in normal-phase high-performance liquid chromatography, Scott [1] analyzed the previously obtained data [10]. Sample solutes were chromatographed at different concentrations of polar solvents (e.g., isopropanol) in a dispersion medium (heptane) on a Partisil 10 column. Assuming the activity of the stationary silica gel phase to be constant and independent of the mobile phase composition, one can thus consider the denominator in Eq. (3.8) as a constant. Then, Eq. (3.8) becomes

$$\frac{1}{V'} = A + B c_p$$ (3.9)

where A and B are constants. The reciprocals of the corrected retention volumes are linearly related to the concentration of the polar solvent in the mobile phase given in terms of the mass–unit volume of the mobile phase as predicted by Eq. (3.9) (Fig. 3.1).

Similarly, linear relationships have been observed in plots of $1/V'$ versus buffer concentration in ion exchange chromatography. In Fig. 3.2 data are plotted of the separation of a number of nucleotides on a strong anion-exchange-bonded phase column. Solute interactions in the mobile phase occur with the buffer moiety, which in this case is the dihydrogen phosphate ion, and again it is demonstrated that the probability of interaction is conditioned by the concentration of that moiety in the mobile phase.

Equivalent results were reported by Knox and Pryde [11] in their work on the separation of organic acids using a bonded-phase anion exchanger.

In his paper Scott [1] also gave evidence for the general validity of Eq. (3.9) in the case of thin-layer and gas chromatographic separations. As a proof of the correctness of Eq. (3.9), Scott [1] discussed the results obtained by Laub and Purnell [12]. These authors measured the distribution coefficient of a number of solutes chromatographed on a series of stationary phase mixtures containing squalane (S) and dinonyl phthalate (A) and proved the following relationship:

$$K_R = K_{R(A)}^0 V_A + K_{R(S)}^0 V_S$$ (3.10)

where K_R is the solute distribution coefficient for the mixture, $K_{R(A)}^0$ and $K_{R(S)}^0$

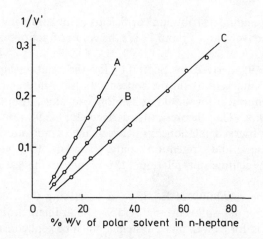

Figure 3.1. Reciprocal of corrected retention volume versus percentage (w/v) of polar solvent in *n*-heptane for (A) benzyl alcohol, (B) 3-phenyl-1-propanol, and (C) deoxycorticosterone alcohol. Polar solvents: (A and B) tetrahydrofuran; (C) isopropanol. Column: 25 cm × 4.6 mm., i.d., packed with Partisil 10. (After R. P. W. Scott, *J. Chromatogr.*, **122**, 35, 1976. With permission.)

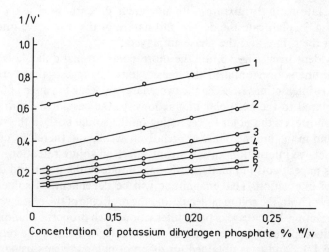

Figure 3.2. Reciprocal of corrected retention volume versus buffer concentration in mobile phase for (1) adenosine 3′,5′-cyclic monophosphate; (2) uridine 5′-monophosphate; (3) cytidine 5′-monophosphate; (4) adenosine 5′-monophosphate; (5) uridine 2′,3′-diphosphate; (6) adenosine 2′,3′-diphosphate; (7) guanosine 5′-monophosphate. Column: 25 cm × 4.6 i.d., packed with Partisil 10 SAX, pH 4.80. (After R. P. W. Scott, *J. Chromatogr.*, **122**, 35, 1976. With permission.)

are the corresponding distribution coefficients of the solute in the pure phases A and S, respectively, and V_A and V_S are the volume fractions of phases A and S, respectively.

Equation (3.9), derived by Scott [1] for the relationship between the probability of interaction and the concentrations of interacting moiety, has been thus confirmed for liquid–solid, ion exchange, thin-layer, and gas chromatography. The theoretically derived Eq. (3.9) resembles the experimentally observed relationship between the composition of chromatographic phases and retention data, similar to the well-known Soczewiński–Wachtmeister relation [13] found in the case of thin-layer chromatography.

To verify the reliability of Eq. (3.8) in describing the role of dispersive interactions in liquid–solid chromatography, polar interactions must be kept constant; that is, for a given phase system c_p must be kept constant. Assuming that dispersive interactions can be changed by varying the density d of the dispersion solvent, an equation analogous to Eq. (3.9) can be derived from Eq. (3.8):

$$\frac{1}{V'} = A' + B'd \tag{3.11}$$

Relationships predicted by Eq. (3.11) were reported [10] for several solutes chromatographed in several mobile phases each comprising 4.51% (w/v) of ethyl acetate in binary mixtures with different hydrocarbons. The $1/V'$ versus d plots (Fig. 3.3) indicate the degree and nature of the dispersive interactions between the solute and the phase employed.

As evident from Fig. 3.3, for the more polar aromatic alcohols the $1/V'$ versus d line is horizontal to the density axis. It suggests that here the net dispersive interactions between the two phases are weak for these polar solutes as compared to the net polar interactions. In the case of the antraquinone solutes, however, the polar forces acting on the solute both in the stationary phase and in the mobile phase are relatively weak, and therefore, changes in the dispersive interaction have a strong effect on solute retention.

In his paper Scott [1] analyzed the magnitude of the polar and dispersive forces of interaction. That magnitude will be dependent on some physical property of both the solute molecule and the molecule of the phase with which it is interacting. In the case of polar interactions such properties, among others, will be incorporated in the constant B in Eq. (3.9). Thus, if the relationship between $1/V'$, and c_p is obtained for different polar solvents using the same solute, different slopes B will be found. These values of B reflect the physical properties of the solvent that condition the magnitude of the interacting polar forces. Similarly, the values of B obtained when a series of different solutes are

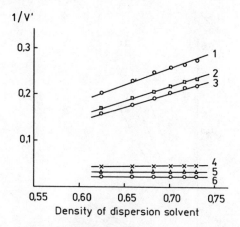

Figure 3.3. Plots of reciprocal of corrected retention volume versus density of mobile phase (4.51% w/v ethyl acetate in different hydrocarbons): (1) 2-ethyl antraquinone; (2) 2-methyl antraquinone; (3) antraquinone; (4) phenyl methyl carbinol; (5) benzyl alcohol; (6) 3-phenyl-1-propanol. (After R. P. W. Scott and P. Kucera, *J. Chromatogr.*, **112**, 425, 1975. With permission.)

employed with the same polar solvent will reflect the different properties of the solute that determine the magnitude of interaction.

The plots of $1/V'$ versus c_p for different polar solvents using the same solute are given in Fig. 3.4.

In Fig 3.5 the logarithm of the slopes of the curves given in Fig. 3.4 for each individual solvent is plotted against the polarizability calculated from the dielectric constants. In fact, dioxane deviated originally from the correlation illustrated in Fig. 3.5. Here, the dielectric constant of dioxane was taken as twice that of diethyl ether because it has been argued that the two dipoles in dioxane are in opposition and, therefore, cancel each other, providing the dielectric constants that do not reflect the polarity of the solvent. Scott's approach to the polarity of dioxane is an example of structure–retention relationship analysis at a submolecular level.

Similar to Fig. 3.4, the relationship $1/V'$ versus c_p was obtained by Scott [1] for different solutes using the same polar solvent ethyl acetate. Next, the logarithms of the slopes of the curves for each respective solute were plotted against the polarizability for each respective solute (Fig. 3.6).

As may be seen in Fig. 3.6, the correlation is good for the esters, ethers, and alcohols studied but not for the ketonic compounds. Although the general validity of Scott's molecular approach to chromatographic distribution mechanism seems to be convincingly proved, the details still require explanation. Most probably, simple dielectric constant measurement may not be

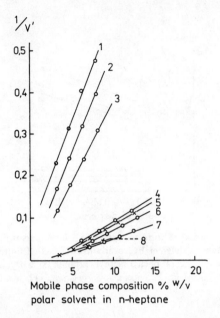

Figure 3.4. Reciprocal of corrected retention volume of phenyl methyl carbinol against composition of mobile phase containing the following polar solvents in *n*-hexane: (1) isopropanol; (2) *n*-butanol; (3) *n*-pentanol; (4) dioxane; (5) tetrahydrofuran; (6) methyl acetate; (7) ethyl acetate; (8) butyl acetate, Column: 25 cm × 4.6 mm i.d., packed with partisil 10. (After R. P. W. Scott, *J. Chromatogr.*, **122**, 35, 1976. With permission.)

suitable for determining the effective chromatographic polarity of substances. To quantify the so-called polar properties of compounds seems to be an especially difficult task of structural chemistry.

The magnitude of the dispersive interactions seems to be clearer. Dispersive interactions have commonly been related to the molecular weight for series of homologous (or closely congeneric) compounds. It should be noted, however, that for a homologous series the molecular weight is linearly related to many other physical properties of each member of the series, and so the apparent significance of molecular weight in dispersive interactions could well be misleading.

Scott's theory aroused considerable controversy [14] due to its apparent failure to account for retention in reversed-phase liquid chromatographic systems. Scott and co-workers [9] argued that the failure was due to incorrectly assuming that the solvent mixture comprising water and methanol was a binary system. Assuming that because of the water–methanol association, the system is in fact ternary, these authors [9] extended the basic theory

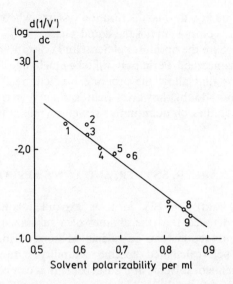

Figure 3.5. Log $d(1/V')/dc_p$ as determined from data given in Fig. 3.4 against solvent polarizability for phenyl methyl carbinol eluted from silica employing the following solvents: (1) butyl acetate; (2) propyl acetate; (3) ethyl acetate; (4) methyl acetate; (5) tetrahydrofuran; (6) dioxane; (7) n-pentanol; (8) n-butanol; (9) isopropanol. (After R. P. W. Scott, *J. Chromatogr.*, **122**, 35, 1976. With permission.)

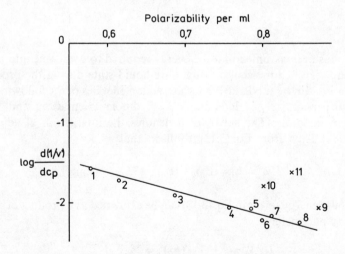

Figure 3.6. Log $d(1/V')/dc_p$ against solute polarizability for (1) benzyl acetate; (2) methyl acetate; (3) tetrahydrofuran; (4) n-octanol; (5) phenyl ethanol; (6) benzyl alcohol; (7) n-pentanol; (8) n-butanol; (9) acetone; (10) methyl ethyl ketone; (11) acetophenone. (After R. P. W. Scott, *J. Chromatogr.*, **122**, 35, 1976. With permission.)

33

to liquid–liquid and liquid–solid distribution systems, where association takes place between the components of the liquid phase.

One need not share the optimism of Scott and co-workers [9] as far as the goal of theory is concerned. Perhaps it will take much time and effort to derive explicit equations that allow the accurate prediction of chromatographic retention. Without doubt, however, Scott's theory, eventually helped to rationally guide studies on quantitative structure–retention relationships.

3.2. KARGER, SNYDER, AND EON'S APPROACH

Karger, Snyder, and Eon [2], in their expanded solubility parameter treatment, have detailed how the abilities of a substance to participate in intermolecular interactions can be calculated from various nonchromatographic data. Such calculated specific solubility parameters provide a classification scheme of solvents and adsorbents that can be used to estimate selectivity in chromatography.

These authors [2] assumed after Hildebrand and Scott [5] that the so-called solubility parameter δ can be determined from the energy of vaporization per mole E_i of a given compound i and its molar volume V_i

$$\delta = \left(\frac{E_{ii}^{\text{v}}}{V_i}\right)^{1/2} \tag{3.12}$$

Next, it has been assumed that the energy required to overcome interactions between adjacent i molecules in the pure liquid state during the process of evaporation can be subdivided into interaction energies of the following type: $(E_{ii})_{\text{d}}^{\text{v}}$, dispersion; $(E_{ii})_{\text{in}}^{\text{v}}$, induction; $(E_{ii})_{\text{o}}^{\text{v}}$, dipole orientation; and $(E_{ii})_{\text{hb}}^{\text{v}}$, hydrogen bonding. (The subscript ii denotes the interaction of adjacent i molecules.) Thus, from Eq. (3.12) it follows that

$$V_i \delta^2 = E_i^{\text{v}} = (E_{ii})_{\text{d}}^{\text{v}} + (E_{ii})_{\text{in}}^{\text{v}} + (E_{ii})_{\text{o}}^{\text{v}} + (E_{ii})_{\text{hb}}^{\text{v}} \tag{3.13}$$

Each of the energy terms in Eq. (3.13) can be expressed as a product of specific solubility parameters:

$$(E_{ii})_{\text{d}}^{\text{v}} = \delta_{\text{d}}^2 V \qquad (E_{ii})_{\text{in}}^{\text{v}} = 2\delta_{\text{in}}\delta_{\text{d}} V$$
$$(E_{ii})_{\text{o}}^{\text{v}} = \delta_{\text{o}}^2 V \qquad (E_{ii})_{\text{hb}}^{\text{v}} = 2\delta_{\text{a}}\delta_{\text{b}} V \tag{3.14}$$

Thus, δ_{d} indicates the ability of a substance to participate in dispersive interactions; δ_o is a measure of the ability of a species to participate in dipole orientation interactions; δ_{in} denotes the ability of that species to induce a dipole moment in surrounding molecules; and δ_a and δ_b characterize the ability of that species to function as a proton donor or acceptor, respectively. Considering orientation, induction, and hydrogen-bonding interactions together as polar interactions, the solubility parameter δ may be described as a sum of polar, δ_p, and dispersion, δ_d, solubility parameter contributions:

$$\delta^2 = \delta_p^2 + \delta_d^2 \qquad (3.15)$$

where δ_p is a composite quantity.

Karger et al. [2] assumed that δ_d does not change significantly from compound to compound in a homologous series. Thus, having δ calculated from experimental energies of vaporization, they could evaluate δ_p from Eq. (3.15).

For a homologous series of a given functional group, the plots of log δ_p versus log V_i are linear. The lines are parallel for different monofunctional homologous series (Fig. 3.7). Included in Fig. 3.7 is the log δ_p versus log V_i plot for n-alcohols. The self-association possible in this case leads to $f(V_i)$ functionality deviating from the parallelism.

The relationship, as illustrated in Fig. 3.7, is as expected from the equation

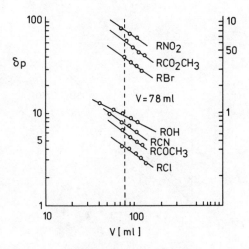

Figure 3.7. Logarithmic plot of polar solubility parameter δ_p against molar volume V_i for homologous series of monofunctional group compounds. (After B. L. Karger, L. R. Snyder, and C. Eon, *J. Chromatogr.*, **125**, 71, 1976. With permission.)

derived by Karger et al. [2] for the polar solubility parameter δ_p:

$$\delta_p^2 = C\mu^2 f(V_i) \qquad (3.16)$$

where C is a constant. The permanent dipole moment or bond dipole μ may also be assumed to be approximately constant for a homologous series of a given functional group. Thus, if there are no hydrogen-bonding interactions involved, the δ_p for the solutes from different series but having the same molar volume should be directly proportional to μ, as illustrated in Fig. 3.8. The molar volume $V_i = 78$ mL was chosen in Fig. 3.7 because for the solutes considered, the contributions of induction and orientation to the polar solubility parameter are equal when $V_i = 78$ mL.

It is evident from Fig. 3.8 that the ester, RCO_2CH_3, is more polar than predicted from its dipole moment. As pointed out by Karger et al [2], this results from the fact that the overall dipole moment of the ester is the sum of two dipole vectors that are partially in opposite directions, causing some cancellation. Yet in solution, single (bond) dipoles from two molecules interact at close range with each other, and therefore, these are the individual bond dipoles that interact effectively.

A more detailed analysis allowed Karger et al. [2] to derive relationships describing the ability of chemical compounds to participate in individual intermolecular interactions. Thus, for quantitation of a dispersive solubility parameter the authors applied the classical Lorentz–Lorenz expression. For dipole induction the interaction energy is the product of induction and

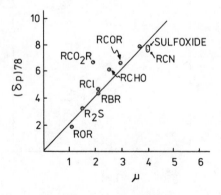

Figure 3.8. Plot of polar solubility parameter normalized to molar volume 78 ml, $(\delta_p)_{78}$, against gas-phase dipole moment μ for monofunctional group compounds. (After B. L. Karger, L. R. Snyder, and C. Eon, *J. Chromatogr.*, **125**, 71, 1976. With permission.)

dispersion solubility parameters, $(E_{ii})_{in}^v = 2\delta_{in}\delta_d V_i$, where δ_{in} is given by

$$\delta_{in} = \frac{C_{in}\mu^2}{V_i} \tag{3.17}$$

where C_{in} is a constant. For dipole orientation $(E_{ii})_o^v = \delta_o^2 V_i$, where

$$\delta_o^2 = \frac{C_o^2 \mu^2}{V_i^2} \tag{3.18}$$

C_o is a constant.

The proton-donor solubility parameter δ_a and proton-acceptor solubility parameter δ_b have been determined, after several assumptions are made, from the enthalpy of hydrogen bonding. Such values, however, must be considered approximate.

An analysis of the dependence of the orientation interaction energy $(E_{ii})_o^v$ on dipole moment μ^2 [Eq. (3.18)] allowed Karger et al. [2] to explain the linear relationship of the gas chromatographic retention index difference $\Delta(RI)$ of a solute between a polar and a nonpolar stationary phase. In the original paper a plot of $\Delta(RI)$ versus μ_i is given for monofunctional solutes when the polar phase n-hexadecylnitrile and the nonpolar phase is n-hexadecane. This relationship has previously been noted by Kováts and Weiss [15].

To illustrate the relative magnitude of the individual solubility parameters of a variety of chromatographic solvents, Table 3.1 provides numerical data after Karger et al. [2].

3.3. TIJSSEN, BILLIET, AND SCHOENMAKER'S APPROACH

Tijssen, Billiet, and Schoenmaker [7] concluded that the total solubility parameter δ_T should be evaluated from internal pressures rather than from the usual vaporization data. The internal pressure p_i is defined by the thermodynamic equation of state, which follows from the first and second laws of thermodynamics:

$$p_i = \left(\frac{\partial u}{\partial u}\right)_T \tag{3.19}$$

Chromatographic distribution is governed by several intermolecular interactions. Tijssen et al. [7] consider three types of interactions: dispersion, orientation, and the so-called acid–base interactions. The abilities of individual chemical compounds to participate in the respective interaction is reflected by its specific partial solubility parameter. The differences in all possible partial polarities or partial solubility parameters determine chromatographic retention rather than the differences in total polarities. In conclusion, Tijssen et al. [7] defined a four-parameter solubility parameter model:

$$\delta_T^2 = \delta_d^2 + \delta_o^2 + 2\delta_a\delta_b \tag{3.20}$$

To calculate the dispersion solubility parameter δ_d^2, Tijssen et al. [7] proposed the formula

$$\delta_d^2 = 228 \times \frac{n^2 - 1}{n^2 + 2} + 5.2 \tag{3.21}$$

where n is the refractive index.

To calculate the orientation solubility parameter δ_o^2, a formula is used derived after Böttcher [16]:

$$\delta_o^2 = \frac{1.2099 \times 10^4}{V^2} \frac{(\varepsilon - 1)(n^2 + 2)}{(2\varepsilon + n^2)} \mu^2 \approx \frac{RT(\varepsilon - 1)(\varepsilon - n^2)}{V} \frac{(\varepsilon - 1)(\varepsilon - n^2)}{(n^2 + 2)\varepsilon} \tag{3.22}$$

where R is a gas constant; T is temperature; V is molar volume (in cm^3/mol); ε is a dielectric constant; n is the refractive index; and μ is dipole moment.

The parameters δ_a and δ_b are related to acid strength and base strength, respectively, or similarly, they can be related to the proton-donating and proton-accepting properties. The product $2\delta_a\delta_b = \delta_h^2$ has a physical meaning (contrary to the separate δ_a and δ_b parameters) and characterizes generally acid–base interactions for polar compounds. The δ_h^2 values may be derived from spectroscopic studies on hydrogen bonding and association enthalpies. To calculate the specific δ_a and δ_b parameters, the ratio of the two quantities must be known along with the δ_h^2 data. The detailed procedure to calculate δ_a and δ_b is given in the original article [7], but the acid–base parameters obtained are uncertain.

The total solubility parameter and the partial solubility parameters calculated by Tijssen et al. [7] are given in Table 3.2.

Having the specific solubility parameters determined for the solute i, stationary phase s, and the mobile phase m, the partition coefficient K_i can be calculated according to the following equation proposed by Tijssen et al. [7]:

$$\ln K_i = \frac{V_i}{RT}\left[(\delta_{T,m}^2 - \delta_{T,s}^2) + 2\delta_{d,i}(\delta_{d,s} - \delta_{d,m}) + 2\delta_{o,i}(\delta_{o,s} - \delta_{o,m}) + 2\delta_{a,i}(\delta_{b,s} - \delta_{b,m}) \right.$$

$$\left. + 2\delta_{b,i}(\delta_{a,s} - \delta_{a,m}) \right] + V_i\left(\frac{1}{V_s} - \frac{1}{V_m}\right) \tag{3.23}$$

Table 3.1. Extended Table of Solubility Parameters $[(cal/mL)^{1/2}]^a$

Solvent	δ_T	δ_d	δ_o	δ_{in}	δ_a	δ_b	V
Perfluoroalkanes	~6.0	~6.0	—	—	1.0	—	—
n-Alkanes	~8.0	~8.0	—	—	—	—	—
Diisopropyl ether	7.1	6.9	1.0	0.1	—	3.0	102
Diethyl ether	7.5	6.7	2.4	0.5	—	3.0	105
Triethylamine	7.5	7.5	—	—	—	4.5	140
Cyclohexane	8.2	8.2	—	—	—	—	108
Propyl chloride	8.4	7.3	2.9	0.6	—	0.7	88
Carbon tetrachloride	8.6	8.6	—	—	—	0.5	97
Diethyl sulfide	8.6	8.2	1.7	0.25	—	2.6	108
Ethyl acetate	8.9	7.0	4.0	1.0	—	2.7	98
Propylamine	8.9	7.3	1.7	0.2	1.8	5.5	82
Ethyl bromide	8.9	7.8	3.1	0.6	—	0.8	77
Toluene	8.9	8.9	—	—	—	0.6	107
Tetrahydrofuran	9.1	7.6	3.5	0.8	—	3.7	82
Benzene	9.2	9.2	—	—	—	0.6	89
Chloroform	9.3	8.1	3.0	0.5	6.5	0.5	81
Ethyl methyl ketone	9.5	7.1	4.7	1.2	—	3.2	90
Acetone	9.6	6.8	5.1	1.5	—	3.0	74
1,2-Dichloroethane	9.7	8.2	4.2	0.5	—	0.7	79
Anisole	9.7	9.1	2.1	0.4	—	1.7	109
Chlorobenzene	9.7	9.2	1.9	0.3	—	1.0	102
Bromobenzene	9.9	9.6	1.5	0.2	—	1.0	105
Methyl iodide	9.9	9.3	2.5	0.3	—	0.7	62
Dioxane	10.1	7.8	5.2	1.0	—	4.6	86
Hexamethylphosphoramide	10.5	8.4	3.4	1.7	—	4.0	176
Pyridine	10.6	9.0	3.8	1.0	—	4.9	81
Acetophenone	10.6	9.6	2.7	0.7	—	3.3	117
Benzonitrile	10.7	9.2	3.4	1.0	—	2.3	103
Propionitrile	10.8	6.9	6.6	1.8	—	2.1	71
Quinoline	10.8	10.3	1.8	0.3	—	4.2	118
N,N-Dimethylacetamide	10.8	8.2	4.7	1.6	—	4.5	92
Nitroethane	11.0	7.3	6.0	2.2	—	1.0	71
Nitrobenzene	11.1	9.5	3.6	1.1	—	1.0	103
Tricresyl phosphate	11.3	9.6	2.5	1.5	—	—	316
Dimethylformamide	11.8	7.9	6.2	2.4	—	4.6	77
Propanol	12.0	7.2	2.6	0.4	6.3	6.3	75
Dimethyl sulfoxide	12.0	8.4	6.1	2.1	—	5.2	71
Acetonitrile	12.1	6.5	8.2	2.8	—	3.8	53
Phenol	12.1	9.5	2.3	0.4	9.3	2.3	92
Ethanol	12.7	6.8	3.4	0.5	6.9	6.9	59
Nitromethane	12.9	7.3	8.3	3.0	—	1.2	54
γ-Butyrolactone	12.9	8.0	7.2	3.2	—	—	77

39

Table 3.1. (Continued)

Solvent	δ_T	δ_d	δ_o	δ_{in}	δ_a	δ_b	V
Propylene carbonate	13.3	9.8	5.9	2.4	—	—	85
Diethylene glycol	14.3	8.2	4.0	0.6	5.3	5.3	96
Methanol	14.5	6.2	4.9	0.8	8.3	8.3	41
Ethylene glycol	17.0	8.0	6.8	1.1	6.1	6.1	56
Formamide	19.2	8.3	—	—	Large	Large	40
Water	23.4	6.3	—	—	Large	Large	18

[a] Symbols: δ_T, total solubility parameter from vaporization; δ_d, dispersion solubility parameter; δ_o, orientation solubility parameter; δ_{in}, induction solubility parameter; δ_a, proton–donor solubility parameter; δ_b, proton–acceptor solubility parameter; V(mL/mol), molar volume. From Karger et al. [2]. With permission.

where the V's are the respective molar volumes. In gas chromatography relative retention $\alpha_{2,1}$ of a solute 2 with respect to the standard solute 1 can even be calculated from Eq. (3.24), resembling the known Rohrschneider's expression [17]:

$$\ln \alpha_{2,1} = \ln \frac{V_1}{V_2} + (V_1 - V_2)\left[\frac{\delta_s^2}{RT} - \frac{1}{V_s}\right] - 2\frac{\delta_s}{RT}(V_1\delta_1 - V_2\delta_2) \qquad (3.24)$$

In Eq. (3.24) the δ's are the total solubility parameters for solutes 1 and 2 and the stationary phases, respectively. Equation (3.24) was found by Tijssen et al. [7] to fail when the compounds under consideration did not form a closely congeneric group.

Although the approach by Tijssen et al. [7] is theoretically interesting, the authors themselves found its predictive properties much worse than, for example, factor analysis [18]. According to Tijssen et al. [7], the failure in predicting liquid chromatographic retention results from the present inability to correct the solubility parameters for the influence of elevated pressures. Also, the attempt by the same authors to predict partition in an octanol–water system may not be considered a success.

3.4. HORVÁTH, MELANDER, AND MOLNAR'S APPROACH

The majority of successful QSRR studies concerns the data obtained from liquid chromatography with nonpolar stationary phases [8]. Commonly, the separation of solutes in reversed-phase chromatography is explained in terms of the so-called hydrophobic (solvophobic) effect. A theoretical framework that accounts for the governing factors in that chromatographic mode was

Table 3.2. Solubility Parameters[a]

Compounds	δ_T	δ_d	δ_o	δ_h	δ_a	δ_b	δ
Acetone	10.51	7.44	6.41	5.49	0.90	16.6	9.93
Acetonitrile	13.15	7.33	10.27	3.70[b]	0.39	17.6	12.50
Acetophenone[b]	10.79	8.74	3.80	2.67	0.27	13.1	10.44
Aniline	12.21	9.04	2.30	7.88[b]	3.49	8.9	11.81
Benzaldehyde[b]	11.34	8.80	4.38	5.65	1.15	13.9	10.95
Benzene	9.71	8.51	0.16	1.95	0.18	10.4	9.19
Benzonitrile	11.38	8.69	5.42	4.96[b]	0.94	13.1	11.00
Benzophenone[b]	10.87	9.15	2.79	5.16	1.11	12.0	10.62
1-Butanol	11.47	7.77	2.64	7.74	3.15	9.1	11.05
tert-Butanol	10.53	7.65	2.47	5.47	2.51	6.1[b]	10.08
Carbon tetrachloride	9.19	8.22	0.29	1.43	0.34[b]	3.0[b]	8.68
Chloroform	9.87	8.12	1.95	3.02	1.54	3.1	9.31
Cyclohexanone[b]	10.76	8.16	4.57	4.13	0.61	13.9	10.36
Dibutyl ether[b]	8.26	7.77	0.86	2.67	0.30	11.9	7.94
1,2-Dichloroethane	10.64	8.11	3.71	3.35	1.68	3.3	10.12
Dichloromethane	10.68	7.96	3.67	4.07	1.92	4.3	10.02
Diethyl ether	7.92	7.38	1.68	2.33	0.19	14.0	7.37
N,N-Dimethylaniline[b]	10.29	8.87	1.43	5.02	1.58	8.0	9.95
Dimethyl sulfoxide	13.45[b]	8.34	9.49	7.51	1.32	21.4	12.98[b]
1,4-Dioxane	10.65	7.95	0.43	7.07[b]	1.47	17.1	10.16
Ethanol	13.65	7.46	4.29	10.81	5.17	11.3	13.09
Ethyl acetate	9.57	7.55	2.32	5.40	0.26	13.5	9.09
Ethylbenzene	9.35	8.47	0.34	1.80	0.15	10.6	8.96
n-Heptane	7.91	7.91	0.03[b]	0	0	0	7.52
n-Hexane	7.74	7.74	0.02	0	0	0	7.29
Methanol	15.85	7.18	6.72	13.53	7.18	12.9	15.15
Methyl acetate	10.19	7.44	2.83	3.90	0.51	15.0	9.63
N-Methylaniline[b]	11.69	8.93	1.85	7.31	3.28	8.2	11.34
Methyl ethyl ketone	9.96	7.60	5.12	4.18	0.58	15.0	9.46
Nitromethane	13.83	7.63	9.58	6.42[b]	—	—	13.23
n-Octanol[b]	9.13	8.00	1.49	4.14	1.85	4.6	8.82
n-Pentane	7.65	7.65	0.03	0	0	0	7.14
Phenol	12.76	8.33	2.52	8.86[b]	5.20	8.7	12.37
1-Propanol	12.27	7.65	3.54	9.04	4.12	10.0	11.78
2-Propanol	12.37	7.59	3.12	9.27	4.21	10.2	11.90
Pyridine	11.12[b]	8.56	3.94	5.90[b]	0.91	19.1	10.62[b]
Tetrahydrofuran	9.88	7.96	2.97	5.04[b]	0.78	16.3	9.32
Toluene	9.53	8.47	0.32	2.45	0.29	10.2	9.09
2,2,4-Trimethylpentane	7.33	7.33	0.04	0	0	0	6.96
Water	25.52	7.22	15.35	19.07	16.65	10.9	24.55

[a] Symbols: δ_T, total solubility parameter from internal pressure; δ, calculated total solubility parameter. All other symbols as in Table 3.1. From Tijssen et al. [7]. With permission.
[b] Estimated data.

given by Horváth et al. [8] in 1976. It acquired a wide assent, and the paper is one of the most often quoted in the chromatographic literature.

Horváth et al. [8] treated the interaction between the solute and the stationary phase in solvophobic chromatography as a reversible association of the solute molecules S with the hydrocarbonaceous ligand L, yielding the complex SL. The equlibrium constant for the process, K, is defined by

$$K = \frac{[SL]}{[S][L]}$$
(3.25)

To evaluate K, Horváth et al. [8] employed the formalism of solvophobic theory put forward by Sinanoğlu and co-workers [19–21].

According to the theory, molecular associations in solution can be conceptually separated into two processes. One would be the interaction of the species S and L to yield SL in a hypothetical gas phase without any influence of the solvent. Another process entails the interactions of the associating molecules and the complex individually with the solvent. The association in the gas phase is assumed to occur by van der Waals forces only. The free-energy change of the process is denoted by $\Delta F_{vdW, assoc}$. The standard unitary free-energy change resulting from placing the species into the solution is equal to the difference between the free-energy change required for the creation of a cavity to accommodate the species j in the solvent, $\Delta F_{c, j}$, and that arising from the interaction between the species and the surrounding solvent molecules, $\Delta F_{i, j}$. Additionally, the entropy change arising from the change in the so-called free volume is taken into account. Thus, the total standard unitary free energy change ΔF_j^0 of the process entailing the interactions of S, L, and SL with the solvent is given for each species j by

$$\Delta F_j^0 = \Delta F_{c, j} + \Delta F_{i, j} + RT \ln(RT/P_0 V)$$
(3.26)

where R is a gas constant, T temperature, V the mole volume of the solvent, and P_0 the atmospheric pressure. The overall standard unitary free-energy change for the association in solution, ΔF_{assoc}^0, is then given by

$$\Delta F_{assoc}^0 = \Delta F_{vdW, assoc} + (\Delta F_{c, SL} + \Delta F_{i, SL}) - (\Delta F_{c, S} + \Delta F_{i, S})$$
$$- (\Delta F_{c, L} + \Delta F_{i, L}) - RT \ln(RT/P_0 V)$$
(3.27)

The thermodynamic equilibrium constant K is related to the free-energy change by

$$\ln K = -\frac{\Delta F_{assoc}^0}{RT}$$
(3.28)

The measured chromatographic parameter is the capacity factor k', which is related to K according to

$$\ln k' = \ln K + \varphi \tag{3.29}$$

where φ is a constant for a given column (φ is the logarithm of the so-called phase ratio). Thus, if one is able to determine precisely the terms on the right side of Eq. (3.27), the prediction of retention in reversed-phase liquid chromatography is feasible.

Based on the solvophobic theory of Sinanoğlu, Horváth et al. [8] gave individual relationships, allowing evaluation of the respective energetic terms in Eq. (3.27). With all the assumptions, approximations, and limitations involved in deriving the components of Eq. (3.27), the resulting general equation for the capacity factor in liquid chromatography has the form

$$\ln k' = \varphi - \frac{\Delta F_{\text{vdW, assoc}}}{RT} + \frac{\Delta F_{\text{vdW, S}}}{RT} - \frac{N(\lambda - 1)}{2\lambda} \frac{\mu_S^2}{V_S RT} DP$$

$$+ \frac{N \Delta A \gamma}{RT} + \frac{4.836 N^{1/3} (\kappa^e - 1) V^{2/3} \gamma}{RT} + \ln \frac{RT}{P_0 V} \tag{3.30}$$

where $\Delta F_{\text{vdW, S}}$ is the van der Waals component of the interaction of the solute with the solvent. This term is calculated according to Sinanoğlu by a rather complex procedure from the following molecular parameters of the solute and the solvent: ionization potential, refractive index, and a function of molecular diameter. Other symbols in Eq. (3.30) are N, the Avogadro number; λ, the proportionality factor reflecting the volume change occurring upon the binding of the solute to the ligand of a stationary phase; μ_S, the solute static dipole moment; V_S, the molecular volume of the solute; ΔA, the contact surface area of the associated species S and L; γ, the surface tension of the solvent; κ^e, a parameter related to the internal energy change associated with the vaporization of the solvent; D, a function of the static dielectric constant of the solvent; and P, a function of the polarizability of the species. The remaining symbols have been explained earlier.

The canonical expression for the capacity factor as given in Eq. (3.30) can be further simplified in certain special cases. Assuming that the solute and ligand properties as well as the $\Delta F_{\text{vdW, assoc}}$ term are invariant for a given solute when the solvent composition is changed at constant temperature and flow rate with a fixed column, Eq. (3.30) can be written as

$$\ln k' = a + bD + c\gamma + d(\kappa^e - 1) V^{2/3} \gamma + e + \ln RT/P_0 V \tag{3.31}$$

where a and c may be regarded as constant and can be determined experimentally; b, d, e, and the last term are also constant and can be estimated by the relations given in the original work [8].

When the same eluent and column are used, the capacity factor of different solutes may be obtained at a fixed temperature and flow rate from

$$\ln k' = a' + b' \frac{1-\lambda}{2} \frac{\mu_S^2}{V_S} \frac{1}{1-(\alpha_S/V_S)} + c' \Delta A \qquad (3.32)$$

where a', b', and c' are constants; α_S is the polarizability of the solute, and the remaining symbols are as before.

Equations (3.31) and (3.32) well illustrate the importance of specific molecular features of the solute and the solvent for distribution in liquid chromatography. Unfortunately, to verify experimentally the approximate equations (3.31) and (3.32) still further simplifying assumptions and approximations had to be introduced by the authors [8].

Equation (3.32) has been obtained from Eq. (3.30) assuming (after some simplification) that

$$a' = \varphi - \frac{\Delta F_{vdW, assoc}}{RT} + \frac{\Delta F_{vdW, S}}{RT} + \frac{4.836 N^{1/3}(\kappa^e - 1) V^{2/3}\gamma}{RT} + \ln \frac{RT}{P_0 V} \qquad (3.33)$$

Thus, returning to Eq. (3.32), one notes that the two van der Waals terms have been lumped together in the leading constant a'. For closely related solutes the two van der Waals terms are assumed to vary to the same extent, and the changes in the two terms cancel because of their opposite signs. In other words, a' is considered constant for solutes having commensurable molecular dimensions. The second term in Eq. (3.32) entails the dipole moment, the molecular volume, and the polarizability of the solute as well as the volume change occurring upon binding the solute to the ligand. Since the c' constant is defined as $c' = N\gamma/RT$, the third term in Eq. (3.32) is essentially the product of the contact surface area in the complex SL and the surface tension of the eluent.

Horváth et al. [8] determined capacity factor k' values with a neat aqueous phosphate buffer at the same conditions for three groups of related aromatic solutes: carboxylic acids, amino acids, and amines. The straight line of a similar slope was obtained when logarithm capacity factors were plotted against the hydrocarbonaceous surfaces area of the solutes (Fig. 3.9). It may be assumed that the contact area ΔA in Eq. (3.32) is directly proportional to the hydrocarbonaceous surface given on the abscissa in Fig. 3.9. Consequently, the second term in Eq. (3.32) is closely similar for all solutes of a given group. Thus, for closely related substances Eq. (3.32) can be simplified to

$$\ln k' = a'' + c' \Delta A = a'' + \frac{N}{RT}\gamma \Delta A \qquad (3.34)$$

where a'' is the sum of a' and the second term in Eq. (3.32) and $c' = N\gamma/RT$. Since the surface tension in the experiments was constant, Eq. (3.34) predicts the linear behavior shown in Fig. 3.9. The relative displacement of the lines for different groups of solutes observed in Fig. 3.9 can be accounted for by the differences mainly in the second term in Eq. (3.32), which expresses the electrostatic effect. Horváth et al. [8] explain the displacement of the individual lines in Fig. 3.9 by the difference in the average dipole moment for the groups of solutes.

For a given solute the effect of changing eluent composition has been expressed by Eq. (3.31). When only the surface tension changes, Eq. (3.31) can be simplified to

$$\ln k' = a''' + b''\gamma \tag{3.35}$$

where a'' is the sum of all terms that do not contain the surface tension and b'' is given by

$$b'' = \frac{N\,\Delta A + 4.836 N^{1/3}(\kappa^e - 1)V^{2/3}}{RT} \tag{3.36}$$

According to Eq. (3.35), log k' depends linearly on the surface tension of the

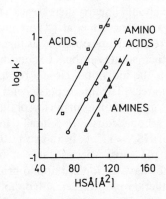

Figure 3.9. Relationship between logarithm of capacity factor, log k', and hydrocarbonaceous surface area, HSA, for biogenic aromatic solutes: (1) carboxylic acids; (2) amino acids; and (3) amines. Column: Partisil 1025 ODS; eluent: 1.0 M Na$_2$SO$_4$ in 0.1 M phosphate buffer pH 2.05; inlet pressure 1000 psi. (After Cs. Horváth, W. Melander, and I. Molnár, *J. Chromatogr.*, **125**, 129, 1976. With permission.)

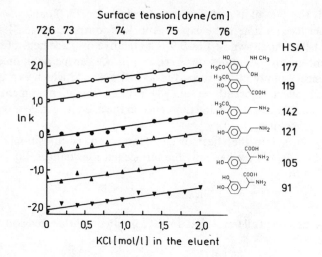

Figure 3.10. Plots of logarithm capacity factor against surface tension or salt (KCl) concentration in eluent (0.05 M KH$_2$PO$_4$ solution). The hydrocarbonaceous surface area, HSA, is also given for the solute molecules. Column: Partsisl 1025 ODS; flow rate: 1 mL/min. (After Cs. Horváth, W. Melander, and I. Molnár, *J. Chromatogr.*, **125**, 129, 1976. With permission.)

solvent for a given solute provided, κ^e and the mole volume of the eluent remains constant. The surface tension of inorganic salt solutions in water is to a good approximation a linear function of the salt concentration. According to Eq. (3.35), the log k' values should linearly increase with the salt concentration. The experimental findings by Horváth et al. [8] are given in Fig. 3.10. These authors [8] consider the relationships obtained as linear; however, it may be questioned for several of their solutes at the KCl concentration range studied. The slopes of the lines for the individual closely related solutes are similar, indicating similar contact areas. The intercepts depend on the van der Waals terms and the electrostatic interaction term in view of Eq. (3.33). The intercepts decrease with the hydrocarbonaceous surface area (HSA) of the solutes and only homovanillic acid of HSA = 119A^2 is an exception. It should be noted here that at the eluent pH most species carry a net charge, and only homovanillic acid does not have an ionized amino group at the pH studied.

Horváth et al. [22] further developed their solvophobic theory of reversed-phase liquid chromatography. There are several other theoretical descriptions of chromatographic behavior in reversed-phase systems, (e.g. the approach Martire and Boehm [23] based on molecular statistical theory). Unfortunately, like the theoretical approaches discussed here, the complex nonempirical descriptions of chromatrographic processes require knowledge of a number of physiochemical parameters that are mostly not available for individual chemical compounds. In fact, the experimental verification of the theories published so far is practically possible only assuming that all the simplifications and approximations applied really hold. As far as the retention of a given group of chemically defined solutes is concerned with respect to their structure, the general conclusion may be drawn from all the approaches discussed, namely, that the differences observed in retention are determined by differences in the polar and nonspecific (mass, bulk) properties of the solutes. That conclusion resembles the suggestion by Kováts and Weiss [15] put forward more than 20 years ago about the importance of dipole moment and molecular mass for gas chromatographic retention. In fact, the term *polar* is still recognized intuitively rather than being defined precisely. The polar interactions are complex, and to characterize them by means of a single parameter is difficult (if possible) for diverse sets of solutes. The so-called nonspecific properties reflect most probably the abilities of the compounds to participate in dispersive interactions and seem to be unequivocally determinable. Studies on the role of the individual molecular and submolecular structural features of the substances engaged in the chromatographic process by the method of QSRR should afford information that in turn would form the basis of a general comprehensible chromatographic theory.

References

1. R. P. W. Scott, The role of molecular interactions in chromatography, *J. Chromatogr.*, **122**, 35, 1976.

2. B. L. Karger, L. R. Snyder, and C. Econ, An expanded solubility parameter treatment for classification and use of chromatographic solvents and adsorbents. Parameters for dispersion, dipole and hydrogen bonding interactions, *J. Chromatogr.*, **125**, 71, 1976.

3. L. R. Snyder, The role of dispersion interactions in adsorption chromatography, *J. Chromatogr.*, **36**, 455, 1968.

4. L. R. Snyder and J. J. Kirkland, *Introduction to Modern Liquid Chromatography*, Wiley, New York, 1974, p. 216.

5. J. H. Hildebrand and R. L. Scott, *The Solubility of Nonelectrolytes*, Dover, New York, 1964.

6. E. sz. Kováts, Zu Fragen der Polarität. Die Methode der Linearkombination der Wechselwirkungskräfte (LKWW), *Chimia*, **22**, 459, 1968.

7. R. Tijssen, H. A. H. Billiet, and P. J. Schoenmakers, Use of the solubility parameter for predicting selectivity and retention in chromatography, *J. Chromatogr.*, **122**, 185, 1976.

8. Cs. Horváth, W. Melander, and J. Molnár, Solvophobic interactions in liquid chromatography with nonpolar stationary phases, *J. Chromatogr.*, **125**, 129, 1976.

9. E. D. Katz, K. Ogan, and R. P. W. Scott, Distribution of a solute between two phases. The basic theory and its application to the prediction of chromatographic retention, *J. Chromatogr.*, **352**, 67, 1986.

10. R. P. W. Scott and P. Kucera, Solute interactions with the mobile and stationary phases in liquid–solid chromatography, *J. Chromatogr.*, **112**, 425, 1975.

11. J. H. Knox and A. Pryde, Performance and selected applications of a new range of chemically bonded packing materials in high-performance liquid chromatography, *J. Chromatogr.*, **112**, 171, 1975.

12. R. J. Laub and J. H. Purnell, Criteria for the use of mixed solvents in gas–liquid chromatography, *J. Chromatogr.*, **112**, 71, 1975.

13. E. Soczewiński and C. A. Wachtmeister, The relation between the composition of certain ternary two-phase solvent systems and R_M values, *J. Chromatogr.*, **7**, 311, 1962.

14. L. R. Snyder and H. Poppe, Mechanism of solute retention in liquid–solid chromatography and the role of the mobile phase in affecting separation. Competition versus "sorption," *J. Chromatogr.*, **184**, 363, 1980.

15. E. sz. Kováts and P. B. Weiss, Über den Retentionsindex und seine Vervendung zur Aufstellung einer Polaritätsskala für Losungsmittel, *Ber. Bunsenges. Phys. Chem.*, **69**, 812, 1965.

16. C. J. F. Bötcher, *Theory of Electric Polarization*, Elsevier, New York, 1975.

17. L. Rohrschneider, Der Lösungsmitteleinfluss auf die gas-chromatographische Retention gelöster Stoffe, *Fortschr. Chem. Forsch.* **11**, 146, 1968.

18. P. T. Funke, E. R. Malinowski, D. E. Martire, and L. Z. Pollara, Application of factor analysis to the prediction of activity coefficients of nonelectrolytes, *Separ. Sci.*, **1**, 661, 1966.

19. O. Sinanoğlu and S. Abdulnur, Effect of water and other solvents on the structure of biopolymers, *Fed. Proce. Suppl. 15*, **24**(2), 12, 1965.

20. T. Halicioğlu, and O. Sinanoğlu, Solvent effects on cis–trans azobenzene isomerization: A detailed application of a theory of solvent effects on molecular association, *Ann. N.Y. Acad. Sci.*, **158**, 308, 1969.

21. O. Sinanoğlu, The C-potential surface for predicting conformations of molecules in solution, *Theor. Chim. Acta*, **33**, 279, 1974.

22. Cs. Horváth, W. Melander and I. Molnár, Liquid chromatography of ionogenic substances with nonpolar stationary phases, *Anal. Chem.*, **49**, 142, 1977.

23. D. E. Martire and R. E. Boehm, Molecular theory of liquid adsorption chromatography, *J. Liq. Chromatogr.*, **3**, 753, 1980.

CHAPTER

4

DETERMINATION OF CHROMATOGRAPHIC RETENTION DATA FOR QUANTITATIVE STRUCTURE–RETENTION RELATIONSHIP STUDIES

The starting condition for derivation of meaningful quantitative relationships between retention data and molecular structure descriptors is that the chromatographic data considered are reliable, reproducible, and precise.

As pointed out in Chapter 1, the retention parameters used in QSRR studies are assumed to be proportional to the free-energy change associated with the chromatographic distribution process. The quantitative structure–retention relationships are a manifestation of linear free-energy relationships (LFER). For LFER to be found between real and model systems, the so-called enthalpy–entropy compensation should be observed with a family of compounds. The method of verification of enthalpy–entropy compensation in reversed-phase HPLC has been described in detail in the chromatographic and related literature [1–3]. However, the procedure is time consuming, and the temperature adjustment must be feasible with the equipment used. In such a situation the enthalpy–entropy compensation is analyzed very occasionally only, and LFER are assumed a priori.

QSRR are derived statistically, and thus the error in determination of the quantities correlated may either produce relationships statistically more significant or conceal the true interdependences. Actually, the most difficult problem for QSRR students is the quantitation of chemical structural parameters of the solutes concerned. To achieve final reliable, measurable (calculable) molecular descriptors is the main task of all structure–activity relationship studies. Nonetheless, the dependent variables (i.e., retention parameter, bioactivity data, etc.) have to be determined as accurately as possible. Certainly, chromatographic retention can be quantified much better than, generally, biological activity. On the other hand, there is no reason to believe that in a living organism the drug–receptor interactions involve a special type of intermolecular interaction, different than those observed in, for example, chromatography. If QSRR are unequivocally derived, the other more complex QSA/activity/R will subsequently be formulated.

The retention parameters derived by different chromatographic techniques and modes are now discussed from the point of view of their usefulness for QSRR studies.

4.1. GAS CHROMATOGRAPHY

Soon after introduction of gas chromatography (GC) in the early 1950s [4] it became apparent that a uniform system of data presentation was necessary to give the accuracy required for the comparison of data between laboratories. This requirement is especially important in the case of QSRR studies.

In order to overcome the problem of the uniform reporting of retention data, Kováts [5] proposed a system using the n-alkanes as a series of standards. The retention of other substances has been expressed relative to these standards. Confirming a suggestion by Ettre [6] that the originally proposed retention volumes could be replaced by adjusted retention times, Kováts [7] introduced his widely known formula

$$I_i(T) = 100 \times \frac{\log t'_{Ri} - \log t'_{Rz}}{\log t'_{R(z+1)} - \log t'_{Rz}} + 100z \qquad (4.1)$$

In Eq. (4.1) $I_i(T)$ is the Kováts retention index characteristic of a substance i chromatographed on a given column (stationary phase) at a definite temperature; t'_{Rz} is the adjusted retention time of a homologue standard with carbon number z; $t'_{R(z+1)}$ is an analogous parameter for another standard with carbon number $z + 1$; and t'_{Ri} is the adjusted retention time of a substance i.

In GC the property directly measured is the total retention time t_R. This value is a sum of two factors: (1) dead time t_M, which is dependent on the system flow rate as well as on the void volume of the apparatus, and (2) the adjusted retention time t'_R, which is independent of the equipment used and characterizes the separation process

$$t_R = t_M + t'_R \qquad (4.2)$$

Thus, the adjusted retention time t'_R is the quantity of interest that could be related to solute structure. However, although independent of the equipment used, adjusted retention times are still dependent on several variables such as column temperature, flow rate, pressure drop, and liquid phase. When adjusted retention times are used to calculate Kováts indices, the retention parameters are obtained, which depend only on the column temperature and stationary phase used.

According to Eq. (4.2), the adjusted retention time can be calculated from the measured retention time after subtraction of the column dead time. Thus, the dead time t_M must be known for a given column. There are many papers published in the chromatographic literature that deal with various methods for determining dead time. The problem is discussed extensively in more recent reviews by Ettre [8] and Haken and co-workers [9, 10]. Here, the updated discussion by Haken et al. [10] is utilized as the complete one. It should be added here that there are a few theoretical methods [11] of calculating retention indices that do not require a knowledge of the adjusted retention times. These methods have not been applied in practice, however.

James and Martin [4] designated the air peak as the dead time and measured the retention times of substances being analyzed as the distances from this peak. Later a discussion started concerning the proper substance to use for dead-time measurement. The problem has been especially acute in the case of flame ionization detectors, which are insensitive to air or an inert gas. For such detectors McReynolds [12] has suggested methane as the t_M marker. However, the use of air or methane causes experimental problems, and moreover, some net retention is observed that is especially significant with methane. As found in an extensive study by Kováts et al. [13] on a series of substances at temperatures between 30 and 230°C, neon had the smallest gross retention volume. The retention volumes of nitrogen, hydrogen, and argon, were slightly greater than for neon, while the light hydrocarbons tested had much larger gross retention volumes. Keeping the above in mind, Haken and co-workers [10] conclude that the use of methane under conditions of low retention for the calculation of Kováts indices above 400 is unlikely to introduce large errors. The retention of methane is larger at low temperatures, in gas–solid chromatography, and at high pressures. Thus, the use of methane at low temperatures on porous polymers may introduce significant error for all but very strongly retained substances. In the gas chromatography–mass spectrometry system multiple analyses using neon should generate most reliable and accurate dead times. Haken et al. [10] also conclude that mathematical methods of dead-time calculation should be used when flame ionization detectors are applied, due to their insensitivity toward permanent gases such as neon and the error resulting from methane retention.

Various mathematical methods of determination of adjusted retention times rely on the accepted linear relationship between ln t'_{Rz} and retention index I. Under isothermal conditions Kováts indices can be calculated by interpolation within the linear relationship

$$\ln t'_{Rz} = bI + c \qquad (I = 100z) \qquad (4.3)$$

where b and c are constants and the Kováts retention indices for n-alkanes are defined as 100 times the carbon number for each liquid phase at all temperatures.

The method of t_M calculation recommended by Haken et al. [10] is that presented by Guardino et al. [14]. In the method an iteration is carried out on t_M, with b and c [Eq. (4.3)] calculated using a least-squares fit. A flow chart of the method is shown in Fig. 4.1.

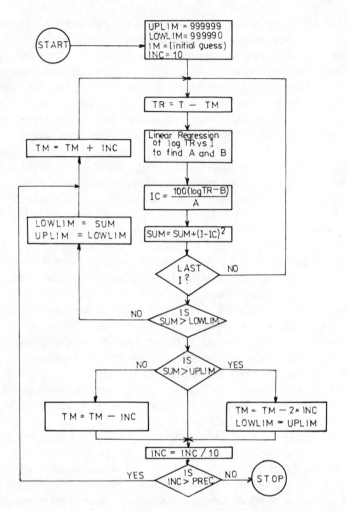

Figure 4.1. Flow chart for calculation of dead time. UPLIM and LOWLIM are upper and lower limits, respectively, of sum of squares of deviation; TM is dead time; INC is increment in deadtime; IC is calculated Kováts index; SUM is sum of squares of deviations; TR is unadjusted retention time of homologues; I is known Kováts index ($I = 100z$); PREC is precision required. (After X. Guardino, J. Albaigés, G. Firpo, R. Rodriguez-Viñals, and M. Gassiot, *J. Chromatogr.*, **118**, 13, 1976. With permission.)

The initial estimate of the dead time (which must be less than the true dead time) is used to determine adjusted retention times. A linear regression then allows b and c to be calculated, and thus retention indices can be determined.

Subtracting these from the known values gives a sum of differences that is compared to the upper and lower limits. If the estimate of t_M is below the lower limit, the limits are reduced and the estimate of t_M is increased. When this estimate increases above the lower limit, it is decreased and the increment is lowered by a factor 10. The procedure is repeated until the increment is less than the precision required.

Although Eq. (4.3) is assumed to hold generally over the entire range of carbon numbers, nonlinearity has been observed experimentally [15] and predicted thermodynamically [16].

Heeg et al. [17] suggested a cubic relationship between adjusted retention time and carbon number. The methods of determination of retention index that are designed to fit cubic and higher degree polynomials are relatively complex and suffer from round-off and subtraction cancellation errors [10, 18].

Certainly, the reproductivity of Kováts indices depends not only on the mathematical processing of the raw chromatographic data but also on the original data. With a well-designed experimental technique and an accurate timing mechanism (nearest tenth of a second), an interlaboratory reproducibility of one unit for larger values of Kováts indices and two units for indices below about 400 is possible [10, 19].

- The temperature dependence of Kováts retention indices has been discussed extensively by Novak et al. [20] and Budahegyi et al. [21]. As observed by Wehrli and Kováts [23] in 1959, a temperature dependence of less than one unit per degree applies in most instances. However, the dependence of retention indices on column temperature can be correctly described by an Antoine-type hyperbolic curve:

$$I(T) = A + \frac{B}{T + C} \tag{4.4}$$

where T is the column temperature, and A, B, and C are constants. In the case of nonpolar solutes chromatographed on nonpolar stationary phases, the retention indices show an almost completely linear dependence on column temperature. Also, in other systems the curve according to Eq. (4.4) has a significant linear portion. Its length depends on the polarity of a solute and on the stationary phase applied. In practice, three retention indices $(I_1, I_2,$ and $I_3)$ determined at temperatures T_1, T_2, and T_3 are used for the calculation of the constants in Eq. (4.4) [22]:

$$C = \frac{(T_3 - T_1)(I_3 T_3 - I_1 T_1) + (T_3 - T_1)(I_1 T_1 - I_2 T_2)}{(T_3 - T_1)(I_2 - I_1) - (T_2 - T_1)(I_3 - I_1)} \tag{4.5}$$

$$A = \frac{I_2 T_2 - I_1 T_1 + C(I_2 - I_1)}{T_2 - T_1} \qquad (4.6)$$

$$B = (I_2 - A)(T_2 + C) \qquad (4.7)$$

A parameter from GC that is often considered in QSRR studies is the difference of Kováts indices determined for a given solute on two stationary phases at identical temperatures. This parameter ΔI is usually derived using the phases of highly different polarity.

As is evident from the material discussed in this chapter, determination of gas chromatographic data designed for QSRR studies would require substantial effort.

According to Budahegyi et al. [21], there exist the following sources of error and errors in retention index determination: (1) errors in the determination of retention times (measurement errors, influence of the amount of sample, operator errors); (2) errors in the measurement or calculation of the gas holdup time (e.g., calculation is carried out in the nonlinear range of n-alkanes); (3) the support is not perfectly coated, or it is active; there are wall effects in the open-tubular (capillary) column; (4) an inhomogenous stationary phase; and (5) fluctuations of instrumental and/or gas chromatographic parameters. According to the same authors [21], the determination of the retention indices of highly polar substances on nonpolar stationary phases and the determination of the retention indices of nonpolar substances, including n-alkanes, on highly polar stationary phases are problematic, especially with respect to accuracy and reproducibility.

The QSRRs reported so far (as discussed subsequently) concern the GC data routinely determined. Because the correlations obtained are often highly significant, there is a chance that they can be improved if precise chromatographic data are collected. With proper retention data the QSRRs derived from GC can be especially useful for characterization of intermolecular interactions.

4.2. HIGH-PERFORMANCE LIQUID CHROMATOGRAPHY

The measure of the degree of retention of a solute in HPLC is the so-called phase capacity ratio or capacity factor k'. A scale of retention indices, resembling the Kováts indices in GC, has also been proposed [24] for the calibration of reversed-phase systems and will be briefly described later in this section.

The capacity factor k' is defined by

$$k' = \frac{t_R - t_M}{t_M} = \frac{V_R - V_M}{V_M} \tag{4.8}$$

where t_R and V_R are the retention time and retention volume, respectively, of the solute chromatographed; t_M and V_M are the retention time and retention volume, respectively, of a so-called unretained solute. The logarithm of the capacity factor, $\log k'$, can be considered as a quantity analogous to the well-known thin-layer chromatography parameter R_M as defined by Bate-Smith and Westall [25], that is, $R_M = \log(1/R_F - 1)$.

So that the formula for the capacity ratio [Eq. (4.8)] can be used for QSRR purposes and data compared between laboratories, the dead volume V_M should be readily determined with adequate precision. However, determination of the dead volume of a column in liquid chromatography poses some problems. In a broad sense, V_M is very often stated to be the elution volume of an *unretained* and *unexcluded* solute. Unfortunately, this statement begs many questions and has led to much confusion, especially in regard to how V_M should be measured. The problem is to decide which, if any, solute is both unretained and unexcluded. Various experimental techniques of determination of the dead volume in HPLC were reviewed by Berendsen et al. [26] and by Wells and Clark [27]. The problem has also been a subject of a recent publication by Knox and Kaliszan [28], where the following were listed as recipes for the determination of V_M found in the literature: (1) V_M is the elution volume of a solvent disturbance or system peak obtained by injecting an eluent component; (2) V_M is the elution volume of an un-ionized solute that gives the lowest retention volume and is small enough not to be sterically excluded; (3) V_M is the elution volume of an isotopically labeled component of eluent, for example, ^2H_2O, in the case of a reversed-phase packing material; (4) V_M is the elution volume of salt or ion, usually a UV-absorbing ion; (5) V_M is the volume of liquid the column contains (obtained, e.g., by weighing the column full of liquid and then empty) less the volume of any adsorbed eluent components; and (6) V_M is the volume that, when subtracted from the elution volumes V_{Rn} of a series of homologues, provides a linear dependence of $\log(V_{Rn} - V_M)$ against n, the number of carbon atoms in the homologues.

Unfortunately, none of the above statements give an acceptable definition of V_M. Assuming that the thermodynamic dead volume V_M is the total volume of all eluent components within the column bed, Knox and Kaliszan [28, 29] introduced a precise method of measurement of that quantity. The method consists in the measurements of the retention volumes V_R of isotopically labeled samples of *all* eluent components and calculating V_M using the formula

$$V_M = V_A* x_A + V_B* x_B + \cdots \qquad (4.9)$$

where x_A, x_B, $\cdots$ are the volume fractions of components A, B, $\cdots$ in the eluent and V_A*, V_B*, $\cdots$ are the retention volumes of radiolabeled samples of A, B, $\cdots$.

The procedure can be illustrated as follows. A mixture of acetonitride (A) and water (B) (75:25 v/v) is fed into a column packed with a chemically bonded octadecylsilica material. The column is equilibrated. This mixture is suddenly replaced by a mixture of identical composition but containing trace quantities of isotopically labeled components, that is, $^{14}CH_3CN$ and ^3H_2O, denoted by A* and B*. It is assumed that the distribution coefficients for molecules of the labeled components are identical to those of the unlabeled materials. The breakthrough volumes V_A* and V_B* of the labeled eluent fronts can be detected by a scintillation counter (Fig. 4.2).

A theoretical expression for the breakthrough volume of each labeled solute is obtained by considering the volume of eluent fed to the column between the time when labeled eluent first meets the column and the appropriate front emerges from the end of the column. Since the amount of any labeled

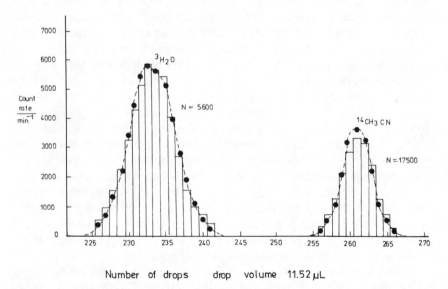

Figure 4.2. Histograms of count rates for individual drops from elution of radiolabeled water (^3H_2O) and acetonitrile ($^{14}CH_3CN$) from a 250 × 5-mm-i.d. column packed with ODS Hypersil. Eluent: acetonitrile–water (75:25, v/v). Flow rate: 1.06 mL/min. N value is number of plates to which column is equivalent. (After J. H. Knox and R. Kaliszan, *J. Chromatogr.*, **349**, 211, 1985. With permission.)

component X^* must be conserved, it can be stated with complete certainty that the amount of X^* fed to the column between the time when the labeled eluent first enters the column and when its front leaves the column equals the amount of X^* that will be found within the column at equilibrium. Thus, it can be written for each labeled component that

$$V_A * x_A = V_M y_A \qquad V_B * x_B = V_M y_B \cdots \qquad (4.10)$$

where $y_A, y_B, \cdots$ are the volume fractions of components A, B, $\cdots$ of the eluent within the packed bed after the column is fully equilibrated. Although y_A, $y_B, \cdots$ will in general differ from $x_A, x_B, \cdots$ (see Fig. 4.3), both x and y are volume fractions, and

$$x_A + x_B + \cdots = 1 \qquad y_A + y_B + \cdots 1 \qquad (4.11)$$

Thus, from Eqs. (4.10) and (4.11) the formula for determining V_M [Eq. (4.9)] may be obtained.

In the work by Knox and Kaliszan [28] it has been established experimentally that, over a wide range of composition and with a range of eluent components, Eq. (4.9) gives essentially identical values of V_M. In practice, it would be simplest to determine V_M by flushing the column with a one-component eluent and determining the simple $V_R *$ of an isotopically labeled sample.

The determination of V_M by the isotope labeling method is time consuming and requires the use of a specific detector. In such a situation a subsidiary standard is desirable. Such a subsidiary standard could be a solvent disturbance peak. However, as documented by Knox and Kaliszan [28], special conditions must be fulfilled. It has been demonstrated that the elution volume of the solvent disturbance peak does not provide a correct value of V_M, except by chance. However, if the eluent components are more or less equally sorbed by the packing, the solvent disturbance peak volume may not be seriously in error. If one uses the solvent disturbance peak as a subsidiary standard for V_M, the absolute retention volume of this peak should be measured directly as a volume for each eluent composition and related to the true V_M in a special experiment.

To calculate reliable capacity coefficients k' in liquid chromatography, not only the column dead volume V_M must be known, but also a solute retention volume V_R should be precisely determined. Chromatographic data for QSRR analysis of a series of solutes are usually determined at identical conditions. However, the chromatographic behavior of individual solutes from the series may be different than the others at the chromatographic system applied. There is a possibility that some solutes may be excluded from a porous matrix on the

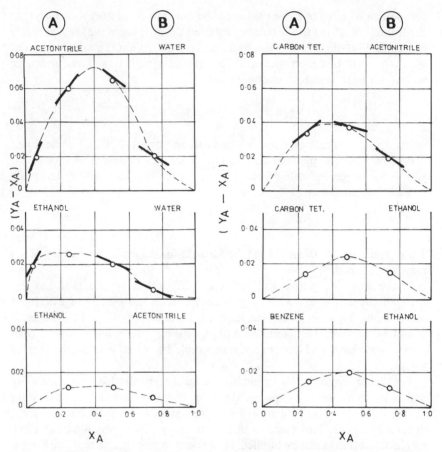

Figure 4.3. Partition isotherms for binary mixtures on ODS Hypersil. Points and broken lines derived from retention volumes of radiolabeled components. Heavy, short lines indicate gradients, calculated from retention volumes of solvent disturbance peaks. (Reprinted from J. H. Knox and R. Kaliszan, *J. Chromatogr.*, **349**, 211, 1985. Courtesy of Marcel Dekker, Inc.)

basis of unfavorable enthalpy changes; that is, they may be eluted before V_M, and thus their capacity ratio k' may be negative.

Two forms of enthalpic exclusion have been demonstrated by Knox et al. [29]. The first is the exclusion of nonpolar solutes, such as pentane, from a polar adsorbent (Corning porous glass) when eluted with a highly polar eluent such as ethanol. The degree of exclusion closely follows the eluotropic series as established by Snyder [30] (Fig. 4.4). The second type of exclusion is that of anionic solutes eluted from a reversed-phase packing material by water–ethanol mixtures. That type of exclusion strongly depends on the ionic strength of the solution (Fig. 4.5).

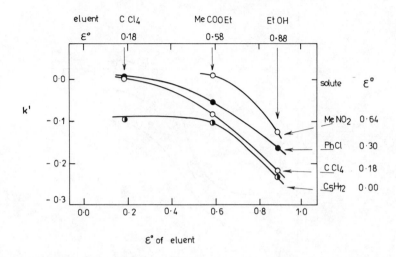

Figure 4.4. Exclusion of nonpolar solutes eluted by polar eluents from 44–53-μm chips of Corning porous glass CPG 75. Values of ε^0 are taken from ref. 30. (After J. H. Knox, R. Kaliszan, and G. F. Kennedy, *J. Chem. Soc., Faraday Symp.*, **15**, 113, 1980. With permission.)

Exclusion of the first type arises because of the difficulty a nonpolar solute finds in displacing a strongly adsorbed polar molecule from the surface of the polar adsorbent. Exclusion of the second type arises from the presence of fixed negative charges, probably due to $\equiv$Si–O$^-$ groups within the matrix of the packing material. The presence of these groups causes the cationic solutes to be strongly retained.

Any general model of QSRR should consider the influence of the mobile phase composition and temperature. Soczewiński et al. [31, 32] have found the following equation to be frequently obeyed in paper chromatography:

$$\log k' = \log k'_{H_2O} x_{H_2O} + \log k'_{org} x_{org} \qquad (4.12)$$

where k' is the capacity factor obtained with a given mixed aqueous–organic mobile phase; k'_{H_2O} and k'_{org} are capacity factors obtained with pure water and a pure organic mobile phase, respectively. The terms x_{H_2O} and x_{org} are volume fractions of water and organic solvent, respectively. Such linear behavior for similar groups of substances of $\log k'$ versus x_{H_2O} over a limited range of concentrations of water in a binary mobile phase has been commonly observed in HPLC [33]. However, this dependence is nonlinear over the full composition range of the mobile phase. In any liquid–solid chromatographic

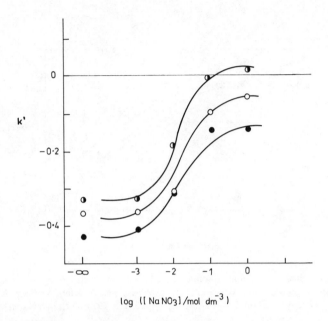

Figure 4.5. Dependence on ionic strength of capacity ratios, k', for benzoic acid (●), salicylic acid (○), and sulfanilic acid (●) eluted from 5 μm ODS Hypersil (Shandon Southern Products, U.K.). Eluent: water–ethanol (70:30 v/v). (After J. H. Knox, R. Kaliszan, and G. F. Kennedy, *J. Chem. Soc., Faraday Symp.*, **15**, 113, 1980. With permission.)

system utilizing silica gel as the stationary phase and a mobile phase consisting of a binary mixture of a nonpolar and a polar solvent, Scott and Lawrence [34] and Maggs [35] observed a nonlinearity of log k' versus the volume fraction of the polar solvent at small concentrations of the polar solvent. Linearity has not been observed before a polar solvent is in excess of 3% w/v, that is, when the concentration of polar solvent is sufficient to deactivate the silica gel.

There have been numerous attempts to relate quantitatively capacity factors and mobile phase composition, especially with regard to reversed-phase HPLC using binary aqueous–organic mobile phases. The best known are the recent approaches by Jandera et al. [36], Schoenmakers et al. [37], and Horváth et al. [38]. For illustration the results obtained by Schoenmakers et al. [37] are given in Fig. 4.6. The analytical relationship found recently by Horváth et al. [38] has the form

$$\log k' = A_1 x(1 - T_c/T) + A_2/T + A_3 + A_4 f(x, T) \qquad (4.13)$$

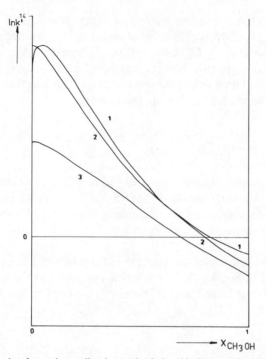

Figure 4.6. Example of experimentally observed relationship between logarithm of reversed-phase HPLC capacity factor, ln k', and volume fraction x_{CH_3OH}, of methanol in methanol–water system. Solutes: (1) naphthalene; (2) benzophenone; (3) p-chlorophenol. (After P. J. Schoenmakers, H. A. H. Billiet, and L. de Galan, *J. Chromatogr.*, **282**, 107, 1983. With permission.)

where x is the volume fraction of the organic co-solvent in the hydroorganic mobile phase; T is absolute temperature and T_c is the so-called compensation temperature derived by taking advantage of enthalpy–entropy compensation [39]. The function $f(x, T)$ may be appropriate either to members of an eluite class (e.g., anions) or to all eluites. Horváth et al. [38] tested various forms of the function $f(x, t)$ [e.g., $f(x, T) = 0$, $f(x, T) = (1 - 0.5x)^{-1}$, $f(x, T) = (x/T)$. Compensation temperature T_c was assumed constant in reversed-phase chromatography and equals 625 K. Coefficients A_1–A_4 of Eq. (4.13) have been obtained by linear regression of $\log k'$ versus the appropriate functions of the mobile phase composition and temperature. The relative error was 6% with use of Eq. (4.13) as compared to 7.8% when a simplified equation was used comprising only the first three terms of Eq. (4.13).

To unify the comparison of reversed-phase HPLC data produced at different chromatographic conditions for compounds of varying polarity,

Baker [40–42] has introduced a retention index scale similar to that used in GC according to Kováts. The scale is based on the relative retention of a homologous series of C_3–C_{23} 2-keto alkanes. The retention index I_i of a given solute i is calculated from its observed capacity factor k_i', the capacity factor for a 2-keto alkane eluting before the test compound, k_N', and the capacity factor of a 2-keto alkane eluting after the test compound, k_{N+1}':

$$I_i = 100 \times \frac{\log k_i' - \log k_N'}{\log k_{N+1}' - \log k_N'} + 100\,N \qquad (4.14)$$

The retention index of a given 2-keto alkane standard is by definition equal to 100 times the carbon number N in the formula. It was found that the retention index of a given compound remains nearly constant, although its retention time may vary significantly with the changes in the composition of the eluent (Fig. 4.7).

A slight modification of the retention index scale introduced by Baker has been proposed by Smith [43, 44]. In Smith's modification the terminal methyl group of the 2-keto alkane series of standards has been replaced by a terminal

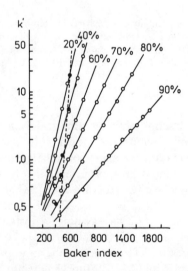

Figure 4.7. Effect of methanol content in aqueous solvent on capacity factor k' for ($\bigcirc$) 2-keto alkanes standards and ($\bullet$) phenacetin. Bake retention index is nearly independent of eluent changes. Column: μ Bondapak C_{18} (Waters Assoc., Milford, MA), pH 7.0; 0.025 M NaH$_2$PO$_4$ aqueous buffer with varying methanol content. (After J. K. Baker and C. -Y. Ma, *J. Chromatogr.*, **169**, 107, 1979. With permission.)

phenyl, which makes the UV detection of the standards somewhat easier. Both HPLC retention indices have been used for QSRR studies.

4.3. THIN-LAYER CHROMATOGRAPHY AND PAPER CHROMATOGRAPHY

Thin-layer chromatography (TLC) and paper chromatography (PC) require considerably less laboratory equipment than column chromatography. Besides, many solutes can be simultaneously analyzed. On the other hand, however, reproducibility of the R_f values is limited. The situation with TLC is relatively better than in PC. The use of well-defined small-diameter particles of stationary phases and a better knowledge of the parameters that determine the efficiency of the chromatographic system have given rise to the term high-performance TLC (HPTLC), by analogy with HPLC, although TLC is not capable of performances comparable to those of column chromatography [45–47].

The conditions usual in normal TLC lead to the characteristic elongated and irregular spots. HPTLC works under such conditions that molecular diffusion determines the spot diameter. Thus, spots are circular, and in chromatograms all spots tend to have similar diameters unless diffusion coefficients vary widely from one compound to another [46].

At first only thin-layer plates precoated with silica were commercially available. As is the case with silica material, if low concentrations of a polar solvent are employed in the mobile phase (less than 3% w/v), a significant proportion of the solvent is adsorbed by the silica gel in the deactivation process. Thus, the concentration of polar solvent in the solvent front is significantly reduced, producing a quasi-gradient elution effect [48]. In fact, this should only occur for low concentrations of polar solvent, and its effect is only significant on fast-moving peaks that follow closely the solvent front for a significant proportion of the development. However, the silica-coated plates are generally not suitable for the analysis of very polar solutes. For such an analysis usually complex mixtures of solvents, impossible to reproduce in practice, have to be used [47]. The development of TLC plates by such mixtures gives rise to the phenomenon of demixing. The chromatograms of polar compounds then exhibit strong tailing and irregular spot shapes. Moreover, whereas standard solutes appear well separated, the resolution of complex mixtures is often poor.

Several of the above disadvantages of the precoated HPTLC plates have been eliminated or reduced since TLC plates precoated with nonpolar chemically bonded phases became commercially available. Obviously, HPTLC is still much less reproducible and less sensitive than HPLC. As there may be a selective adsorption of mobile phase components in TLC, a quasi-gradient elution may be observed for less retained solutes. The same is true for buffer components. In contrast, HPLC provides precise control of the pH and ionic strength during the separation process; temperature control is also easier in HPLC than in TLC.

The retention parameter derived from TLC and PC is the R_M value defined by Bate-Smith and Westall [25]:

$$R_M = \log(1/R_f - 1) \tag{4.15}$$

where R_f is the ratio of a distance passed by the solute to that attained by the solvent front. Accuracy in TLC and PC demands $0.2 < R_f < 0.8$. This gives a range of R_M of less than one and a half decades, as compared to three to four decades in the case of HPLC [49]. In such a situation, it is often necessary to change the composition of the mobile phase for an individual solute in order to obtain reliable R_M values when one deals with a series of compounds of various polarities. This is the case especially in hydrophobicity determinations by the reversed-phase HPTLC with aqueous–organic mobile phases. Then, there arises the problem of linearity of the relationship between R_M and volume fraction of organic component in a water–organic eluent.

Much controversy has been evoked concerning the R_M values extrapolated to the pure water mobile phase (i.e., the R_M^0 parameter). One would expect the R_M^0 value to be identical on a given stationary phase, independent of the water–organic system used for the R_M determinations and the subsequent extrapolation to R_M^0. However, this is not always the case in practice. Certainly, it is extremely difficult to control all the factors that can influence TLC and PC separations. Certainly, it is difficult to interpret the relationship between the mobile phase composition and R_M values as illustrated in Fig. 4.8 [50]. More difficult is the decision of how R_M^0 should be determined from such a relationship.

Reversed-phase PC and especially TLC techniques are very often used for evaluation of the organic–water partitioning properties of solutes. A separate chapter is devoted to the quantitative relationships between chromatographic retention data and both experimentally and theoretically derived measures of solute hydrophobicity. Because relationships of that type are basically derived experimentally, the details of the procedures applied as well as their advantages and limitations will be described in the same chapter. Here, the advantages of TLC listed by Stahl [51], the inventor of the technique, are given because QSRR studies involving data derived by this method are still flourishing. According to Stahl [51], TLC offers the greatest freedom in the easy choice of stationary and mobile phases; it offers a large number of different development techniques; it offers the largest number of methods of detection; in TLC it is

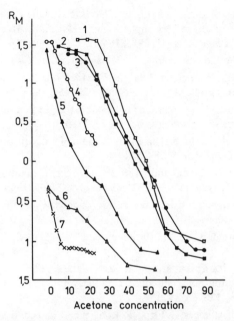

Figure 4.8. Relationship between R_M values and acetone concentration in acetone–water mobile phase. TLC silica plates impregnated with silicone oil. Solutes: (1) testosterone propionate; (2) prazepam; (3) 1-(pyridin-2-azo)-2-naphthol; (4) dicloxacillin; (5) dermorphin; (6) 1,3-dihydroxy-benzene; (7) carbenicillin. (After M. C. Pietrogrande, C. Bighi, P. A. Borea, A. M. Barbaro, M. C. Guerra, and G. L. Biagi, *J. Liq. Chromatogr.*, **8**, 1711, 1985. With permission.)

possible to separate and visualize many samples and reference mixtures simultaneously; and TLC is the simplest and most economical chromatographic technique for fast separation and visual evaluation.

References

1. W. Melander, D. E. Campbell, and Cs. Horváth, Enthalpy–entropy compensation in reversed-phase chromatography, *J. Chromatogr.*, **158**, 213, 1978.

2. E. Tomlinson, H. Poppe, and J. C. Kraak, Thermodynamics of functional groups in reversed-phase high-performance liquid–solid chromatography, *Int. J. Pharm.*, **7**, 225, 1981.

3. Gy. Vigh and Z. Varga-Puchony, Influence of temperature on the retention behaviour of members of homologous series in reversed-phase high-performance liquid chromatography, *J. Chromatogr.*, **196**, 1, 1980.

4. A. T. James and A. J. P. Martin, Gas–liquid partition chromatography: The separation and micro-estimation of volatile fatty acids from formic acid to dodecanoic acid, *Biochem. J.*, **50**, 679, 1952.

5. E. Kováts, Gas-chromatographische Charakterisierung organischer Verbindungen. Teil 1: Retentionsindices aliphatischer Halogenide, Alkohole, Aldehyde und Ketone, *Helv. Chim. Acta*, **41**, 1915, 1958.

6. L. S. Ettre, The Kováts retention index system, *Anal. Chem.*, **36**(8), 31A, 1964.

7. E. sz. Kováts, Zu Fragen der Polarität. Die Methode der Linearkombination der Wechselwirkungskräfte (LKWW), *Chimia*, **22**, 459, 1968.

8. L. S. Ettre, Generalized equations to evaluate the gas hold-up time of chromatographic systems, *Chromatographia*, **13**, 73, 1980.

9. M. S. Wainwright, and J. K. Haken, Evaluation of procedures for the estimation of dead time, *J. Chromatogr.*, **184**, 1, 1980.

10. R. J. Smith, J. K. Haken, and M. S. Wainwright, Estimation of dead time and calculation of Kováts indices, *J. Chromatogr.*, **334**, 95, 1985.

11. G. Tarján, Contribution to the question of precalculation of retention indices according to Takács, *J. Chromatogr. Sci.*, **14**, 309, 1976.

12. W. O. McReynolds, Characterization of some liquid phases, *J. Chromatogr. Sci.*, **8**, 685, 1970.

13. F. Riedo, D. Fritz, G. Tarján, and E. sz. Kováts, A tailor-made C_{87} hydrocarbon as a possible non-polar standard stationary phase for gas chromatography, *J. Chromatogr.*, **126**, 63, 1976.

14. X. Guardino, J. Albaiges, G. Firpo, R. Rodriguez-Viñals, and M. Gassiot, Accuracy in the determination of the Kováts retention index. Mathematical dead time, *J. Chromatogr.*, **118**, 13, 1976.

15. L. Rohrschneider, Zur Frage der Linearität der *n*-Alkan-Geraden, *Chromatographia*, **2**, 437, 1969.

16. R.-W. Bach, E. Dötsch, H. A. Friedrichs, and L. Marx, Zur Thermodynamischen Begründung der Kováts'schen Retentionsindizes, *Chromatographia*, **4**, 459, 1971.

17. F. J. Heeg, R. Zinburg, H. J. Neu, and K. Ballschmiter, Berechnung von Retentionsindices auf der Grundlage eines nichtlinearen Zusammenhanges zwischen Nettoretentionszeit und Kohlenstoffzahl von *n*-Alkanen, *Chromatographia*, **12**, 451, 1979.

18. R. J. Smith, J. K. Haken, and M. S. Wainwright, A problem with Kováts' retention index equation, *J. Chromatogr.*, **331**, 389, 1985.

19. R. J. Smith, J. K. Haken, M. S. Wainwright, and B. G. Madden, Comparison of mathematical methods for the calculation of Kováts indices, *J. Chromatogr.*, **328**, 11, 1985.

20. J. Novák, J. Vejrosta, M. Roth, and J. Jánák, Correlation of gas chromatographic specific retention volumes of homologous compounds with temperature and methylene number, *J. Chromatogr.*, **199**, 209, 1980.

21. M. V. Budahegyi, E. R. Lombosi, T. S. Lombosi, S. Y. Mészáros, Sz. Nyiredy, G. Tarján, I. Timár, and J. Takács, Twenty-fifth anniversary of the retention index system in gas–liquid chromatography, *J. Chromatogr.*, **271**, 213, 1983.

22. J. Takács, M. Rockenbauer, and J. Olácsi, Determination of the relationship between retention index and column temperature in gas chromatography through

the temperature-dependence of the net retention volume, *J. Chromatogr.*, **42**, 19, 1969.

23. A. Wehrli and E. Kováts, Gas-chromatographische Charakterisierung organischer Verbindungen. Teil 3: Berechnung der Retentionsindices aliphatischer, alicyclischer und aromatischer Verbindungen, *Helv. Chim. Acta*, **42**, 2709, 1959.

24. J. K. Baker, Estimation of the high pressure liquid chromatographic retention indices, *Anal. Chem.*, **51**, 1693, 1979.

25. E. C. Bate-Smith and R. G. Westall, Chromatographic behavior and chemical structure. I. Some naturally occurring phenolic substances, *Biochim. Biophys. Acta* **4**, 427, 1950.

26. G. E. Berendsen, P. J. Schoenmakers, L. de Galan, G. Vigh, Z. Varga Puchony, and J. Inczedy, On the determination of the hold-up time in reversed phase liquid chromatography, *J. Liq. Chromatogr.*, **3**, 1669, 1980.

27. M. J. M. Wells and C. R. Clark, Liquid chromatographic elution characteristics of some solutes used to measure column void volume on C_{18} bonded phases, *Anal. Chem.*, **53**, 1341, 1981.

28. J. H. Knox and R. Kaliszan, Theory of solvent disturbance peaks and experimental determination of thermodynamic dead-volume in column liquid chromatography, *J. Chromatogr.*, **349**, 211, 1985.

29. J. H. Knox, R. Kaliszan, and G. J. Kennedy, Enthalpic exclusion chromatography, *J. Chem. Soc., Faraday Symp.*, **15**, 113, 1980.

30. L. R. Snyder, *Principles of Adsorption Chromatography*, Marcel Dekker, New York, 1968.

31. E. Soczewiński and C. A. Wachtmeister, The relation between the composition of certain ternary two-phase solvent systems and R_M values, *J. Chromatogr.*, **7**, 311, 1962.

32. E. Soczewiński and G. Matysik, Two types of R_M–composition relationships in liquid–liquid partition chromatography, *J. Chromatogr.*, **32**, 458, 1968.

33. P. Jandera, Reversed-phase liquid chromatography of homologous series. A general method for prediction of retention, *J. Chromatogr.*, **314**, 13, 1984.

34. R. P. W. Scott and J. G. Lawrence, Gradient elution in liquid chromatography under conditions of axial equilibrium, *J. Chromatogr. Sci.*, **8**, 619, 1970.

35. R. E. Maggs, The role of temperature in liquid–solid chromatography: Some practical considerations, *J. Chromatogr. Sci.*, **7**, 145, 1969.

36. H. Colin, G. Guiochon, and P. Jandera, Interaction indexes and solvent effects in reversed-phase liquid chromatography, *Anal. Chem.*, **55**, 442, 1983.

37. P. J. Schoenmakers, H. A. H. Billiet, and L. de Galan, Description of solute retention over the full range of mobile phase compositions in reversed-phase liquid chromatography, *J. Chromatogr.*, **282**, 107, 1983.

38. W. R. Melander, B.-K. Chen, and Cs. Horváth, Mobile phase effects in reversed-phase chromatography VII. Dependence of retention on mobile phase composition and column temperature, *J. Chromatogr.*, **318**, 1, 1985.

39. W. Melander, D. E. Campbell, and Cs. Horváth, Enthalpy–entropy compensation

in reversed-phase chromatography, *J. Chromatogr.*, **158**, 213, 1978.

40. J. K. Baker, Estimation of high pressure liquid chromatographic retention indices, *Anal. Chem.*, **51**, 1693, 1979.

41. J. K. Baker and C.-Y. Ma, Retention index scale for liquid–liquid chromatography, *J. Chromatogr.*, **169**, 107, 1979.

42. J. K. Baker, G. J. Hite, M. Reamer, and P. Salva, Effect of stereochemistry on the high-performance liquid chromatographic retention index of azabicycloalkanes, *Anal. Chem.*, **56**, 2932, 1984.

43. R. M. Smith, Arylalkylketones as a retention index scale in liquid chromatography, *J. Chromatogr.*, **236**, 313, 1982.

44. R. M. Smith, Characterization of reversed-phase liquid chromatography columns with retention indexes of standards based on an alkyl aryl ketone scale, *Anal. Chem.*, **56**, 256, 1984.

45. A. Zlatkis and R. E. Kaiser (Eds.), *High Performance Thin-Layer Chromatography*, Elsevier, Amsterdam, 1977.

46. A. Siouffi, G. Guichon, H. Engelhardt, and I. Halasz, Study of the performance of thin-layer chromatography I. A phenomenological approach, *J. Chromatogr. Sci.*, **16**, 152, 1978.

47. A. M. Siouffi, T. Wawrzynowicz, F. Bressolle, and G. Guiochon, Problems and applications of reversed-phase thin-layer chromatography, *J. Chromatogr.*, **186**, 563, 1979.

48. R. P. W. Scott, The role of molecular interactions in chromatography, *J. Chromatogr.*, **122**, 35, 1976.

49. M. S. Mirrlees, S. J. Moulton, C. T. Murphy, and P. J. Taylor, Direct measurement of octanol–water partition coefficient by high-pressure liquid chromatography, *J. Med. Chem.*, **19**, 615, 1976.

50. M. C. Pietrogrande, C. Bighi, P. A. Borea, A. M. Barbaro, M. C. Guerra, and G. L. Biagi, Relationship between log k' values of benzodiazepines and composition of the mobile phase, *J. Liq. Chromatogr.*, **8**, 1711, 1985.

51. E. Stahl, A quarter of a century of thin-layer chromatography. An interim report, in H. Kalasz and L. S. Ettre (Eds.), *Chromatography, the State of the Art*, Akadémiai Kiadó, Budapest, 1985, p. 497.

CHAPTER

5

METHODOLOGY OF QUANTITATIVE STRUCTURE–RETENTION RELATIONSHIP ANALYSIS

Chromatographically determined physicochemical data must be some function of the chemical structure of the three entities involved in intermolecular interactions, that is, the solute, the stationary phase, and the mobile phase. Certainly, the conditions under which chromatography is carried out are decisive for the process, but these are assumed constant for the present. As has been discussed in the preceding chapters, it is practically impossible to derive a strict, thermodynamically well-founded, and unequivocally verifiable canonical equation relating retention to the three main chromatographic variables, that is, the structure of the solute, the stationary phase, and the mobile phase. Perhaps, the reason is that at present chemists are still unable to describe molecular structures in precise quantitative terms.

Despite the lack of a general formula relating retention to structure, attempts have been undertaken since the early 1950s to detect any existing regularities of behavior among selected subgroups of solutes chromatographed at fixed conditions. Of the three main variables mentioned, only solute structure was changed, whereas the stationary and mobile phases applied in a given chromatographic technique remained constant. As expected, a smooth change of retention with carbon number was found for a series of homologues. For a more diverse sets of solutes, relationships between empirical or theoretical molecular structural parameters have usually not been unequivocal and required statistical evaluation in order to check the significance of the resulting correlations.

As long as the simple linear regression analysis of the relationship between the retention parameter and one of the structural parameters as an independent variable has been performed, the calculations required can be done manually. With a large number of independent structural variables determining retention data, the calculations become very tedious. Further progress in all types of property–structure analyses had not been possible until computers became commonly available (i.e., the early 1960s). Soon such statistical techniques came into regular use like, for example, multiparameter regression analysis, nonparameter de novo methods, and principal component and factor analyses. Interestingly, some of these methods have originally been elaborated to solve problems in econometrics, psychology, and weather forecasting.

The above-mentioned advanced statistical methods first found their application in studies of quantitative relationships of chemical structure and pharmacological activity (QSAR). Until the late 1970s these methods have been only occasionally applied to structure–retention relationship (QSRR) studies. Since then, however, the number of QSRR publications employing statistical procedures exceeded 200 and is still increasing [1], due to the easy access to computers (often personal computers suffice) and the commerical supply of the programs required [2].

There is certainly no need at present to write computer programs for the standard statistical methods of QSRR calculations. However, the programs available are sometimes used rather mechanically, and the results obtained (and even reported occasionally) are insignificant or incorrectly interpreted. Thus, it seems worthwhile to discuss briefly the main features, and especially the limitations, of the most commonly used statistical procedures applied in QSRR studies.

5.1. MULTIPARAMETER REGRESSION ANALYSIS

Multiple linear regression [3, 4] was the first, and foremost, statistical method applied in QSAR studies. This method has been popularized by Hansch [5] to relate bioactivity data to the measures of lipophilic, electronic, and steric properties of a series of derivatives. At present it is the statistical method most frequently used in QSRR studies.

Assuming the useful formalism of linear free-energy relationships (LFER), a given chromatographic retention parameter R may quantitatively be described by a set of experimentally or theoretically derived molecular parameters, numerically expressed, x_i:

$$R = f(ax_1, \cdots, zx_i) \tag{5.1}$$

The coefficients at the individual x_i parameters are calculated by multiple linear regression. A number of statistics are derived in conjunction with such a calculation that allow the statistical significance of the resulting correlation to be assessed. The most important among them are the following:

1. R (or r for simple linear regression), the correlation coefficient; its square, R^2 (or r^2), is the coefficient of determination or percentage of data variance accounted for by the regression equation.
2. s, the standard error of the estimate, often called the standard deviation from regression.

3. F, a statistic for assessing the overall significance of the derived equation; the F value can be calculated by the formula [3, 4]

$$F = \frac{(n-k-1)R^2}{k(1-R^2)} \qquad (5.2)$$

where R is the correlation coefficient, n is the number of data points (retention data) for the dependent variable, and k is the number of independent variables (structural parameters) in regression equation. The F values are compared with statistics tables that normally are given in statistics textbooks. Occasionally, significance level p is calculated for the equation and the individual parameters.

4. Confidence intervals (usually 95%) for the individual regression coefficients in the equation or the t-test values.

5. The cross-correlation coefficients r_{ij} between the independent variables in the equation. These must be low to assure true independence of the variables. The independence or orthogonality of the variables is a necessary condition for meaningful results.

Recently, the Committee on Minimal Standards for Reporting Results of Regression Analysis of the International Group for Correlation Analysis in Organic Chemistry has published an announcement entitled "Recommendations for Reporting the Results of Correlation Analysis in Chemistry using Regression Analysis" [6]. According to the publication, the following requirements should be met:

1. All of the data points (i.e., values of the dependent variable, the quantity to be correlated) in a set should be reported either in the paper or in supplementary material. If values are excluded from the correlation, it should be clearly indicated and the reason for exclusion stated.

2. All of the independent variables (i.e., explanatory variables) included in the correlation analysis at any stage should be reported even if they are not present in the final correlation. This requirement is meant to apply to multiple linear regression analyses in which no a priori model is available and correlation is carried out with a collection of independent variables some of which are dropped when they fail to contribute significantly to the correlation equation. Values of the independent variables used in the correlation analysis should be given in the paper or in supplementary material whenever possible. Otherwise, their sources should be reported. If values are estimated, the method of estimation and its source should be given.

3. The statistics reported should include at least the following:
   (a) Standard error of the estimate.
   (b) Standard errors of the regression coefficients.
   (c) At least two of the following:
       (i) R or $100R^2$ (or for simple linear regression, r or $100r^2$);
       (ii) F (with confidence levels); and
       (iii) f or ψ—the former is intended for use with linear regression through the origin, the latter for linear regression with an intercept.
   (d) A correlation matrix or table of correlation coefficients r_{ij} for the correlation of the ith independent variable with the jth independent variable. They are called zeroth-order correlation coefficients.

As optional but recommended are the following:

1. Values of R_i for the correlation of the ith independent variable on all of the other independent variables. In the case of four independent variables, for example, the R_i are $R_{1,234}$, $R_{2,134}$, $R_{3,124}$, and $R_{4,123}$. Values of R for the ith independent variable on all other possible combinations of the other independent variables are also of interest. The R_i are sufficient, however, for the detection of collinearity among the independent variables.
2. Graphs:
   (a) In simple regression a plot of y versus x.
   (b) In multiple regression a plot of y calculated versus y observed. Alternatively, a residual plot may be used. A plot of the first two principal components of the x matrix and information about the variance explained would also be of use. The number of degrees of freedom should be given on the graph or in its caption.

To illustrate the meaning of individual statistical parameters, an example is given from the author's recent publication [7]. In Table 5.1 the correlated data are given for a set of 12 aromatic solutes. The dependent variable b_j is the intercept of the relationship

$$\log k_j' = a_j X + b_j \qquad (5.3)$$

where k_j' is the capacity factor for a solute j determined in reversed-phase HPLC with a water–methanol mobile phase; X is the mole fraction of water in the binary solvent. Independent variables (i.e., explanatory variables) are quantum chemically calculated by the CNDO/2 method the total energy $E_{T,j}$ and maximum excess charge difference in a given molecule, Δ_j. The data in Table 5.1 are extracted from a larger set of data for illustrative purpose. The

Table 5.1. Exemplary Multiparameter QSRR Data[a]

Solute	b_j Determined	$E_{T,j}$ atomic units	Δ_j electrons	b_j Calculated
1	0.6639	−65.5548	0.4328	0.7447
2	1.1786	−81.1756	0.5088	1.2084
3	1.1596	−94.8446	0.7774	1.1290
4	1.6051	−99.6184	0.6836	1.5517
5	1.0646	−74.2375	0.4265	1.1190
6	1.5152	−82.6639	0.4275	1.4652
7	2.0059	−91.3479	0.4270	1.8257
8	2.3567	−100.1758	0.4246	2.1966
9	0.5849	−59.5473	0.3737	0.6381
10	1.0879	−68.2305	0.3564	1.0388
11	1.7990	−96.6241	0.5028	1.8620
12	2.2945	−112.3142	0.4922	2.5365

[a] RP HPLC capacity factors extrapolated to pure methanol in mobile phase, $k'_i = b_j$. Structural parameters: $E_{T,j}$, total energy; Δ_j maximum excess charge difference for set of aromatic solutes [7]. With permission.

QSRR derived will be discussed in detail in the discussion concerning the application of quantum chemical parameters in structure–retention correlation studies in Section 9.4.

The regression equation derived from the data given in Table 5.1 has the form

$$b_j = -0.0414(\pm 0.0066)\, E_{T,j} - 2.4010(\pm 0.8535)\, \Delta_j - 0.9280(\pm 0.5027) \qquad (5.4)$$
$$n = 12; \quad s = 0.1335; \quad R = 0.9810; \quad F_{2,9} = 115; \quad F_{2,9,\,\alpha=0.05} = 4.26$$

The regression coefficients are accompanied by the 95% confidence intervals according to the t-test. These intervals are the products of standard errors of regression coefficients times a value of a multiplier from statistics tables, specified for $n - k - 1 = 12 - 2 - 1 = 9$ degrees of freedom, where n is the number of the dependent variable data points, and k is the number of independent variables considered. The F-test value is calculated using Eq. (5.2) for two independent variables and 9 degrees of freedom. For Eq. (5.4) this value is larger than the value tabulated for significance level $\alpha = 0.05$, which means that Eq. (5.4) is significant at least at the 95% significance level. The value of F for the 99% significance level is $F_{2,9,\,\alpha=0.01} = 8.02$. Thus, Eq. (5.4) is significant at the 99.9% significance level. With the help of a computer the precise significance levels may easily be calculated for the regression equations obtained as well as for the individual regression coefficients. In the case of the regression equation [Eq. (5.4)], the significance level calculated is $p = 4.8 \times 10^{-6}$, whereas the coefficients at $E_{T,j}$, Δ_j, and the free term are significant at the levels 0.0002, 0.0016, and 0.0022, respectively.

In searching for the "best" regression equation, (i.e., the equation describing most precisely the observed variance in retention among the solutes studied), some workers have aimed at attaining as high a correlation coefficient as possible. Occasionally, due to introduction of several structural parameters, the standard deviations calculated are lower than the experimental error of the retention parameter measurements. Such overfitting of the data makes the QSRR obtained dubious.

The easiest way to detect if successive introduction of individual independent variables to the regression is statistically justified is to apply the sequential F-test. The general formula, whether improvement of the correlation due to introduction of an additional independent variable into the regression is statistically significant, is

$$F^* = \frac{(R_1^2 - R_2^2)/p}{(1 - R_1^2)/[n - (k+p) - 1]} \tag{5.5}$$

If R_1 is the multiple correlation obtained when $k + p$ independent variables are used and R_2 is the multiple correlation found when only k of them are retained, Eq. (5.5) provides a test of whether R_1 is significantly greater than R_2. This F^* has p and $n - (k+p) - 1$ degrees of freedom. When $p = 1$, this is a test of whether the addition of a single variable provides a significant increase in the multiple correlation. The F^* values calculated are compared with statistics tables for specific significance level, usually for $\alpha = 0.05$.

In the case of the example regression [Eq. (5.4)], the question arises whether the introduction of the term Δ_j to the regression is statistically justified. The correlation coefficients of the linear relationship

$$b_j = A E_{T,j} + B \tag{5.6}$$

$$b_j = A' \Delta_j + B' \tag{5.7}$$

are $r_2 = 0.8789$ for Eq. (5.6), and $r_2' = 0.0837$ for Eq. (5.7). As stated, the correlation coefficient of the two-parameter Eq. (5.4) is $R_1 = 0.9810$. Thus, the F^*-test of improvement of the correlation by the equation $b_j = f(E_{T,j}, \Delta_j)$ as compared to the equation $b_j = f(E_{T,j})$ equals

$$F^* = \frac{(0.9810^2 - 0.8789^2)/1}{(1 - 0.9810^2)/[12 - (1+1) - 1]} = 45.4 \tag{5.8}$$

This F^* value is compared with the tabulated F-value for, for example, $\alpha = 0.05$. As the value $F^*_{p,\,[n-(k+p)-1]} = F^*_{1,\,9} = 45.4$ is larger than the tabulated value $F_{1,\,9,\,\alpha=0.05} = 5.12$, the introduction of the Δ_j term into regression is justified at least at the 95% significance level. In fact, the F^* value is even larger than the respective tabulated value for the $\alpha = 0.01$ significance level (10.6). As can be easily calculated, introduction of the $E_{T,\,j}$ term in Eq. (5.4) in addition to Δ_j gives an F^*-test value that is even larger than yielded by Eq. (5.8).

The other reason why sometimes the QSRR equations reported as the "best" are without relevance is high intercorrelation among the structural descriptors used as independent variables in regression equations. The two data vectors are fully orthogonal if the intercorrelation coefficient $r = 0$. There are structural parameters highly intercorrelated, that is, the r coefficient attains a value close to 1 if two such parameters are compared for a given set of solutes. Such an intercorrelation is obvious if one compares molecular mass and carbon number in a series of homologues. However, in some instances the intercorrelation takes place with parameters that are generally not related to one another. For a given series of solutes, however, a peculiar combination of substituents or some other structural features may give rise to a high collinearity of the molecular descriptors. Thus, the intercorrelation of the parameters used in regression must be analyzed, and the resulting data should be given.

In the case of Eq. (5.4), the intercorrelation of the $E_{T,\,j}$ and Δ_j variables is $r = 0.4879$. Taking the square of this value, $r^2 \times 100\%$, it may be stated that less than 25% of the structural information may be mutually contained in both the $E_{T,\,j}$ and Δ_j parameters. It would certainly be desirable for more meaningful QSRR results to find equally significant but less collinear independent variables.

One cannot rely on statistics only to find independent variables for QSRRs. It is very important, but also difficult, to properly evaluate that any statistically derived equation make good chemical sense.

It is an important assumption of regression analysis that the independent variables have minimal error [3]. In structure–biological activity correlations the error in determining structural descriptors can generally be expected to be low as compared to the error of biological data. Certainly, the precision of retention data is much higher than the majority pharmacological data. Unfortunately, the effect of error in the independent variable on the reliability of the regression results is only mentioned in the QSAR literature [8]. Also, the potential for chance correlation when too many variables are surveyed to correlate too few data has been pointed out by Topliss and co-workers [9, 10]. In QSAR studies at least four to five biological data (compounds) should fall per independent variable. According to Goodford [8], the final number of degrees of freedom $(n - k - 1)$ should be at least 10.

As a final remark, the simplicity principle should be observed, that is, if several correlation equations are equivalent, the simplest one should be chosen [11].

5.2. DE NOVO NONPARAMETER METHOD OF CORRELATION ANALYSIS

In 1964 Free and Wilson [12] demonstrated a general mathematical method both for assessing the occurrence of additive substituent effects on bioactivity in an analog series and for quantitatively estimating their magnitude. According to this method, the molecules of a drug series are structurally decomposed into a common moiety or core that is variously substituted in multiple positions. A series of linear equations are constructed of the form

$$BA_i = \sum_j a_j X_{ij} + \mu \tag{5.9}$$

where BA_i is the biological activity of the ith compound, X_j is the jth substituent with a value of 1 if present and 0 if not, a_j is the contribution of the jth substituent to BA, and μ is the overall average activity. All activity contributions at each position of substitution must sum to zero.

In 1979 an analogous approach to the evaluation of substituent contributions to chromatographic retention was reported by Chen and Horváth [13]. These authors assumed that the difference between the logarithm capacity factor ($\log k'$ or R_M) of a substance i, $\log k'_i (R_{M_i})$, and that of its parent compound, $\log k'_p (R_{M_p})$, is given by

$$\log r_{i,p} = \log k'_i - \log k'_p = \sum_{j=1}^{m} \tau_{i,j} \tag{5.10}$$

where τ_{ij} is a substituent parameter that measures the change in chromatographic retention upon replacing a hydrogen atom by the jth substituent; $r_{i,p}$ is the corresponding relative retention. As the τ_{ij} values for a given substituent in a certain position are the same in different compounds [14], τ_j can be written instead of τ_{ij}. A given substituent in different positions, however, may have different τ_j values depending on the particular molecular environment, which can change the effect of the substituent on retention, and therefore the number of τ_j may be greater than the number of actual substituents. Chen and Horváth [13] assumed that in a given liquid chromatographic system the logarithms of the capacity factors ($\log k'$ or R_M) of n congeners containing m possible substituents can be expressed by a set of linear equations

$$\log k'_1 = \log k'_p + I_{11}\tau_1 + I_{12}\tau_2 + \cdots + I_{1m}\tau_m$$

$$\log k'_2 = \log k'_p + I_{21}\tau_1 + I_{22}\tau_2 + \cdots + I_{2m}\tau_m$$

$$\vdots \qquad\qquad\qquad\qquad\qquad (5.11)$$

$$\log k'_n = \log k'_p + I_{n1}\tau_1 + I_{n2}\tau_2 + \cdots + I_{nm}\tau_m$$

where the coefficient I_{ij} is the indicator variable, which is set to unity when a τ_j value is assigned to a substituent in compound i and to zero otherwise. Thus, for the parent compound, all I variables are zero, and for the congeners only those that correspond to a position without a substituent are zero.

Equation (5.11) can be rewritten in matrix form and simplified as follows:

$$[I_{ij}]_{n \times m}[\tau_j]_{m \times i} + [\log k'_p]_{n \times i} = [\log k'_i]_{n \times i} \qquad (5.12)$$

Subtracting $\log k'_p$ from each array of $\log k'_i$ in Eq. (5.12), an expression is obtained containing the relative retention:

$$[I_{ij}]_{n \times m}[\tau_j]_{m \times i} = [\log k'_i]_{n \times i} - [\log k'_p]_{n \times i} = [\log r_{i,\,p}]_{n \times i} \qquad (5.13)$$

If the number of substances is equal to the number of retention increments to be determined, the τ_j values can be obtained by simple matrix operation. In practice, however, the number of congeners should exceed the number of τ_j values, and each substituent should appear more than once at a position in different combinations with substituents at other positions. By the method of multiple-regression analysis, the matrix τ_j is then solved so that the sum of squared deviations between the regressed value $\log r_{i,\,p}$ and the values observed is minimized.

For a meaningful de novo analysis a substantial number of compounds with varying substituent combinations is required. One should also be cognizant that the model will break down if nonlinear dependence on substituent properties is important or if there are interactions between the substituents [15].

In fact, the de novo approach is not frequently applied in QSRR studies. The reason is that the method is rather intended for predicting retention of a new derivative, and it supplies no direct information connected with the inter-molecular interactions governing the chromatographic distribution. With this method one type of chromatographic data is related to another, that is, retention of a compound to its component parts contributions. As the QSRRs are only occasionally derived to predict retention, the popularity of the method in chromatography is much limited.

5.3. PRINCIPAL COMPONENTS AND FACTOR ANALYSIS

Factor analysis is applied in chemistry to determine the "intrinsic dimensionality" of certain experimentally determined chemical properties, that is, the number of "fundamental factors" required to account for the variance [16].

As already discussed, the intercorrelation among independent variables affects the reliability of the multiple-regression analysis. The cross correlation between the independent variables can be assessed by examination of the correlation matrix of the parameters. If the dependent variable is included, the variance–covariance matrix is constructed. Such matrices may be transformed by the prescribed methods [17] of linear algebra into new matrices containing nonzero elements only on the diagonal. These are the so-called eigenvalues of the matrix. With each eigenvalue of the matrix thus obtained, an eigenvector is associated that is a linear combination of the original set of variables. The characteristic feature of the eigenvectors is that the correlation coefficient between them is exactly zero, that is, they are completely orthogonal, as opposed to the original set of variables. If there is significant covariance in the set of original variables, most of the total variance will be accounted for by a number of eigenvectors equal to a fraction of the original number of variables. A reduced set containing only the major eigenvectors (principal components) may then be examined or used in various ways. Factor analysis involves other manipulations of the eigenvectors and is aimed more at gaining insight into the structure of a multidimensional data set. The appropriate computer calculation programs modified by Weiner et al. [18] are now commerically available [2].

It should be noted here, however, that the determination of how many eigenvectors may be reasonably ignored in the principal components analysis is a subjective decision. In any case, if only the principal components are considered, new orthogonal variables can be constructed from the eigenvectors, and hence the dimensionality of the parameter space can be reduced while most of the information in the original variable set is retained [15].

Usually, attempts are undertaken to give a physicochemical meaning to the individual abstract factors derived mathematically. To this end, the multiple-regression analysis is applied with various structural descriptors as independent variables.

Although factor analysis has not been frequently used in QSRR studies, computer-assisted multivariate techniques become increasingly more popular for evaluating large sets of retention data.

References

1. R. Kaliszan, Quantitative relationships between molecular structure and chromatographic retention. Implications in physical, analytical, and medicinal chemistry, *Crit. Rev. Anal. Chem.*, **16**, 323, 1980.

2. For example, sets of programs written for personal computers are distributed by Dr. F. Darvas, Compudrug Ltd., Fürst Sándor u.5, H-1136 Budapest, Hungary.

3. C. Daniel and F. S. Wood, *Fitting Equations to Data*, Wiley, New York, 1971.

4. N. R. Draper and H. Smith, *Applied Regression Analysis*, Wiley, New York, 1966.

5. C. Hansch, Quantitative structure–activity relationships in drug design, in E. J. Ariëns (Ed.), *Drug Design*, Vol. 1, Academic press, New York, 1971, p. 271.

6. M. Charton, S. Clementi, S. Ehrenson, O. Exner, J. Shorter, and S. Wold, Recommendations for reporting the results of correlation analysis in chemistry using regression analysis, *Quant. Struct. Act. Relat.*, **4**, 29, 1985.

7. R. Kaliszan, K. Ośmialowski, S. A. Tomellini, S.-H. Hsu, S. D. Fazio, and R. A. Hartwick, Quantitative retention relationships as a function of mobile and C_{18} stationary phase composition for non-cogeneric solutes, *J. Chromatogr.*, **352**, 141, 1986.

8. P. J. Goodford, Prediction of pharmacological activity by the method of physicochemical–activity relations, *Adv. Pharmacol. Chemother.*, **11**, 51, 1973.

9. J. G. Topliss and R. J. Costello, Chance correlations in structure–activity studies using multiple regression analysis, *J. Med. Chem.*, **15**, 1066, 1972.

10. J. G. Topliss and R. P. Edwards, Chance factor in studies of quantitative structure–activity relationships, *J. Med. Chem.*, **22**, 1238, 1979.

11. J. K. Seydel and K.-J. Schaper, *Chemische Struktur und Biologische Aktivität von Wirkstoffen. Methoden der Quantitativen Struktur-Wirkung-Analyse*, Verlag Chemie, Weinheim, New York, 1979.

12. S. M. Free, Jr. and J. W. Wilson, Mathematical contribution to structure–activity studies, *J. Med. Chem.*, **7**, 395, 1964.

13. B.-K. Chen and Cs. Horváth, Evaluation of substituent contributions to chromatographic retention: Quantitative structure–retention relationships, *J. Chromatogr.*, **171**, 15, 1979.

14. I. Molnár and Cs. Horváth, Catecholamines and related compounds. Effect of substituent on retention in reversed-phase chromatography, *J. Chromatogr.*, **145**, 371, 1978.

15. C. J. Blankley, Introduction: A review of QSAR methodology, in *Quantitative Structure: Activity Relationships of Drugs*, Academic, New York, 1983.

16. E. R. Malinowski and G. D. Howery, *Factor Analysis in Chemistry*, Wiley, New York, 1980.

17. R. D. Cramer, III, BC/DEF/Parameters, 1 and 2, *J. Am. Chem. Soc.*, **102**, 1837, 1849, 1980.

18. P. H. Weiner, E. R. Malinowski, and A. R. Levinstone, Factor analysis of solvent shifts in proton magnetic resonance, *J. Phys. Chem.*, **74**, 4537, 1970.

6

ADDITIVE STRUCTURAL PARAMETERS IN QUANTITATIVE STRUCTURE–RETENTION RELATIONSHIP STUDIES

This chapter gives a detailed review of the reported relationships between retention parameters and molecular descriptors, which, as a general rule, can be derived from the structural formula of the solutes chromatographed. A separate chapter will be devoted to the QSRRs employing experimentally determined physicochemical properties as the molecular parameters, for example, boiling point, refraction index, and especially n-octanol–water partition coefficient. Certainly, the QSRR relating retention to any arduously determinable physicochemical property can be of practical analytical value. An example may be the utilization of the reversed-phase liquid chromatography for the evaluation of partitioning properties in a given group of compounds. Nonetheless, for the fundamental chemical studies the more informative could be the QSRRs in which retention would be predicted based on the a priori determinable molecular structural parameters. First, the molecular structural parameters are considered, which are determined as a simple sum of individual subunits forming any given molecule.

6.1. CARBON NUMBER

The linear relationships between the retention data and carbon number for different homologous series have long been observed in various types and modes of chromatographic separation. In GC, the linearity between the isothermal relative retention data and carbon numbers of homologous hydrocarbons has given rise to the designing of the retention index system by Kováts. However, for the lower hydrocarbon homologues the deviations from linearity observed cause some complications in the retention index system. This has been pointed out in chapter 4 in the discussion of GC retention quantitation.

Nevertheless, in GC the linear relationship between retention indices and carbon number is generally valid among the classes of substances. Basing their work on this observation, Ladon and Sandler [1] were able to derive regression equations relating the solute carbon number to its retention time on each of the three selected stationary phases. This relation holds true for classes of compounds containing the same kind and number of functional groups such as alkanes, cycloalkanes, alkenes, cycloalkenes, and alkadienes. On the other hand, the position of the hydroxyl group in primary and secondary straight-chain alcohols was found by Castello and D'Amato [2] to affect the retention of homologous series of that class of solutes. Plots of Kováts retention indices versus carbon number of related solutes are given in the paper by Hoshikawa et al. [3].

Carbon-number-versus-retention relationships in a series of homologues have extensively been studied from the point of view of the reliability of the retention index systems. A detailed literature review concerning this problem may be found in papers by Haken and co-workers [4] and Takács and co-workers [5].

For the sake of illustration of generally observed regularities, the retention-versus-carbon-number plots are given here after Haken and Korhonen [6] for four homologous series (Fig. 6.1). The experimental data were determined by the isothermal capillary GC.

As evident from Fig. 6.1, starting from chain lengths of four to five carbon atoms, the retention index increments for methylene groups at a given temperature on a given stationary phase are constant.

Saura-Calixto and Garcia-Raso [7] derived two-parameter regression equations relating Kováts indices of homologues to both carbon number and column temperature (°C).

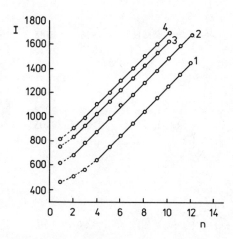

Figure 6.1. Retention index I versus carbon number n in alkyl chain of four homologous series. Column: vitreous silica SE-30 wall coated open tubular, 25 m × 0.33 mm i.d.; temperature 120°C. Solutes: (1) n-alkanols; (2) n-propanoates; (3) 2-chloropropanoates; (4) 3-chloropropanoates. (After J. K. Haken I. O. O. Korhonen, *J. Chromatogr.*, **319**, 131, 1985. With permission.)

The results obtained by Garcia-Raso et al. [8] in their studies on bicyclo[n.4.0]alken-2-ones are interesting. Retention indices of the solutes, determined on a 2-m × $\frac{1}{4}$-in.-i.d. column packed with 10% UCC (Supelco) on Chromosorb P AW DMCS at three temperatures, are given in Table 6.1 along with the boiling points.

A general equation was derived that represents the chromatographic behavior of these solutes at the temperature range studied:

$$I = 627.83 + 103.93N + 0.868t + P$$

$$n = 24; \quad R = 1.000; \quad s = 8.60; \quad F = 25,423 \tag{6.1}$$

Table 6.1. GC Retention Indices for Two Homologous Groups of Bicyclo[n.4.0] alken-2-ones Determined at Three Temperatures and Boiling Points of Solutes[a]

Compound Number	Compound		Retention Index (220°C)	(230°C)	(240°C)	Boiling Point (°C)
1		R = H	1338	1346	1354	90
2		R = CH$_3$	1380	1388	1396	90–92
3		R = H	1451	1461	1470	98–100
4		R = CH$_3$	1489	1498	1507	98–100
5		R = H	1536	1545	1553	110
6		R = CH$_3$	1570	1579	1588	115
7		R = H	2069	2077	2085	160
8		R = CH$_3$	2105	2105	2124	180
9		R = H	1545	1554	1563	—

[a] According to Garcia-Raso et al. [8]. With permission.

where I is retention index of an individual solute at a given temperature; N is number of carbon atoms in ring A; t is column temperature (°C); and P is a structural parameter equal to zero for $R = H$ and 37.8 for $R = CH_3$. The statistical symbols are as explained in the preceding paragraph. The variable P is the indicator variable that allows two homologous series to join in one equation. However, if a methyl substituent is attached to the ring A, the retention index of such a solute deviates from Eq. (6.1).

Kováts-index-versus-carbon-number relationships for homologues hold for various chemical groups. Examples for organosilicon derivatives have also been reported [9, 10].

For estimation of Kováts indices of homologues, Golovnya and Grigor'eva [11] proposed an equation of the form

$$I_{R_nX}^{st.phase(T)} = \alpha + \beta n + \gamma \frac{\log n}{n} + \delta \frac{1}{(n-2)^2 + 0.1} \qquad (6.2)$$

where $I_{R_nX}^{st.\,phase}$ is the Kováts index of a solute with n carbons in the aliphatic chain belonging to a homologous group with a functionality X chromatographed at temperature T on a given stationary phase; α, β, γ, and δ are experimentally determined coefficients. For example, for amines of the series R_nNH_2 chromatographed at 100°C on Apiezon L, the coefficients α, β, γ, and δ are 281.09, 97.97, -213.66, and 1.31, respectively. For the same solutes on Carbowax 1000 the corresponding values are 623.18, 98.25, -409.03, and 1.02, respectively.

Equation (6.2) accounts for the nonlinearities usually observed in retention-versus-carbon-number plots. However, statistical significance of the last two terms on right side of Eq. (6.2) has not been determined. In any case, the last term is meaningless except for carbon number 2.

There is a vast number of publications reporting regular changes of GC retention with carbon atom number in homologous series starting from carbon number 4. In fact, the prevailing majority of these works, which are still regularly published, add little to the rule established in the fundamental work of James and Martin [12]. It should be noted here, perhaps, that similarly to carbon number, linear correlations between retention values and number of other atoms have been found; for example, retention was correlated with the number of chlorine atoms in solutes [13–15].

Analogously, linearities have been reported between GC retention parameters and number of atoms of silicon, sulfur, nitrogen, and oxygen as well as the number of phenyl groups and double bonds [14].

To account for deviations from linearities, especially evident for polar solute series chromatographed on polar phases, various combinations of carbon (or other atom) number and its transformations (reciprocals, squares, logarithms) are occasionally introduced based on experimental data for often much limited series [11].

In various forms of liquid chromatography the linear relationship between retention data and carbon number is also valid for homologous series of solutes. This relationship has been studied extensively in partitioning chromatographic systems, usually employing the reversed-phase mode. Thus, in reversed-phase paper chromatography and reversed-phase TLC smooth linear relationships have been long since reported between R_M data and carbon atom number in alkyl chains of various chemical series. Best known are numerous publications by Green and co-workers [e.g., 16] and Clifford et al. [17].

Very important historically is the paper by Boyce and Milborrow [18] reporting a quadratic relationship between chromatographic retention data and a biological activity measure. This publication opened the way to wide application of chromatography in QSAR. In the paper, linear relationships are described between carbon atom number in the alkyl chain of N-n-alkyltrityl-amines and R_M values from reversed-phase TLC. Boyce and Milborrow [18] reported that the linearity has been observed for C_2–C_8 homologues at different mobile phase compositions. However, the C_1 compound apparently deviated from linearity.

In cation exchange chromatography of aliphatic amines, β-phenylethyl-amine, and aminoalcohols Murakami [19] obtained curvilinear plots of the selectivity coefficients K_M of the amines for the counterion M against carbon number for individual subgroups of solutes (Fig. 6.2).

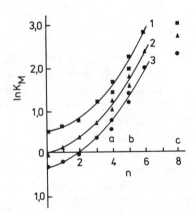

Figure 6.2. Correlation between carbon number n of alkylamines and logarithm of selectivity coefficient, ln K_M, in cation exchange chromatography. Column: $1 = K^+$ form; $2 = Na^+$ form; $3 = Li^+$ form. (a) Isobutylamine; (b) isoamylamine; (c) β-phenylethylamine. (After F. Murakami, J. Chromatogr., **198**, 241, 1980. With permission.)

Comparing aliphatic amines and aminoalcohols, the same author observed that the introduction of a hydroxyl group into the aliphatic amine molecule caused a decrease in adsorption, corresponding to a decrease in carbon number of about 2.

Of all chromatographic methods, reversed-phase HPLC is able to generate the most precise and reproducible numerical retention data. The relationships between the logarithms of the reversed-phase HPLC capacity factors and carbon numbers for solutes that are closely related structurally (i.e., congeneric) are still analyzed [e.g., 20–22]. The plots of log k' versus carbon number are usually linear, but at some limited range of the aliphatic chain length. The reported range of linearity does not exceed six to eight carbon atoms, depending on the functional groups present in the series. Deviations from linearities of the retention data obtained in reversed-phase systems, when plotted versus carbon number in the alkyl chain of the homologues, may result from differences in shielding a polar functional group present in the series of alkyl chains of different length. Apparent nonlinearity of the log k' versus carbon number plot for alkylbenzoates is illustrated in Fig. 6.3 after Tomlinson et al. [20].

The relationship between log k' and carbon number holds only for a homologous series. For the compounds belonging to closely related but different groups, the relation is no longer valid. An example of such incompatibility for alkylbenzenes and polyaromatics is given in a paper by Koopmans and Rekker [22].

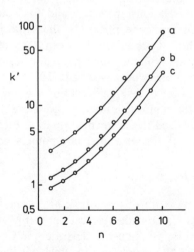

Figure 6.3. Relationship between capacity factor, log k', and carbon number of alkyl chain, n, for homologous series of (a) 3-Cl-; (b) 2-OCH$_3$-, and (c) 2-NO$_2$-substituted alkylbenzoates. (After E. Tomlinson, H. Poppe, and J. C. Kraak, *Int. J. Pharm.* **7**, 225, 1981. With permission.)

6.2. MOLECULAR MASS AND PARACHOR

Where specific interactions within a particular group of compounds contribute to retention to the same extent, a linear correlation between retention parameters and molecular mass is observed for solutes that are not homologues. Such relationships have occasionally been reported in GC. For example, Bortnikov et al. [23] reported that the retention volumes of organo-element (M) substances of $(C_2H_5)_3MH$ and $(C_2H_5)_4M$ series tend to increase when the atomic weight of the element M increases. The element M in the experiments was silicon, germanium, or lead.

A linear relationship between logarithm of GC retention volume and molecular mass was found by Preisler [15] for polydimethylcyclosiloxane, methyl hydrocyclopolysiloxane, and methylethylhydrocyclopolysiloxane series. Kochetov et al. [24] observed a similar relationship for nitrogen-containing organosilicon compounds.

Retention indices of a series of N-substituted saturated nitrogen-containing heterocycles were shown to be linearly related to molecular mass [25, 26]. Similar regularity has been observed in the case of GC retention indices of methyl (2-furyl)silanes, -germanes, -stannanes, and -plumbanes of the formula $(CH_3)_{4-n}M(furyl)_n$, where M is silicon, germanium, tin, or lead. [27].

Some correlation between GC retention data and solute molecular mass has been discussed by Castello and D'Amato [2].

Gudzinowicz and Martin [28] have described the proportionality between logarithm retention volume and a quantity $(M_w - 170)^{1/3}$, where M_w is molecular mass. This relationship was found for compounds chromatographed on a nonpolar silicone phase. In the same paper retention was correlated to parachor. Correlation between GC retention parameters (logarithms of retention volume) and parachor for related solutes was analyzed by Wurst in a series of papers [e.g., 29].

Molecular mass has been occasionally employed as one of several structural parameters used jointly in multiparameter regressions. For example, Dimov and Boneva [30] proposed the following equation to describe relative GC retention, R_i, of seven 2-nitroalkanol-1 compounds on a Carbowax 20M phase:

$$R_i = -9.70 + 0.044T_b + 0.493M_w/(W + h_{NO_2}) \qquad (6.3)$$

where T_b is the boiling point of the solute; M_w its molecular mass; W is a topological parameter, the so-called Wiener number [31] (topological indices as molecular descriptors are discussed separately in Chapter 8); and h_{NO_2} is defined as a measure of NO_2 hindrance to which a value of 20 units has been assigned for tertiary NO_2 groups and 10 units otherwise.

It is difficult to determine the significance of the M_w parameter in Eq. (6.3), even if one assumes that there is no chance correlation, obtained with four independent variables and only seven data points. The reason is that it is rather impossible to assign any physical meaning to the absolute sum of different molecular descriptors (e.g., $W + h_{NO_2}$).

In conclusion, molecular mass provides structural information of low value for QSRR studies. That parameter is completely nonspecific and may actually be quantitatively related to retention only in homologous series, yielding the same correlations as carbon number.

6.3. MOLAR VOLUME

Molar volume is an easily calculable molecular structural parameter often used in QSAR and QSRR studies. Due to the application of this parameter, some diversity of the solutes comprised by individual QSRR equations is observed; that is, compounds belonging to separate homologous series can be treated jointly. Still, however, the single-parameter retention versus molar volume relationships concern closely similar classes of solutes (e.g., normal-chain and branched hydrocarbons), often including those differing in the position of unsaturation.

Molar volume is a molecular descriptor that may be related to the ability of a solute to take part in nonspecific intermolecular interactions of the London (dispersion) type. As such, molar volume is frequently used in multiparameter regression analyses both in medicinal chemistry and chromatography. In chapter 9 applications related to multiparameter QSRR analysis will be discussed. Here, retention–molar volume relationships are briefly surveyed.

Molar volume belongs to a group of "bulk" parameters whose dimensions are volume. The group include molar refractivity as well as parachor. It must be noted here that there is a significant correlation between the Bondi–van der Waals volume V_W and molar refractivity as well as parachor [32]. The correlation between V_W and molecular mass is relatively low ($r = 0.611$). The correlation between molar refractivity and molecular mass ($r = 0.751$) is slightly higher. Since there is a significant intercorrelation between all "bulk" parameters, only one parameter can be used in a regression equation as a measure of solute polarizability, that is, as a measure of solute dispersive properties.

Van der Waals volumes are normally calculated using the group increments after Bondi [33] (Table 6.2).

A linear correlation between logarithms of GC retention volumes and solute molar volumes was demonstrated by Wurst and Churacek [34] for organosilicon compounds. Several authors found high correlations between Kováts retention indices and molar volume for closely related sets of solutes.

Table 6.2. Van der Waals Group Increments[a]

Group	Volume, V_W (cm³/mol)	Total Surface Area, A_W (10^9 cm²/mol)	Radius, R_W (10^{-8} cm)
$-CH_3$	13.67	2.12	2.0
$-C_2H_5$	23.90	3.47	2.0
$-C_3H_7$	34.13	4.82	2.0
$-CH(CH_3)_2$	34.12	4.81	—
$-C(CH_3)_3$	44.34	6.36	—
$-C(CH_3)_2-$	30.67	4.24	—
$-CH_2-$	10.23	1.35	2.0
$>CH-$	6.78	0.57	—
$-\overset{\shortmid}{\underset{\shortmid}{C}}-$	3.33	0	—
$-Cyclo-C_6H_9$	46.5	—	—
$-Cyclo-C_6H_{11}$	56.8	—	—
$=CH_2$	11.94	1.86	—
$=CH-$	8.47	1.08	—
$=C<$	5.01	0.61	—
$=C=$ (all.)	6.96	—	—
$\equiv CH$	11.55	1.74	1.78
$\equiv C-$	8.05	0.98	1.78
$-C\equiv$ (diacetylene)	7.82	0.96	1.78
$\geqslant CH$ (aromatic)	8.06	1.00	1.77
$\geqslant C-$ (aromatic cond.)	4.74	0.21	1.77
$\geqslant C-R$ (aromatic aliphatic)	5.54	0.30	1.77
$-C_6H_5$	45.84	5.33	3.73
Phenylene	43.32	4.65	3.73
Naphthyl	71.45	7.76	—
$-NH_2$ (aromatic aliphatic)	10.54[b]	1.74	1.75
$-NH-$ (aliphatic)	8.08[b]	0.99	1.65
$>N-$ (aliphatic)	4.33	0.23	1.55
$\geqslant N$ (heteroaromatic)	5.2	—	1.6
4-Pyridyl	43.0	—	—
$-N=N-$	—	—	1.60
$-CN$ (aromatic aliphatic)	14.7	2.19	1.78
$-NO_2$ (aliphatic, mono)	16.8	2.55	—
$-PH_2$ (aliphatic, mono)	16.6	—	—
$-PH-$ (aliphatic)	13.53	—	—
$>P-$ (aliphatic)	10.44	—	1.80
$(-O)_3P=O$ (aromatic aliphatic)	25.8	—	—
$-O-$ (aliphatic)	3.7	0.60	1.45
$-O-$ (aromatic)	3.2	0.54	1.45
$-OH$ (aromatic aliphatic)	8.04	1.46	—
$>C=O$ (aliphatic)	11.70	1.60	—
$-CH=O$ (aromatic aliphatic)	15.14	2.37	1.65

[a] According to Bondi [33].
[b] Correction for neighboring carbonyl or amide group is -0.22.

Muchtarova and Dimov [35] described the relative retention of a set of aromatic hydrocarbons, aliphatic alcohols, and acetates in terms of solute boiling point and the reciprocal of molecular volume. However, the statistics has not been given, and the significance of a rather unusual $1/V$ term is unknown.

For homologous or closely related series of solutes chromatographed on stationary phases of different polarities, linear relationships between van der Waals volume and Kováts indices have been reported by Saura Calixto and Garcia Raso [36–38].

In the case of nonpolar solutes, alkylbenzenes, the linear correlations between retention indices and molar volume (calculated from molecular mass and density), van der Waals volume, molar refractivity, and index of refraction are significant [39]. For 61 alkylbenzenes, correlation coefficients r of indices obtained on a nonpolar stationary phase (squalane) against van der Waals volume was 0.94; in the case of molar volume, molar refractivity, and refraction index, the r values were 0.88, 0.92, and 0.23, respectively. The corresponding correlations obtained in the case of Kováts indices determined on a polar phase (Carbowax 20M) were significantly lower (except the refraction index) and amounted to 0.88, 0.77, 0.84, and 0.40, respectively. Experimentally determined molar volume has also been linearly related to retention of alkines on nonpolar phases, but the correlation dropped if retention data obtained on polar phases were considered [40, 41].

The lower correlations generally obtained on polar phases, as compared to the nonpolar ones, result from the relatively higher contribution of specific "polar" interactions with solutes on the polar phases. These polar interactions are not accounted for when bulk molecular parameters are the regressors. Bermejo et al. [39] derived also a two-parameter regression equation for a subgroup of 26 derivatives that in their opinion could be used for the identification of components of industrial mixtures. Unfortunately, the subgroup selected is not defined precisely, and the highly possible intercorrelation between the two parameters used in the equation (i.e., van der Waals volume and refraction index) is not determined for the data set considered.

In their studies on aliphatic esters, Bermejo and Guillen [42] observed a high correlation between GC retention indices determined on five phases of varying polarity and solute boiling points. The correlation coefficients ranged from 0.9987 on SE-30 to 0.9724 on Carbowax 1000. Having the predictive power of their relations in mind, these authors introduced additional parameters into their regression equations. In effect, they diminished the standard error by introducing to the regression the reciprocal of the van der Waals volume along with the refraction index. However, the problem arises of orthogonality of the data used in multiple regression correlation equation. The correlation matrix is not given in the original paper. The authors have attempted to avoid obvious intercorrelations among the regressors, and instead of the van der Waals volume, they used its reciprocal. It may be noted, however, that the boiling point of the solutes and the reciprocals of the molecular volumes used in the regression equations are still intercorrelated and should not be used together. Recently, Bermejo and co-workers [43] reversed their procedure to some extent, aiming at predicting boiling points based on Kováts indices and van der Waals volumes of a short series of chloro derivatives of 1,4-dimethylbenzene.

Bojarski and Ekiert [44] studied the linear relationships between van der Waals volumes and literature retention data generated by various chromatographic techniques for barbituric acid derivatives. The correlations with PC, TLC, GC, and HPLC retention parameters have been significant, especially in the case of GC and TLC data. The significance level of the relationships observed decreased with increasing diversity of structure of the barbiturates considered. The quality of the correlations obtained with van der Waals volume has been comparable to that observed with topological connectivity indices (see Chapter 8).

Macek and Smolková-Keulemansowa [45] related van der Waals volume to retention data for a group of 16 phenylacetic and phenylpropionic acid derivatives. Retention indices on two stationary phases OV-17 and OV-101 were determined along with the reversed-phase HPLC capacity factors. It has been found that in reversed-phase HPLC molar volume gave lower values of linear correlation coefficients than in GC. The correlation of Kováts indices determined on the OV-101 phase with V_W was higher than in the case of more polar OV-17 phase.

Jinno and Kawasaki [46] studied correlations between the logarithm of reversed-phase HPLC capacity factors and the van der Waals volumes of 19 isomeric alkylbenzenes. Three reversed-phase stationary materials (C-2, C-8, and C-18) of increasing alkyl chain lengths were used for generating retention data. Significant correlations between $\log k'$ and van der Waals volume were found, especially in the case of C-8 and C-18 phases (r of 0.93 and 0.92, respectively). These authors [46] combined the van der Waals volumes, connectivity index, and a hydrophobic parameter ($\log P$) [47] as the supposedly independent variables into one regression equation describing $\log k'$ values. The parameters were defined as follows: the van der Waals volume describes solute size, connectivity index characterizes its shape, and $\log P$ determines its hydrophobic properties. The correlation coefficient of the resulting three-parameter equations increases with respect to that obtained with a single parameter. However, the strong intercorrelations among the molecular descriptors used make the physical significance of the three-parameter equations reported questionable. The information concerning molecular size, shape, and hydrophobicity are contained in each of the individual descriptors used for the set of compounds studied. In other words, the individual descriptors used in the three-parameter equations may be represented by a linear function of one of them. Thus, the physical meaning of the three-parameter equations reduces to that expressed by a respective one-parameter relation.

In another paper [48] Jinno and Kawasaki correlated a $\log k'$ determined on various reversed-phase columns for a set of polycyclic aromatic hydrocarbons (PAH) with several structural descriptors, including van der Waals volume V_W. As expected for such closely congeneric solutes like PAH, high correlation exists between $\log k'$ versus V_W as well as when any other molecular-size-related quantity is used as the regressor. Despite the high correlation, the prediction of elution order of isomers is impossible unless a second molecular-shape-related parameter is included in the regression.

Van der Waals volumes as well as the other "bulky" parameters describe $\log k'$ satisfactorily in the reversed-phase HPLC of closely related nonpolar solutes, like PAH or alkylbenzenes. However, if more diverse sets of solutes are considered, correlations become insignificant. Thus, for PAH chromatographed on a C-18 stationary phase with methanol–water (75 : 25 v/v) solvent, the $\log k'$ versus V_W correlation coefficient is $r = 0.989$; if van der Waals surface area A_W is the independent variable, an analogous correlation is $r = 0.982$. If alkylbenzenes are chromatographed at identical conditions, then for $\log k'$ versus V_W relationships the correlation is $r = 0.992$, and similarly for $\log k'$ versus A_W, $r = 0.991$. However, if the solutes chromatographed at identical conditions are benzene derivatives possesing such functional groups like NH_2, OH, CN, NO_2, $COOCH_3$, and so on, the respective correlations are dropped to $r = 0.351$ for $\log k'$ versus V_W and to $r = 0.336$ for $\log k'$ versus A_W [49]. For a larger set of substituted benzene derivatives chromatographed in a C-18 reversed-phase HPLC system with an aqueous acetonitrile mobile phase, the variables $\log k'$ and V_W are practically orthogonal (i.e., $r = 0.075$). Similarly, correlations with other nonspecific molecular descriptors are low (Tables 6.3 and 6.4) [50].

The results of Jinno and Kawasaki clearly illustrate the predictive value of the van der Waals volume for QSRR analyses. It should be emphasized that V_W may be a valuable descriptor of the abilities of solutes to take part in dispersive interactions with the chromatographic phases, and it is often used in a multiparameter approach to QSRR.

Molecular surface area is a parameter that has a value to QSRR analysis similar to the van der Waals volume. The results of Horváth et al. [52] illustrate this. In Fig. 6.4 plots of $\log k'$ versus surface area, HSA, calculated for the hydrocarbonaceous part of the molecule form separate lines for different classes of substances. The relative displacement of the lines representing the three families of compounds in Fig. 6.4 can be accounted for by the differences

Table 6.3. RP HPLC Capacity Factor k' of Benzene Derivatives Determined on Octadecylsilica Column with Acetonitrile–water 65:35 Eluent with Molecular Descriptors[a]

Compound	k'	F	χ	V_w	A_w	π	HA	HD	HA+HD	HA−HD
Aniline	0.63	3	2.199	56.38	7.07	−1.23	1	1	2	0
N-Methylaniline	0.89	4	2.661	67.59	8.44	−0.47	1	1	2	0
N-Ethylaniline	1.13	5	3.221	77.82	9.79	0.08	1	1	2	0
Benzaldehyde	0.68	3	2.435	60.06	7.61	−0.65	1	0	1	1
Benzonitrile	0.73	3	2.384	60.54	7.52	−0.57	1	0	1	1
Nitrobenzene	0.83	3	2.448	62.64	7.88	−0.28	1	0	1	1
Anisole	0.91	4	2.523	62.71	7.99	−0.02	1	0	1	1
Acetophenone	0.72	4	2.865	71.21	9.05	−0.55	1	0	1	1
Methyl benzoate	0.91	4	2.977	76.73	9.83	−0.01	1	0	1	1
α-Bromotoluene	1.12	4	3.667	70.47	8.76	0.79	0	0	0	0
α-Bromoacetophenone	0.89	4	4.121	82.17	10.36	−0.32	1	0	1	1
o-Nitroaniline	0.66	3	2.653	70.66	8.94	−1.51	2	1	3	1
m-Nitroaniline	0.62	3	2.647	70.66	8.94	−1.51	2	1	3	1
p-Nitroaniline	0.54	3	2.647	70.66	8.94	−1.51	2	1	3	1
o-Chloroaniline	0.78	3	2.718	65.86	8.20	−0.52	1	1	2	0
m-Chloroaniline	0.74	3	2.712	65.86	8.20	−0.52	1	1	2	0
p-Chloroaniline	0.78	3	2.712	65.86	8.20	−0.52	1	1	2	0

p-Ethylaniline	1.05	5	3.171	77.76	9.86	−0.21	1	1	2	0
o-Methoxyaniline	0.68	4	2.728	70.73	9.05	−1.25	2	1	3	1
m-Aminoacetophenone	0.52	4	3.064	79.23	10.11	−1.78	2	1	3	1
p-Aminoacetophenone	0.45	4	3.064	79.23	10.11	−1.78	2	1	3	1
Phthalaldehyde	0.64	3	2.876	71.76	9.21	−1.30	2	0	2	2
Acetylsalicyclic acid	0.19	4	3.552	93.95	12.21	−4.37	2	1	3	1
Dimethyl phthalate	0.71	5	3.960	105.10	13.65	−0.02	2	0	2	2
Diethyl phthalate	1.02	7	5.135	125.56	16.35	1.02	2	0	2	2
o-Bromotoluene	1.74	4	3.319	72.11	8.90	1.42	0	0	0	0
m-Bromotoluene	1.79	4	3.313	72.11	8.90	1.42	0	0	0	0
o-Nitrotoluene	1.01	4	2.865	73.79	9.32	0.28	1	0	1	1
m-Nitrotoluene	1.09	4	2.859	73.79	9.32	0.28	1	0	1	1
Methoxy-3-methylbenzene	1.14	5	2.934	73.86	9.43	0.54	1	0	1	1
α-Bromo-p-nitrotoluene	1.05	4	4.115	84.75	10.63	0.51	1	0	1	1

^{a} From Jinno and Kawasaki [50]. With permission. F = number of double bonds plus number of primary and secondary carbons minus 0.5 for a nonaromatic ring; χ = molecular connectivity [51]; V_w and A_w = van der Walls volume and surface area; π = hydrophobic constant [47]; HA = number of electron acceptor groups; HD = number of electron donor groups.

Table 6.4. Correlations Between log k′ and Molecular Descriptors in Table 6.3[a]

Descriptor	Correlation Coefficient, r
F	0.294
χ	0.176
V_W	−0.075
A_W	−0.110
π	0.962
HA	−0.717
HD	−0.491
HA + HD	−0.731
HA − HD	−0.314

[a]Number of dependent variables $n = 31$. According to Jinno and Kawasaki [50]. With permission.

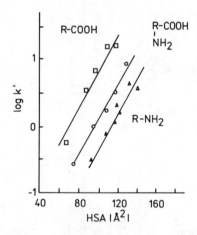

Figure 6.4. Relationship between logarithm of capacity factor, log $k′$, and hydrocarbonaceous surface area of different classes of solutes. (After Cs. Horváth, W. Melander, and I. Molnár, *J. Chromatogr.*, **125**, 129, 1976. With permission.)

in the abilities of solutes to participate in chemically specific polar interactions with the chromatographic phases.

An interesting approach to QSRR was presented by Hanai et al. [53], who determined log $k′$ in a reversed-phase HPLC system for alkanes, alkyl alcoholes, alkylbenzenes, halogenated benzenes, and PAH at six temperatures using acetonitrile–water (80 : 20 and 70 : 30 v/v) as the mobile phase. Thus,

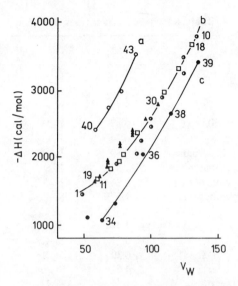

Figure 6.5. Relation between enthalpy values, ΔH, calculated from log k' values and van der Waals volumes, V_W, for (a) aliphatic hydrocarbons, (b) aromatic hydrocarbons and their alkyl and chloro derivatives, and (c) aliphatic alcohols. (After T. Hanai, A. Jukurogi, and J. Hubert, *Chromatographia*, **19**, 266, 1984. With permission.)

enthalpy values, ΔH, of the solutes could be determined. Such chromatographically derived enthalpies are not a simple linear function of molar volume, as illustrated in Fig. 6.5.

The curves in Fig. 6.5 join separate classes of solutes. In the case of a relatively diverse set of aromatic hydrocarbons and their chloro and alkyl derivatives, the deviations from linearity are evident.

6.4. MOLAR REFRACTIVITY AND POLARIZABILITY

Molar refractivity and molecular polarizability are other "bulk" parameters that may be related to solute ability to undergo dispersive interactions with chromatographic phase components. For any given molecular formula refractivity may easily be calculated by summation of the atomic, the group, or the bond-type increments. Independent of the method applied for calculation, the refractivity values obtained are highly interrelated. Summation of the refractivities assigned to individual bonds in the molecule is a convenient calculation. The appropriate data are given in Table 6.5 [54, 55].

As expected in GC, linearity between retention parameters and solute molar

Table 6.5. Bond Refractivities, $MR_B{}^a$

Bond Type	MR_B	Bond Type	MR_B
C–H	1.676	N=N	4.12
C–C	1.296	P–C	3.575
C=C	4.17	P–Cl	8.856
C≡C (terminal)	5.87	P–H	4.010
C≡C (other)	6.24	P–O	3.102
C–C (in cyclopropane)	1.49	P→O	−1.032
C–C (in cyclobutane)	1.37	P–S	7.583
C–C (in cyclopentane)	1.26	P=S	6.866
C–C (in cyclohexane)	1.27	Si–C	2.52
C_{arom}–C_{arom}	2.688	Si–C_{arom}	2.93
C–F	1.44	Si–F	1.7
C–Cl	6.51	Si–Cl	7.11
C–Br	0.39	Si–Br	10.24
C–I	14.61	Si–O	1.80
C–O (in ethers)	1.54	Si–H	3.17
C–O (in acetols)	1.46	Si–S	6.14
C=O	3.32	Si–N	2.16
C=O (in methyl ketones)	3.49	Ge–C	3.05
C–S	4.61	Ge–Cl	7.6
C=S	11.91	Ge–Br	11.1
C–N	1.57	Ge–I	16.7
C=N	3.76	Ge–F	1.3
C≡N	4.82	Ge–O	2.47
O–H (in alcohols)	1.66	Ge–S	7.02
O–H (in acids)	1.80	Ge–N	2.33
S–H	4.80	Pb–C	5.26
S–S	8.11	Hg–C	7.21
S–O	4.94	Sn–C	4.16
S→O	−0.20	Sn–C_{arom}	3.78
N–H	1.76	Sn–Cl	8.91
N–O	2.43	Sn–Br	12.00
N→O	1.78	S–I	17.92
N=O	4.00	Sn–Sn	10.77
N–N	1.99	Sn–O	3.84

^{a} According to Vogel [54, 55] as collected by Iophphe [56]. Data for sodium D line.

refractivity was often observed, especially when congeneric solutes were chromatographed on nonpolar stationary phases. The importance of solute refractivity for its retention has been suggested in early papers by Kováts. Basing their work on refraction–retention relationships, Ellrén et al. [57] proposed the *refraction number*, which is analogous to the Kováts retention index. Linearity of logarithms of retention parameters plotted against molar refraction was reported for several homologous series by Wurst [29, 58, 59]. At the same time Putman and Pu [60] found correlations between molar refraction and Kováts indices measured on apiezon for a series of tetraalkyltin compounds. Other early examples of such correlations for several classes of organic substances chromatographed on nonpolar stationary phases were published by Vorob'ev [61], Iwata et al. [62], and Araki [63].

Radecki et al. [64] and Grzybowski et al. [65] connected the Kováts indices for a series of phenols determined on stationary phases of different polarity by means of molar refractivity. They obtained highly significant regression equations of the general form

$$I_P = k_1 I_{NP} - k_2(MR) + k_3 \tag{6.4}$$

where I_P is the Kováts index determined for a given solute on a polar phase (e.g., polyneopentyl glycol adipate, NGA); I_{NP} is the corresponding index from a nonpolar phase (e.g., dimethylpolysiloxane, SE-30); MR is the molar refractivity calculated by summation of group increments taken from a compilation by Hansch et al. [66] and k_1, k_2, and k_3 are the regression coefficients. These authors [66] misinterpreted their equations of the form of Eq. (6.4), suggesting that the larger the MR value of a compound, the lower the retention index, which certainly is not true. Their observations, however, gave rise to detailed analysis of the relationship by Kaliszan and Höltje [67]. To explain relationships of the type of Eq. (6.4), the following argument was proposed [67, 68].

If one has retention indices for a given set of compounds on a polar phase, I_P, and on a nonpolar phase, I_{NP}, according to the most general rules discussed in the preceding chapters, one can write

$$I_P = a_P P + b_P D + c_P \tag{6.5}$$

$$I_{NP} = a_{NP} P + b_{NP} D + c_{NP} \tag{6.6}$$

where a, b, and c are constants depending on the properties of the individual phases; P and D are solute polarity and dispersive descriptors, respectively. From Eq. (6.6),

$$P = (I_{NP} - b_{NP}D - c_{NP})/a_{NP} \tag{6.7}$$

Equation (6.5) can be rewritten as

$$I_P = a_P(I_{NP} - b_{NP}D - c_{NP})/a_{NP} + b_P D + c_P \qquad (6.8)$$

Rearranging,

$$I_P = \frac{a_P}{a_{NP}} I_{NP} - \left(\frac{a_P}{a_{NP}} b_{NP} - b_P \right) D + \text{const} \qquad (6.9)$$

or

$$I_P = k_1 I_{NP} - k_2 D + k_3 \qquad (6.10)$$

where k_1, k_2, and k_3 are new constants yielded by the constants of Eq. (6.9). Thus, Eq. (6.10) explains the experimentally observed relationships between retention indices determined on GC phases of different polarities. Instead of MR, other dispersive property descriptors may be used. Recently, a relationship of the type of Eq. (6.10) has been reported by Halkiewicz et al. [69]. Here, a quantity reflecting the size of the alkyl substituent in dialkyldithiocarbamates complexes with copper and nickel was used as the bulky parameter.

Bermejo and co-workers [39, 42] found MR to be as good a descriptor of solute polarizability as the van der Waals volume V_M in a series of alkylbenzenes chromatographed on squalane and on Carbowax 20 M. The objections can be applied to the multiple correlations similar to those already discussed (see page 89) involving the boiling point and the reciprocal of molar refraction.

The results recently obtained by Kuchař and co-workers [70, 71] seem valuable. These authors analyzed GC data for nonhomologous, although related, series of solutes. In the case of variously substituted dibenzo[b, f]thiepins [70] the Kováts index determined on a nonpolar phase (Dexsil 300) has been significantly related to MR: $r = 0.845$, F-test value 79.6. However, seven solutes of total number 41 did not fit the regression and were omitted. The outliers were the derivatives possessing trifluoromethyl and nitro substituents. The other group of solutes analyzed by Kuchař et al. [71] formed arylalkoxy- and alkoxyarylaliphatic acids. This group has certainly been more congeneric than the former. As expected, correlation between Kováts indices determined on a nonpolar phase (Apolane 87) and MR was relatively high: $r = 0.941$, $F = 271.8$. Correlations can be further improved by separating the solutes in two more congeneric series (i.e., arylalkoxy and alkoxy derivatives).

That molar refractivity is a valuable descriptor of nonspecific chromatographic interactions for related solutes (not only for homologues) is evident from the paper by Macek and Smolkova-Keulemansova [45]. The solutes,

Table 6.6. GC Retention Indices (I) Determined on Stationary Phases OV-101 and OV-17, Reversed-Phase HPLC Capacity Factors of Undissociated Forms of Acids Studied and Molar Refraction[a]

Compound (acid)	$I_{\text{OV}-101}$	$I_{\text{OV}-17}$	k'	Molar Refraction (cm^3/mol)
2-Phenylpropionic	1299.7 ± 1.2	1453.0 ± 0.9	1.179	72.24
2-(4-Methylphenyl)propionic	1385.6 ± 0.7	1547.8 ± 0.9	1.832	81.33
2-(4-Ethylphenyl)propionic	1473.1 ± 0.8	1634.2 ± 0.4	2.810	89.06
2-(4-Isobutylphenyl)propionic	1614.3 ± 0.3	1751.9 ± 0.8	7.551	104.59
2-(3-Chlorophenyl)propionic	1455.4 ± 0.3	1567.6 ± 0.4	1.885	82.38
2-(4-Bromophenyl)propionic	1574.0 ± 0.6	1751.3 ± 0.7	2.152	87.70
2-(4-Methoxyphenyl)propionic	1528.6 ± 0.2	1735.2 ± 0.3	0.926	84.19
2-(4-Phenoxyphenyl)propionic	2066.7 ± 0.9	2270.7 ± 1.3	4.992	120.00
Phenylacetic.	1269.4 ± 0.3	1442.8 ± 0.3	0.737	64.45
4-Ethylphenylacetic	1452.2 ± 0.6	1571.2 ± 0.4	1.890	81.27
4-Isopropylphenylacetic	1514.4 ± 0.5	1680.8 ± 0.7	2.683	89.06
4-Cyclohexylphenylacetic	1939.5 ± 1.2	2142.9 ± 0.9	9.448	109.57
4-(n-Octyloxy)-3-chlorophenylacetic	2305.0 ± 1.7	1552.8 ± 1.4	78.01	141.24
4-Isopropoxy-3-chlorophenylacetic	1766.5 ± 0.5	1966.6 ± 0.9	2.770	102.05
4-Benzyloxy-3-chlorophenylacetic	2206.8 ± 1.4	2618.7 ± 1.5	5.696	130.07
4-Allyloxy-3-chlorophenylacetic	1837.0 ± 1.0	2080.6 ± 0.9	2.273	101.97

[a] Molar refraction according to Gladstone and Dale. Data compiled from Macek et al. [45]. With permission.

along with retention and structural data, are given in Table 6.6. The correlation for the OV-101 phase, which is characterized by $r = 0.977$, is given in Fig. 6.6.

For the more polar phase OV-17, the analogous correlation is lower, and $r = 0.960$.

The correlation obtained in the same paper of the ln k' from reversed-phase HPLC against MR is markedly lower ($r = 0.88$) but is still significant. The capacity factor k' determined by Macek and Smolková-Keulemansowa [45] is a quantity corrected for ionization of the acids in the column and thus adequately reflects the properties of the undissociated form of the solute, for which molar refraction is calculated. Such corrections of retention data are, unfortunately, not always done prior to QSRR studies.

The relationships between GC retention data and molecular polarizabilities of solutes have been reported for organosilicon compounds [27, 72].

In their studies on the GC retention indices of 10 polycyclic aromatic hydrocarbons (PAH) on five stationary phases, Lamparczyk et al. [73] found linear relationships with average molecular polarizabilities, calculated after Miller and Savchik [74]. The discussion of the results in the original paper is somewhat obscured by the incorrect assumption of the identity of the dimensionless Kováts index and the term in the classical Debye equation related to inductive solute–stationary phase interactions. In any case, linear relationships between retention indices and polarizability data were found for that closely related group of solutes. The percentage error of polarizability calculated from the regression equations ranged from 2 to 4% when compared with experimental values applied for deriving the relationships.

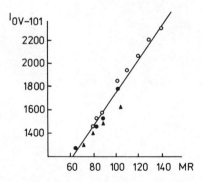

Figure 6.6. Correlation between retention indices on OV-101 phase I_{OV-101}, and molar refraction according to Gladstone and Dale, MR (see Table 6.6). △, Alkyl derivatives of phenylpropionic acid; ●, alkyl derivatives of phenylacetic acid; ○, other derivatives. (After J. Macek and E. Smolková-Keulemansová, *J. Chromatogr.*, **333**, 309, 1985. With permission.)

In a later paper, Lamparczyk and Radecki [75] attempted to substantiate the polarizability-versus-retention relationship previously observed for PAH. They started from the equation for the potential energy of van der Waals interactions, E:

$$E = -\frac{1}{r^6}\left[\frac{2}{3kT}\mu_a^2\mu_b^2 + \alpha_a\mu_b^2 + \alpha_b\mu_a^2 + \frac{3I_aI_b}{2(I_a+I_b)}\alpha_a\alpha_b\right] \quad (6.11)$$

where the subscripts a and b denote the interacting molecules; μ is the dipole moment; T is the absolute temperature; r is the distance of interaction; α is the molecular polarizability; I is the first ionization potential; and k is the Boltzmann constant. Equation (6.11) has previously been analyzed by Gassiot-Matas and Firpo-Pamies [76], but these authors omitted the second and third terms in the equation, assuming that the induction effect is never important in interactions between neutral molecules.

In considering closely related compounds, Lamparczyk and Radecki [75] assumed that the ionization potential was constant among the series. At any given phase and temperature conditions, the terms r, T, μ_{phase}, and α_{phase} can also be considered constant. Thus, after simple calculations, one obtains the relation

$$I = f(E) = a\mu_{solute}^2 + b\alpha_{solute} + c \quad (6.12)$$

where I is the retention index of the individual solute of the series determined at given conditions, and a, b, and c are constants. Thus, in QSRR studies, the practically verifiable form of Eq. (6.11) is of the same type as that suggested in the late 1950s by Kováts and applied practically in QSRR by Gassiot-Matas and Firpo-Pamies [76] and Kaliszan and Höltje [67].

Lamparczyk and co-workers [75, 77] further simplified Eq. (6.12) assuming for PAH the dipole moment $\mu_{solute} = 0$, and thus they derived a description of Kováts indices determined for these compounds on various phases as a linear function of their polarizability. That correlation allowed Lamparczyk [77] to propose the chromatographic interaction index I_1, defined as

$$I_1 = (\alpha_{s2} - \alpha_{s1})\frac{\log X - \log X_{s1}}{\log X_{s2} - \log X_{s1}} + \alpha_{s1} \quad (6.13)$$

where α_1 and α_2 denote two standard solutes, having all properties equal except polarizability; X is the retention parameter depending on the system applied, that is, the Kováts index in GC, $\log k'$ in HPLC, and $\log(1/R_f)$ in TLC.

Good correlations between the indices I_1 and the polarizabilities of PAH were found in the case of Kováts indices determined on nonpolar stationary phases as well as for logarithms of capacity factors determined in reversed-phase HPLC systems. Correlation with the data from partitioning TLC (straight-phase system) was lower, but still significant. However, no correlation ($r = 0.37$) was found when the TLC data from adsorption chromatography on acetylated cellulose were considered.

References

1. A. Ladon and S. Sandler, Gas-chromatographic retention and molecular structure, An extension of the James–Martin rule, *Anal. Chem.*, **45**, 921, 1973.

2. G. Castello and G. D'Amato, The correlation between the physical properties and structure of alcohols and their gas chromatographic behaviour on polar and nonpolar stationary phases. I. C_1–C_8 straight-chain alcohols, *J. Chromatogr.*, **131**, 41, 1977.

3. Y. Hoshikawa, K. Koike, and T. Kuriyama, Estimation of retention indexes of higher dimethyl paraffins by a graphical method, *Bunseki Kagaku*, **21**, 307, 1972.

4. R. J. Smith, J. K. Haken, and M. S. Wainwright, Estimation of dead time and calculation of Kováts indices, *J. Chromatogr.*, **334**, 95, 1985.

5. M. V. Budahegyi, E. R. Lombosi, T. S. Lombosi, S. Y. Mészáros, Sz. Nyiredy, G. Tarján, I. Timár, and J. M. Takács, Twenty-fifth anniversary of the retention index system in gas–liquid chromatography, *J. Chromatogr.*, **271**, 213, 1983.

6. J. K. Haken and I. O. O. Korhonen, Gas Chromatography of homologous esters. XXVII. Retention increments of C_1–C_{18} primary alkanols and their 2-chloropropanoyl and 3-chloropropanoyl derivatives on SE-30 and OV-351 capillary columns, *J. Chromatogr.*, **319**, 131, 1985.

7. F. Saura Calixto and A. Garcia-Raso, Relationship between gas chromatographic retention indices of homologous compounds and the number of carbon atoms and temperature, *J. Chromatogr.*, **216**, 326, 1981.

8. A. Garcia-Raso, M. A. Vázquez, and P. Ballester, Gas chromatography of bicyclo[n.4.0]alken-2-ones, *J. Chromatogr.*, **331**, 406, 1985.

9. V. A. Chernoplekova, A. N. Korol', K. I. Sakodynskii, U. S. Lopatina, and K. A. Kocheshkov, Primenenie termodinamicheskich phunkcii rastvoreniya dlya identiphikatsii elementoorganicheskikh soedinenii v gazozhidkostnoi khromatographii, *Zh. Anal. Khim.*, **30**, 1285, 1975.

10. O. Ellrén, I.-B. Peetre, and B. E. F. Smith, Gas Chromatographic investigation of organometallic compounds and their carbon analogues. I. Determination, calculation and correlation of Kováts' retention indices for tetraalkoxysilanes, *J. Chromatogr.*, **88**, 295, 1974.

11. R. U. Golovnya and D. N. Grigor'eva, Universalnyi vid uravneniya dlya rascheta udelnykh uderzhivaemykh obemov, indeksov uderzhivaniya i differentsialnykh molnykh svobodnykh energii rostvoreniya chlenov gomologicheskikh ryadov organicheskikh soedinenii, *Zh. Anal. Khim.*, **40**, 316, 1985.

12. A. T. James and A. J. P. Martin, Gas–liquid partition chromatography: The

separation and microestimation of volatile fatty acids from formic acid to dodecanoic acid, *Biochem. J.*, **50**, 679, 1952.

13. E. A. Kirichenko and B. A. Markov, *Tr. Mosk. Khim. Tekhnol. Inst.*, **70**, 143, 1972; after Shatz, V. D., Sturkovich, R. Ya, and Lukevics, E., Gas chromatographic analysis of organosilicon compounds, *J. Chromatogr.*, **165**, 257, 1979.

14. A. N. Korol, Vozmozhnosti predskazaniya velichin uderzhivaniya v gazozhidkostnoi khromatographii, *Uspekhy Khim.*, **60**, 1225, 1982.

15. J. Preisler, Separation of the homologous series of methylcyclopolysiloxanes, methylhydrocyclopolysiloxanes, methylethylhydrocyclopolysiloxanes and linear methylpolysiloxanes by means of gas–liquid chromatography, *Z. Anal. Chem.*, **240**, 389, 1968.

16. J. Green, S. Marcinkiewicz, and D. McHale, Paper chromatography and chemical structure. III. The correlation of complex and simple molecules. The calculation of R_M values for tocopherols, vitamins K, ubiquinones and ubichromenols from R_M phenol. Effect of unsaturation and chain branching, *J. Chromatogr.*, **10**, 158, 1963.

17. D. R. Clifford, D. M. Fieldgate, and D. A. M. Watkins, Chromatography of dinitrophenols, *J. Chromatogr.*, **43**, 110, 1969.

18. C. B. C. Boyce and B. V. Milborrow, A simple assessment of partition data for correlating structure and biological activity using thin-layer chromatography, *Nature (London)*, **208**, 537, 1965.

19. F. Murakami, Quantitative structure–retention relationships in cation-exchange chromatography, *J. Chromatogr.*, **198**, 241, 1980.

20. E. Tomlinson, H. Poppe, and J. C. Kraak, Thermodynamics of functional groups in reversed-phase high performance liquid–solid chromatography, *Int. J. Pharm.*, **7**, 225, 1981.

21. T. Vitali, E. Gaetani, C. F. Laureri, and C. Branca, Cromatografia liquido–liquido ad alta pressione. Correlazione fra tempi di ritenzione ed attivita biologica di derivativi 1,3,5-triazinici, *Il Farmaco (Ed. Sci.)*, **31**, 58, 1976.

22. R. E. Koopmans and R. F. Rekker, High-performance liquid chromatography of alkylbenzenes. Relationship with lipophilicities as determined from octanol–water partition coefficients or calculated from hydrophobic fragmental data and connectivity indices; Lipophilicity predictions for polyaromatics, *J. Chromatogr.*, **285**, 267, 1984.

23. G. N. Bortnikov, N. S. Vyazankin, N. P. Nikulina, and Ya. I. Yashin, Gazovaya khromatographiya etilnykh proizvodnykh kremniya, germaniya i olova, *Izv. Akad. Nauk SSSR Ser. Khim.*, 21, 1973.

24. V. A. Kochetov, V. M. Kopylov, B. A. Markov, M. I. Shkol'nik, E. A. Kirichenko, and K. A. Andrianov, Khromatographicheskii analiz azotsoderzhashchikh kremniiorganicheskikh soedinenii, *Zh. Anal. Khim.*, **33**, 1214, 1978.

25. E. Lukevics, R. Ya. Moskovich, and V. D. Shatz, Azotosoderzhanie kremniiorganicheskie soedineniya. XLIII. Khromatographicheskoe issledovanie 1-zameshchennykh piperidinov i pergidrazepinov, *Zh. Obshch. Khim.*, **44**, 105, 1974.

26. E. Lukevics, R. Moskivich, and V. Shatz, *Izv. Akad. Nauk Latv. SSR, Ser. Khim.*,

53, 1976; after Shatz, V. D., Sturkovich, R. Ya, and Lukevics, E., Gas chromatographic analysis of organisilicon compounds, *J. Chromatogr.*, **165**, 257, 1979.

27. V. D. Shatz, R. Ya. Sturkovich, and E. Lukevics, Gas chromatographic analysis of organosilicon compounds, *J. Chromatogr.*, **165**, 257, 1979.

28. B. J. Gudzinowicz and H. F. Martin, Separation of organo- and organobromoarsenic compounds by gas–liquid chromatography, *Anal. Chem.*, **34**, 648, 1962.

29. M. Wurst, Gaschromatographie. III. Die Abhängigkeit der Elutionsdaten organischer Stoffe vom Parachor, *Microchim. Acta*, 379, 1966.

30. N. Dimov and S. Boneva, Separation and identification of 2-nitroalkanol-1 compounds by gas chromatography, *J. Chromatogr.*, **206**, 549, 1981.

31. H. Wiener, Structural determination of paraffin boiling points, *J. Am. Chem. Soc.*, **69**, 17, 1947.

32. M. Charton, Volume and bulk parameters, in M. Charton and I. Motoc (Eds.), *Steric Effects in Drug Design*, Akademie-Verlag, Berlin, 1983, p. 107.

33. A. Bondi, *Physical Properties of Molecular Crystals, Liquids and Glasses*, Wiley, New York, 1968.

34. M. Wurst and J. Churacek, Gaschromatographie. V. Die Beziehung zwischen den Elutionsdaten und dem Molvolumen, *J. Chromatogr.*, **70**, 1, 1972.

35. M. Muchtarova and N. Dimov, Gas chromatographic identification of some indoor air pollutants using correlation equations, *J. Chromatogr.*, **148**, 269, 1978.

36. F. Saura Calixto and A. Garcia-Raso, Correlation between van der Waals' volume and retention index. General equation applicable to different homologous series, *Chromatographia*, **14**, 596, 1982.

37. F. Saura Calixto and A. Garcia Raso, Retention index, connectivity index and van der Waals' volume of alkanes GLC, *Chromatographia*, **15**, 521, 1982.

38. F. Saura Calixto and A. Garcia Raso, Influence of the van der Waals volume of ethers, esters, carbonyls, and alcohols on retention on gas chromatographic stationary phases of different polarity, *Chromatographia*, **15**, 771, 1982.

39. J. Bermejo, J. S. Canga, O. M. Gayol, and M. D. Guillén, Utilization of physicochemical properties for calculating retention indices of alkylbenzenes, *J. Chromatogr. Sci.*, **22**, 252, 1984.

40. P. G. Robinson and A. L. Odell, A system of standard retention indices and its uses. The characterization of stationary phases and the prediction of retention indices, *J. Chromatogr.*, **57**, 1, 1971.

41. A. F. Shlyakhov, R. I. Koreshkova, and M. S. Telkova, Gas chromatography of izoprenoid alkanes, *J. Chromatogr.*, **104**, 337, 1975.

42. J. Bermejo and M. D. Guillén, A study of Kováts retention indices and their relation to the polarity of the stationary phase, *J. Chromatogr.*, **318**, 187, 1985.

43. J. Bermejo, C. G. Blanco, and M. D. Guillén, Capillary gas chromatography of chloro derivatives of 1,4-dimethylbenzene. Separation, identification and prediction of boiling points, *J. Chromatogr.*, **331**, 237, 1985.

44. J. Bojarski and L. Ekiert, The evaluation of molecular connectivity indices and van

der Waals volumes for correlation of chromatographic parameters, in H. Kalasz (Ed.), *Proceedings of the Symposium on Advanced Liquid Chromatography, Szeged,* 1982, Akadémiai Kiadó, Budapest, 1984.

45. J. Macek and E. Smolková-Keulemansowa, Correlation of the chromatographic retention of some phenylacetic and phenylpropionic acid derivatives with molecular structure, *J. Chromatogr.,* **333,** 309, 1985.

46. K. Jinno and K. Kawasaki, Correlations between retention data of izomeric alkylbenzenes and physical parameters in reversed-phase micro high-performance liquid chromatography, *Chromatographia,* **17,** 337, 1983.

47. C. Hansch and A. Leo, *Substituent Constants for Correlation Analysis in Chemistry and Biology,* Wiley, New York, 1979.

48. K. Jinno and K. Kawasaki, Correlation between the retention data of polycyclic aromatic hydrocarbons and several descriptors in reversed-phase HPLC, *Chromatographia,* **17,** 445, 1983.

49. K. Jinno and K. Kawasaki, Computer-assisted retention prediction system for reversed-phase micro high-performance liquid chromatography, *J. Chromatogr.,* **316,** 1, 1984.

50. K. Jinno and K. Kawasaki, Retention prediction of substituted benzenes in reversed-phase HPLC, *Chromatographia,* **18,** 90, 1984.

51. L. B. Kier and L. H. Hall, *Molecular Connectivity in Chemistry and Drug Research,* Academic, New York, 1976.

52. Cs. Horváth, W. Melander, anion prediction of substituted benzenes in reversed-phase HPLC, *Chromatographia,* **18,** 90, 1984.

53. T. Hanai, A. Jukurogi, and J. Hubert, Retention vs. van der Waals volume, pi-energy and hydrogen-bonding energy effects and enthalpy in liquid chromatography, *Chromatographia,* **19,** 266, 1984.

54. A. I. Vogel, W. T. Cresswell, G. H. Jeffery, and J. Leicester, Physical Properties and chemical constitution. Part XXIV. Aliphatic aldoximes, ketoximes, and ketoxime O-alkyl ethers, N,N-dialkylhydrazines, aliphatic ketazines, mono- and di-alkylaminopropionitriles, alkoxypropionitriles, dialkyl azodiformates, and dialkyl carbonates. Bond parachors bond refractions, and bond-refraction coefficients, *J. Chem. Soc.,* 514, 1952.

55. A. I. Vogel, W. T. Cresswell, and J. Leicester, Bond refractions for tin, silicon, lead, germanium and mercury compounds, *J. Phys. Chem.,* **58,** 174, 1954.

56. B. V. Iophphe, Rephraktometricheskie metody khimii, GNTIKhL, Leningrad, 1960.

57. O. Ellrén, I.-B. Peetre, and B. E. F. Smith, Gas chromatographic investigation of organometallic compounds and their carbon analogues. V. Use of refractive index in conjunction with Kováts' retention index for the identification of organosilicon compounds, *J. Chromatogr.,* **93,** 383, 1974.

58. M. Wurst, Analyse von Organosiliciumverbindungen III. Trennung und Bestimmung linearer und cyclischer Polydimethylsiloxane mittels Gaschromatographie, *Collect. Czech. Chem. Commun.,* **29,** 1458, 1964.

59. M. Wurst, Die gaschromatographische Trennung und Bestimmung linearer und cyclischer Polydimethylsiloxane, *Z. Anal. Chem.*, **211**, 73, 1965.

60. R. C. Putnam and H. Pu, Retention indexes of organotins, *J. Gas Chromatogr.*, **3**, 160, 1965.

61. L. N. Vorob'ev, Gazo-zhidkostnaya khromatographiya; O zavisimosti uderzhivaemykh obemov organicheskikh soedinenii ot reaktsii, *Coll. Czech. Chem. Commun.*, **27**, 1045, 1962.

62. R. Iwata, Y. Makide, and T. Tominaga, The gas-chromatographic behaviour of perhaloalkanes, *Bull. Chem. Soc. Japan*, **47**, 3071, 1974.

63. T. Arahi, The effects of stationary liquids on the separation of the isomers of aromatic hydrocarbons in gas chromatography, *Bull. Chem. Soc. Japan*, **36**, 879, 1963.

64. A. Radecki, J. Grzybowski, H. Lamparczyk, and A. Nasal, Relationship between retention indices and substituent constants of phenols on polar stationary phases, *J. High Resolut. Chromatogr. Chromatogr. Commun.*, **2**, 581, 1979.

65. J. Grzybowski, H. Lamparczyk, A. Nasal, and A. Radecki, Relationship between the retention indices of phenols on polar and non-polar stationary phases, *J. Chromatogr.*, **196**, 217, 1980.

66. C. Hansch, A. Leo, S. H. Unger, K.-H. Kim, D. Nikaitani, and E. J. Lien, "Aromatic" substituent constants for structure–activity correlations, *J. Med. Chem.*, **16**, 1207, 1973.

67. R. Kaliszan and H.-D. Höltje, Gas chromatographic determination of molecular polarity and quantum chemical calculation of dipole moments in a group of substituted phenols, *J. Chromatogr.*, **234**, 303, 1982.

68. R. Kaliszan, Quantitative relationships between molecular structure and chromatographic retention. Implications in physical, analytical, and medicinal chemistry, *Crit. Rev. Anal. Chem.*, **16**, 323, 1986.

69. J. Halkiewicz, H. Lamparczyk, and A. Radecki, Behaviour of copper(II) and nickel(II) dialkyldithiocarbamates on polar and non-polar stationary phases, *Z. Anal. Chem.*, **320**, 577, 1985.

70. H. Tomková, M. Kuchař, V. Rajholec, V. Pacáková, and V. E. Smolková-Keulemansowá, Relationship between the chromatographic behaviour and structure of some substituted dibenzo[b,f]thiepins and their analogues, *J. Chromatogr.*, **329**, 113, 1985.

71. M. Kuchař, H. Tomková, V. Rejholec, and O. Skalická, Relationships between gas–liquid chromatographic behaviour and structure of arylaliphatic acids, *J. Chromatogr.*, **333**, 21, 1985.

72. G. N. Bortnikov, A. N. Egorochkin, N. S. Vyazankin, E. A. Chernyshev, and Ya. I. Yashin, Izuchenie elektronnykh ephphektov kremnezameshchennykh tiophenov metodami gazovoi khromatographii i yadernogo magnitnogo rezonansa, *Izv. Akad. Nauk SSSR Ser. Khim.*, 1402, 1970.

73. H. Lamparczyk, D. Wilczyńska, D. and A. Radecki, Relationship between the average molecular polarizabilities of polycyclic aromatic hydrocarbons and their

retention indices determined on various stationary phases, *Chromatographia*, **17**, 301, 1983.

74. K. J. Miller and J. A. Savchik, A new empirical method to calculate average molecular polarizabilities, *J. Am. Chem. Soc.*, **101**, 7206, 1979.

75. H. Lamparczyk and A. Radecki, The role of electric interactions in the retention index concept; implications in quantitative structure–retention studies, *Chromatographia*, **18**, 615, 1984.

76. M. Gassiot-Matas, and G. Firpo-Pamies, Relationships between gas chromatographic retention index and molecular structure, *J. Chromatogr.*, **187**, 1, 1980.

77. H. Lamparczyk, The role of electric interaction in the retention index concept. Universal interaction indices for GLC, HPLC and TLC, *Chromatographia*, **20**, 283, 1985.

CHAPTER

7

PARAMETERS RELATED TO SPECIFIC
PHYSICOCHEMICAL PROPERTIES OF SOLUTES

In this chapter molecular structural descriptors in QSRR studies that basically are not additive are discussed. The quantities discussed may be related to the electronic properties of a solute as a whole or to some structural fragments of the solute that play a decisive role in chromatographic interactions. Thus, in the following sections the QSRR applications of dipole moments, of electronic substituent constants, and of quantum chemical indices are reviewed. A separate chapter is devoted to the chromatographic applications of some molecular shape descriptors.

7.1. DIPOLE MOMENTS

Early observations of parallel displacements for separate homologous series of the retention-versus-dispersive descriptor relationships suggested the important role of specific solute–chromatograpic phase interactions for retention. These specific, or polar, interactions are related to the properties of functional groups present in individual series. There are various functional groups that do not differ significantly in van der Waals volumes, refractivities, or other "bulk" properties. On the other hand, the differences created by a substituent and concerning the overall electric properties of the solutes are often pronounced. Dipole moment has long since been known as a measurable and/or calculable theoretically quantity characterizing dielectric constants and generally a chemical compound polarity.

To prove quantitatively the contribution of dipole moment to the overall retention, the solutes selected have dispersive properties that are nearly constant in the series, whereas dipole moments differ. Such an approach has successfully been applied by Scott [1] and Karger et al. [2] (see Chapter 3 for details). A valuable finding by these authors is that local dipoles, rather than the overall molecular dipole moment, contribute to retention.

Very recently a paper appeared [3] that provides an illustration of the importance to GC retention of the polar properties of the solutes. In that publication the linear relationship is given between logarithmic retention time and the position of keto group in a series of positional isomers, methyl

108

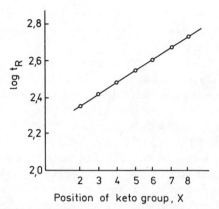

Figure 7.1. Plot of logarithm of retention time, log t_R, against position of keto group, X, of methyl X-ketononanoates. Stationary phase: 10% Carbowax 20 M terminated with terephthalic acid; 150°C. (After B. Abegaz and S. Gerba, *J. Chromatogr.*, **324**, 440, 1985. With permission.)

ketononanoates (Fig. 7.1). Certainly, for the isomers considered, dispersive properties are the same. Polarities, however, are different (e.g., dipole moment increases with the increasing distance of the keto group from the carbomethoxy moiety). Thus, the position of the keto group on the abscissa reflects the solute dipole moment.

Recently, there is an increase in applying dipole moments and other related electronic quantities, due to the popularization among QSRR students of the routine quantum chemical calculations. Certainly, the applications concern multiparameter QSRRs, where the electronic data are considered along with the dispersive properties descriptors.

Specifically, dipole moments have long since been occasionally used in structure–retention relationship studies. Already in 1956 James [4] related the chromatographic behavior of a series of isomeric xilidines to their dipole moments. An attempt to relate the GC retention indices and dipole moments of a series of alkyl- and arylchlorosilanes was undertaken by Ainshtein and Shulyatieva [5]. Another correlation of that type for a series of carbon- and silicon-substituted silatranes was reported by Shatz et al. [6]. The latter authors have claimed obtaining a linear equation, which permits the evaluation of dipole moments with $\pm 5\%$ accuracy using chromatographic data.

Proportionality between GC retention data and the square of the dipole moment of the solutes was reported for a group of aromatic mononitro derivatives; if dinitro derivatives were studied, the proportionality disappeared [7].

As with dipole moments, it was attempted to use the dielectric constants of solutes as descriptors of the ability of molecules to take part in dipole orientation interactions in GC [8].

In more recent years the dipole moment of solutes is from time to time reported as one of several molecular descriptors in QSRR equations. Usually, the quantum chemically calculated dipole moments are applied because the reliable respective experimental data are scanty.

Multiparameter QSRRs will be discussed separately in Chapter 9. Here publications are cited that report QSRR equations containing a statistically significant dipole moment term. The first complex approach to the now "classical" multiparameter QSRR studies, employing experimentally determined total dipole moments, was presented by Gassiot-Matas and Firpo-Pamies [9]. Based on Kováts indices obtained on different stationary phases and connectivity indices (see Chapter 8), these authors proposed a method of determination of the *chromatographic dipole moment*.

Kaliszan and Höltje [10] used dipole moment (either experimental or quantum chemically calculated by the CNDO/2 method) in combination with molar refractivity for QSRR studies of GC data of a series of phenols. These authors also proposed a method of determination of the *chromatographic polarity parameter* based on Kováts indices determined on two phases of different polarities and molar refractivities of the solutes (see Chapter 9 for details). However, Kaliszan and Höltje [10] found rather mediocre correlation between their chromatographic polarity parameter and a square dipole moment calculated quantum chemically; the correlation coefficient was $r = 0.919$, but of the 43 phenols studied, 16 were excluded from regression for various reasons.

Among the numerous multiparameter QSRR equations listed by Bermejo Guillén [11] and describing the Kováts indices of aliphatic esters, two contain either the reciprocal of the dielectric constant, $1/\varepsilon$, or the square of the dipole moment, μ^2, as the independent variables. Introduction of these variables into the regressions, however, is not justified statistically according to the sequential F^*-test [Eq. (5.5)]. The terms $1/\varepsilon$ or μ^2 improve the correlation coefficient (in both cases) from $r = 0.9990$ without these terms to $r = 0.9992$ with the terms. After adding $1/\varepsilon$ or μ^2, the equations considered become four-parameter and comprise 16 derivatives. Thus, in Eq. (5.5) $R_1 = 0.9992$, $R_2 = 0.9990$, $p = 1$, $n = 16$, and $k = 3$. With these quantities $F^*_{1, 11} = 2.75$, whereas $F^*_{1, 11, \alpha = 0.05} = 4.84$. In other words, for GC retention of a congeneric group of aliphatic saturated esters analyzed by Bermejo and Guillén [11], the dipole moment is not a discriminative value. This does not mean that only dispersive interactions of the solutes with stationary phases of different polarity are taking place. The lack of significance of a term related to specific, polar interactions in QSRR equations for a given set of solutes and a chromatographic system suggests that the dipole moment (or generally polar) contributions to retention are either negligible or nearly similar for the individual solutes. Certainly, the polar interaction contributions will be relatively large in the case of polar phases (especially if the solutes are also polar in properties) but may be of secondary importance in the case of nonpolar phases (if both a solute and a phase are nonpolar, only dispersive forces operate).

7.2. ELECTRONIC SUBSTITUENT CONSTANTS

In 1937 Hammett formulated his well-known equation for the calculation of substituent effects on reaction rates and chemical equilibria using the dissociation of meta- or para-substituted benzoic acids in water at 25°C as a reference process:

$$\log \frac{K_X}{K_H} = \rho\sigma \qquad (7.1)$$

where K_X is the dissociation constant of the derivative possessing substituent X in the para or meta position; K_H is the constant for the unsubstituted acid; ρ is the reaction constant ($\rho = 1$ for the standard reaction by definition); σ_p or σ_m are the Hammett constants for substituents in an aromatic system located at the para or meta position, respectively. In the case of the ortho position, steric effects preclude unequivocal determination of the electronic constant for substituents.

Hammett constants are based on the extrathermodynamic assumption of a linear free-energy relationship (LFER). As such, the Hammett constants are additive; that is, for a multiple-substituted aromatic system the Hammett equation is

$$\log \frac{K_R}{K_H} = \rho \sum \sigma \qquad (7.2)$$

where K_R is the dissociation constant for the multiple-substituted derivative, and σ is a given defined Hammett constant.

Since Hammett's fundamental works, much has been done to isolate and describe specific electronic effects, for example, mesomeric, inductive, and field effects. Now one can encounter in literature (especially in biological QSAR) such new substituent constants of σ-Hammett type as σ_p, σ_m, σ_o, σ^+, σ^-, σ^*, σ_I, σ_R, σ^o, E_R, δ, F, and R. Individual numerical values for more common substituents may be found in several compilations [12, 13]. In QSRRs, the

Hammett constants are applied only occasionally.

The relationships between electronic substituent constants and GC retention data have been reported since the 1960s. Retention was related to the substitution effects for several isomeric pairs or groups [14–16]. Attempts to evaluate specific contributions to GC retention have been numerous (e.g., hydrogen-bonding contributions) by means of electronic substituent constants [17–22].

By means of regression analysis Nieudorp et al. [23] derived an equation relating the differences of logarithms of retention volume determined on Apiezon L and Carbowax 2M in the GC of a series of aromatic compounds to Hammett sigmas at three positions of substitution.

Several researchers have used various electronic substituent constants to derive structure–GC retention relationships. For example, Ono [24] applied the ordinary Hammett constant σ, the Taft electronic constant σ_0, and the resonance constant R in a single equation. Other examples can be found in the literature [25–32]. The reliability of the approach may be questioned in many instances, however, because there are too few data points (differing in numerical value), allowing interpretation of the significance of changes in the individual substituent constant considered for retention. Besides, different types of substituent constants are added to one another or treated as fully exchangeable.

For the sake of illustration a report [33] on QSRRs for a series of 26 substituted phenols may be considered. Radecki et al. added classical Hammett substituent constants for the meta and para positions (common practice) together with the Taft constants for the ortho position. The polarity descriptor thus obtained was found to be significant in multiparameter regression equations derived to predict retention of the solutes on polar GC phases. The question arises, however, as to what may be the physical meaning of an arithmetic sum of different substituent constants.

Interesting results have been reported by Steurbaut et al. [34]. These authors determined substituent increments to GC and TLC retention parameters for various phase systems of a group of 11 O-alkyl-O-arylphenyl-phosphonothioates. Next, these substituent increments were correlated with Hammett constants. In TLC the correlations obtained were low, except, perhaps, the correlation ($r = 0.766$) for the relation ΔR_M versus σ, with ΔR_M obtained on polyamide TLC with n-hexane–acetic acid 95:5 as eluent. The respective correlations with GC data, even if obtained on a highly polar (DEGS) column ($r = 0.756$), were lower than expected.

Buydens et al. [34–37] studied the importance of the Hammett sigma derived from the hydrolysis of aliphatic acids for the prediction of GC retention on nonpolar stationary phases. As expected, the electronic constant was the significant parameter in regression equations obtained. Generally, however, when the quantum chemical electronic discriptors are included in the regression, they are preferred to the experimental σ parameter [36].

In TLC, Mokrosz and Ekiert [38, 39] analyzed the effect of substituents in the phenyl ring of 5-arylidenebarbiturates and their N,N-dimethyl analogs on the ability of the solutes to adsorb on silica gel. Apparent correlations were found between R_M values and the electronic substituent constants. The correlations were significant when the compounds studies were divided into subgroups by type and position of a substituent. Although the predictive potency of the relationship is too low for analytical purposes, the regularities found can be of use in discussing the TLC adsorption mechanism.

The Hammett sigma, σ, was applied by Jinno and Kawasaki [40, 41] in studies of QSRRs of phenols chromatographed in various reversed–phase HPLC systems. The logarithms of the capacity factors, $\log k'$, were the dependent variables. Along with the Hammett constant, the hydrophobic substituent constant, π [12], was applied as an independent variable. The σ values were found to reflect differences in polarity among the set of solutes studied. The equations obtained have the form

$$\log k' = k_1 \pi + k_2 \pi \sigma (1 - \pi) + k_3 \qquad (7.3)$$

where k_1, k_2, and k_3 are constants. At the composition of the binary mobile phase of acetonitrile–water 65:35, the correlation coefficient of Eq. (7.3) was $r = 0.954$. Jinno and Kawasaki [40] analyzed the relationship between the constants k_1, k_2, and k_3 and the volume fraction X of the organic modifier in the mobile phase. Here, X ranged from 0.3 to 0.65 acetonitrile. The constant k_2 did not depend on X, whereas k_1 and k_3 were described by a two-parameter regression equation involving the X^2 and X terms. Only five data points were used to perform the multiple–regression analysis of k_1, k_2, and k_3 as functions of the mobile phase composition. Thus, for the set of solutes chromatographed on a given reversed-phase column, one can predict the HPLC retention at any binary solvent composition provided there is no chance correlation between the constants k_1, k_2, and k_3 in Eq. (7.3) and the mobile phase parameters. Based on their findings in studies on alkylbenzenes, polycyclic aromatic hydrocarbons (PAH), variously substituted benzene derivatives, and phenols, Jinno and Kawasaki [42] presented a computer-assisted system for retention prediction and thus for the determination of the optimum separation conditions with two mobile phase systems of aqueous acetonitrile and methanol. Despite all the limitations, the system is of high instructive value for QSRR studies.

The numerous attempts to apply the electronic substituent constants in QSRR studies have met only limited success. One reason is the lack of constants for many chemical moieties. The additivity of different sigmas assumed by several researchers is not substantiated. The effect of the chemical environment on electronic properties of the submolecular fragments considered is often prevailing.

The important objection is that most of the electronic constants (at least those available for larger sets of common substituents) have been derived from the benzene system. The question arises whether such parameters can be extrapolated over aliphatic and heterocyclic systems. In the case of aromatic heterocycles, such an extrapolation is probably substantiated, at least as long as one remains within an individual heterocyclic group of solutes. Such is this author's experience with substituted pyridines [43].

In 1975 Hetnarski and O'Brien [44] proposed a new substituent constant, namely, C_T (charge transfer) constant. Analogous to σ, C_T is defined as

$$\log \frac{K_R}{K_H} = C_T \qquad (7.4)$$

where K_R is the association constant of the electron pair donor–electron pair acceptor (EPD–EPA) complex between tetracyanoethylene as the EPA reagent and various aromatic compounds with substituent R as the EPD reagent; K_H is the association constant for the unsubstituted EPD aromatic compound. There is an intercorrelation between the C_T and σ substituent constants. To this author's knowledge, C_T has not yet been applied in QSRR studies, although it was used in biological QSARs [45].

In conclusion, one may summarize that generally electronic substituent constants have no real advantage over dipole moments and quantum chemical indexes as molecular polarity descriptors. Occasionally (as discussed in Chapter 9, devoted to multiparameter QSRRs), electronic constants, used in addition to hydrophobic or dispersive parameters, may yield relationships comprising two or more related classes of solutes. This is observed when the separation process is performed on polar GC phases or when reversed-phase HPLC is done using relatively polar solutes.

7.3. QUANTUM CHEMICAL INDICES

Looking for a reliable measure of solute polarity, several authors turned their attention to quantum chemistry. Initially, the simple extended Hückel method (EHT) or Del Re's quantum chemical empirical method were employed for calculations of quantum chemical indices. Some applications in multiparameter QSRRs found such quantum chemical data like the π-electron density [46, 47] and the sum of "effective" charges on principal individual atoms in the solute molecule [48–50].

Later, standard computer programs of quantum chemical calculations became commercially available that could be routinely used. At present, one need not be a professional quantum chemist to generate structure data of interest for QSRR studies, and thus, quantum chemical indices are often encountered in QSRR publications.

To calculate molecular parameters of orbitals, the now classical Roothaan method is commonly applied, consisting in the complete neglect of differential overlap (CNDO/2 method). This semiempirical approximate molecular orbital method, in which only the valence electrons are explicitly considered, is recognized as satisfactorily reflecting the electronic properties of solutes. In most cases, the calculations are done for the most stable conformation for molecules in the gaseous phase. The energetically favored conformations can be found by means of quantum chemical calculations for various possible conformers. For the calculations, the lengths and angles of bonds can be taken from X-ray crystallography or standard parameterization is assumed.

One of the first applications of the CNDO/2-calculated molecular descriptors in QSRR studies was specifically concerned with solute dipole moments [10]. As discussed in Section 7.1, the attempt to describe GC retention indices of phenols in terms of experimental dipole moments have met with only limited success. Also, with the CNDO/2-calculated dipole moments the QSRRs obtained are not impressive. The calculated dipole moments represent the molecular conformations in the gas phase, which may well differ from the actual conformation of the solute under chromatographic conditions. Choosing the proper solute conformation for CNDO/2 calculations may be a problem. This can be demonstrated with 4-hydroxyphenol (Fig. 7.2). Among the conformations of the compound, two energetically favored conformers

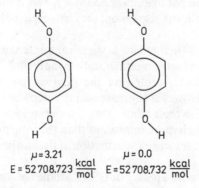

$\mu = 3.21$ $\mu = 0.0$

$E = 52\,708.723\ \frac{kcal}{mol}$ $E = 52\,708.732\ \frac{kcal}{mol}$

Figure 7.2. Dipole moments μ, and conformational energies E for energetically favored conformations of hydroquinone. (After R. Kaliszan and H.-D. Höltje, *J. Chromatogr.*, **234**, 303, 1982.)

exist. In such a situation the calculated dipole moment is obtained as the arithmetic mean of the two conformers and equals $1.61D$. The calculated value is in a good agreement with the experimentally determined dipole moment, μ_{obs}, for the solute at 25°C in benzene solution (i.e., $\mu_{obs} = 1.40$) [51]. However, such an average calculated dipole moment, μ_{calc}, does not correlate with GC retention data. On the other hand, if one takes the μ_{calc} of only one conformer ($\mu_{calc} = 3.21$), there is a very good correlation with chromatographic data. Unfortunately, a priori rationalization of such a choice is impossible.

In 1983 Massart and co-workers [36, 37] published the results of studies on the prediction of GC retention indices for a group of solutes consisting of aliphatic ethers, esters, alcohols, ketones, and aldehydes. The authors used, in multiple-regression analysis, the following CNDO/2-calculated electronic parameters: the magnitude of the dipole moment, DM; the sum of the absolute values of the charges in a given molecule, QT; and the sum of the absolute values of the charges of the atoms constituting the functional group and the atoms in the α position to this functional group, QA. Because in the dipole–dipole interaction the energy of the process is proportional to the square of the dipole moments, the square terms of the individual structural parameters were also included in the regression analysis.

To get meaningful QSRRs, Massart and co-workers [36, 37] had to consider separately the subsets of mono- and bifunctional derivatives. Multiparameter regression equations involving quantum chemical indices, along with several topological indices and substituent constants, have shown the importance of electronic parameters for solute retention on polar stationary phases. Whereas for the nonpolar stationary phases the topological parameters of monofunctional molecules can explain a great part of the total retention variance, for the polar stationary phases a combination of topological and electronic parameters is necessary. When the quantum chemical indices are included in the regression, they are preferred to the experimental Hammett sigma.

As reported [36, 37], there is a significant intercorrelation among the structural parameters used simultaneously in individual regression equations. This makes it difficult to interpret the physical meaning of a particular parameter applied. However controversial the stepwise entry of the variables is, one can agree with the authors that the parameters describing the local polarity (QA) are much more important than the total dipole moment (DM). This is especially evident when the retention indices on polar stationary phases are considered.

CNDO/2 indices were applied in QSRR studies of a series of n-alkenes by Garcia-Raso et al. [52]. These authors analyzed GC retention indices on squalane, previously used by Dubois et al. [53] in their topological analysis of alkenes. Very high correlations were observed between the Kováts index I and the quantum chemically calculated total energy E_T and binding energy E_b. When the subgroup of 62 cis and trans alkenes was separated, the three-parameter equation was derived predicting the retention index with a standard deviation as low as $s = 4.66$. Apart from E_T, the solute–stationary phase interaction factor, ΔE, and the energy of highest occupied molecular orbital, E_{HOMO}, were included in the equation. As Garcia-Raso et al. [52] emphasize, the three-parameter equation accurately represents the chromatographic behavior of cis and trans isomers of homologous alkenes.

Some explanation is required for Garcia-Raso et al.'s [52] statement that the two-parameter equations involving either E_T and ΔE or E_T and E_{HOMO} reduce the standard deviation of the equation I versus $_T$ "slightly", but if they considered these two factors jointly (ΔE and E_{HOMO}), a large reduction in standard deviation resulted. The question arises whether the introduction of individual parameters into regression equations is statistically justified or if a chance correlation takes place. In any case, the quantum chemically calculated total energy (or binding energy) is the parameter of deciding importance for retention of this closely congeneric group of solutes of relatively low and nearly equal polarity. This is the case for retention on nonpolar stationary phases.

Another congeneric set of solutes studied by the same group [54] form homologous series of n-aliphatic esters. Excellent correlation was found between the CNDO/2-calculated total energy E_T and GC retention indices determined on squalane and Carbowax 1540 for an individual homologous series of esters (Fig. 7.3). The well-known shift of the plots of I versus E_T for individual homologous series is evident in Fig. 7.3. Thus, considering the esters of the same E_T value, the most strongly retained were formates, followed by acetates, propionates, and so on. Shifts of plots for respective homologous series were greater on polar Carbowax 1540 phase than on the nonpolar squalane. Saura-Calixto et al. [54] attempted to find the selectivity parameters determining retention of esters with the same number of carbons, thus having very similar E_T values. They considered the orbital coefficients and energy of the highest occupied molecular orbital. However, no two-parameter equation is given relating the retention index to total energy and the selectivity parameter proposed.

Analyzing the reported relationships between retention data and the quantum chemical index total energy E_T within homologous series, one may conclude that E_T reflects differences in the dispersive properties of solutes. In that respect, E_T resembles carbon number, molar volume, refractivity, and other "bulk" structural descriptors. The real advantage of E_T over the other dispersive molecular parameters is that E_T differentiates individual isomers and even conformers. The E_T term has little or no real chemical meaning. It is likely that this total energy term is reflecting the nonspecific properties of the

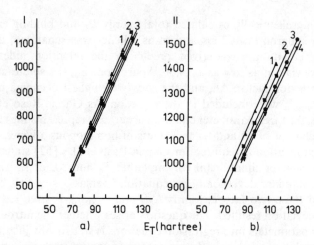

Figure 7.3. Retention index I versus total energy, E_T for homologous series of aliphatic esters: (1) formates; (2) acetates; (3) propionates; (4) butyrates. Columns: (a) squalane; (b) Carbowax 1540; 81°C. (After F. Saura-Calixto, A. Garcia-Raso, and M. A. Raso, *J. Chromatogr. Sci.*, **22**, 22, 1984. With permission.)

solutes indirectly and that other more chemically significant parameters are contained within this term.

Analyzing existing reports on the application of quantum chemical parameters in QSRR studies, one could conclude that there were no quantities proposed that accurately characterized the solute ability to participate in specific polar interactions with chromatographic phases. On the other hand, quantum chemical calculations supply a large amount of information concerning the electronic structure of a solute, however approximate they may be. Undoubtedly, there is a chance that appropriate adaptation of quantum chemical indices would yield a parameter comprising different classes of solutes and stationary phases as well as different chromatographic modes.

Bearing the above in mind, this author's groups undertook QSRR studies in a structurally diverse set of primary, secondary, and tertiary (heterocyclic) amines [55]. GC retention indices were obtained on the methyl silicone (OV-101) phase of low polarity.

Among the several CNDO/2 calculated electronic indices, the total energy E_T was found to be of prevailing effect on retention. Since about 20% of variance in retention data has not been explained by total energy, attempts were undertaken to characterize a solute polarity increment to the retention. However, when the CNDO/2-calculated dipole moment was used along with E_T in a two-parameter regression equation, its significance level was only $p \approx 0.5$. To find a more reliable measure of chromatographic polarity,

Ośmiałowski et al. [55] considered electron-excess charges on individual atoms in a molecule. In effect, a submolecular polarity parameter Δ has been proposed.

To determine Δ, electron densities on each atom in the molecule are calculated and two atoms are found with highest electron excess and deficiency. The difference in electron densities for the two atoms gives Δ (Fig. 7.4).

In the two-parameter regression equation relating the Kováts index of the amines to E_T and Δ, the term Δ is significant at the $p < 0.006$ level. Like the previously mentioned local polarity indices proposed by Massart and co-workers [36], the parameter Δ differentiates solute polarities better than dipole moment.

Recently, the submolecular polarity parameter Δ has been successfully applied in QSRR studies of reversed-phase HPLC capacity factors of a diverse set of substituted benzenes [56, 57]. In that case, again, the parameter Δ gives much better QSRR equations than the CNDO/2-calculated total dipole moment.

The approximate character of polarity parameter Δ is obvious. It characterizes polar interactions with stationary phases better than does dipole moment since, probably, the polar interactions in chromatography are actually the sum of contributions from several local intramolecular dipoles in closest contact

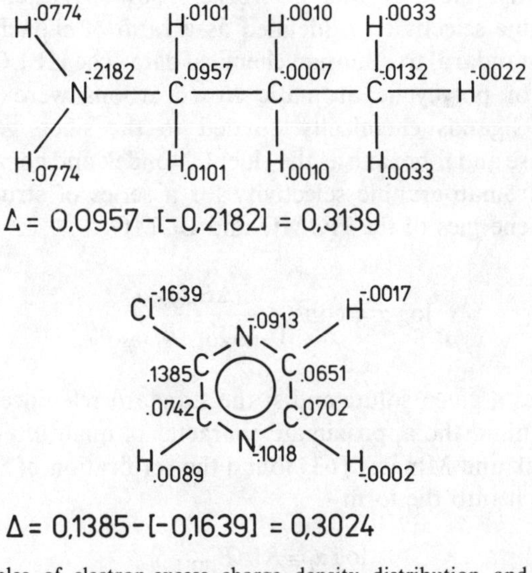

$$\Delta = 0.0957 - [-0.2182] = 0.3139$$

$$\Delta = 0.1385 - [-0.1639] = 0.3024$$

Figure 7.4. Examples of electron-excess charge density distribution and determination of submolecular polarity parameter Δ. (After K. Ośmiałowski, J. Halkiewicz, A. Radecki, and R. Kaliszan, *J. Chromatogr.*, **346**, 53, 1985. With permission.)

with the phase. The Δ is as good as approximation of these interactions as the largest of the acting dipoles, prevailing over the others. In any case, the local polarity measure can be a subsequent step to accommodate the noncongeneric solutes in QSRR.

It must be noted here that the submolecular polarity parameter Δ turned out to be of lower value for characterization of specific interactions with polar GC stationary phases. The same set of primary, secondary, and heterocyclic amines, for which the GC retention indices obtained on the nonpolar OV-101 phase have been described quantitatively in terms of E_T and Δ, did not yield significant QSRRs on polar phases unless Δ was replaced with a new quantum chemical topological index [58].

Quantum chemical data are used occasionally for QSRR studies of HPLC data generated on the electron pair acceptor (EPA) or electron pair donor (EPD) phases. EPD–EPA liquid chromatography has been performed since at least 1964 [59], but the first QSRR studies of such data were probably by Burger and Tomlinson [60]. These authors designed a simple HPLC system by dynamically coating a strong π-electron acceptor, tetracyanoethylene (TCE), on the surface of·silica gel and used mixtures of hexane and 1,2-dichloroethane as eluents. A linear relationship was found between the difference of donor solute capacity factors in TCE-loaded and -nonloaded columns when plotted against the EPD–EPA complex association constant, which was spectroscopically determined.

More recently, Nondek and co-workers [61–63] attempted to relate chromatographic selectivity α_i (defined as a ratio of capacity factors for a solute i and a standard) to quantum chemical data. The HPLC measurements for a group of polycyclic aromatic hydrocarbons were done with the nitroaromatic ligands chemically bonded to the silica gel surface as a stationary phase and n-hexane as the eluent. Nondek and co-workers [61, 63] related the chromatographic selectivity for a series of structurally related solutes to the energies of the HOMO and LUMO:

$$\log \alpha_i = \text{const} \frac{(\Delta E_{\text{HOMO}})_i}{(E_{\text{HOMO}} - E_{\text{LUMO}})_{\text{st}}^2} \tag{7.5}$$

where i denotes a given solute and st the standard reference solute.

Keeping in mind the approximate character of quantum chemical calculations, Nondek and Minárik [63] found the verification of Eq. (7.5) difficult and simplified it into the form

$$\log \alpha_i = K(\Delta E_{\text{HOMO}})_i \tag{7.6}$$

where $(\Delta E_{\text{HOMO}})_i$ is the difference between the energies of HOMOs of solutes

i and the reference standard. The parameter $(\Delta H_{HOMO})_i$ is assumed to reflect the electron pair donation ability of the PAH considered. The discussion of results in the original paper [63] is somewhat obscure, but it seems that "bulk" molecular descriptors, affect HPLC selectivity more than the energy of the HOMO of the solute. Certainly, further efforts are required to quantify the EPD–EPA HPLC retention in structural terms.

An interaction factor ΔE, calculated from energies and coefficients of LUMO of the solutes and HOMO of the stationary phase was applied by Garcia-Raso et al. [52] in their QSRR studies of Kováts indices of alkenes determined on a nonpolar phase (squalene). The term ΔE has been introduced into a three-parameter equation along with E_T and E_{HOMO}. No statistics is given, but the contribution of this term to the Kováts index looks insignificant with respect to E_T.

7.4. MOLECULAR SHAPE DESCRIPTORS IN QSRRs

The general equations describing the distribution coefficient of a solute between chromatographic phases (see Chapter 3) comprise the term that incorporates the probability of the position of contact of the solute molecule with another molecule of a specific type. This term is decided by the size and geometry of the molecules considered. To analyze steric effects in QSSRs, one needs a quantitative description of molecular shape.

To describe molecular shape in quantitative terms is a difficult task, even for relatively rigid molecules. A special difficulty in QSRR studies is that the shape increment to retention is generally of minor importance in comparison to differences resulting from both polar and dispersive interactions. In the case of polar solutes, the molecular shape differences are, at least in part, incorporated in parameters describing the changes in solute polarity. Thus, the separation of shape factors for QSRR purposes may appear questionable.

In biological QSARs, steric effects play, generally, a much more important role than in QSRRs. Medicinal chemists use several parameters related, more or less directly, to steric effects [64]. The first group of parameters is defined from chemical reactivities and is represented by the Taft steric substituent constant, E_s. The role of the steric parameters may also play van der Waals radii of atoms, functional groups, and whole molecules. Some steric parameters are derived from molecular geometry (e.g., Verloop's STERIMOL length parameters) [65]. An alternative to these parameters is the use of the topological methods.

In chromatography, such molecular descriptors as the sum of van der Waals radii of atoms in the molecule are decisive for dispersive interactions and have little importance as the possible shape descriptors. Topological indices, on the other hand, may comprise information related to both steric and polar properties of solutes. As topological indices are extensively applied in QSRR studies, Chapter 8 is devoted to them.

The number of publications concerning QSRRs and involving shape parameters is quite limited. In fact, the single family of solutes for which some evidence is reported of the significance of shape parameters in QSRR equations is the group of PAH. PAH form a unique group of solutes of very low (and practically the same) polarity. As the compounds are actually rigid and coplanar, their shape may be represented by a two-dimensional formula. The group comprises a large variety of isomers for which dispersive properties are also very similar. Thus, the differences in retention data experimentally observed may be considered as reflecting the shape variations of solutes.

No unequivocally successful application in QSRR studies of steric substituent constants has been found. To be sure, Chumakov et al. [66] reported some linearity between logarithms of retention volumes, $\log V_R$, and Taft, Hancock, and Palm steric constants for a small series of alkylpyridines. The retention data were determined by LC on alumina, silica gel, or alumina impregnated with cobalt chloride. A dioxane–isooctane mobile phase was employed. The authors claim that the best correlation is observed between $\log V_R$ and the Palm steric constant for the five solutes under study. However, in the figure included in their paper, the best correlation is actually observed between the log V_R and Taft E_s constant. These authors [66] conclude that steric hindrance determines adsorption in the case of alkylpyridines. This conclusion may be questioned since the series of compounds is closely congeneric, and thus the constants assumed as being steric reflect dispersive properties as well.

Correlations of GC retention increments for substituent with their E_s constants for a series of para-substituted O-alkyl-O-arylphenylphospono-thioates (Fig. 7.5) have been significant at the $p = 0.05$ significance level if

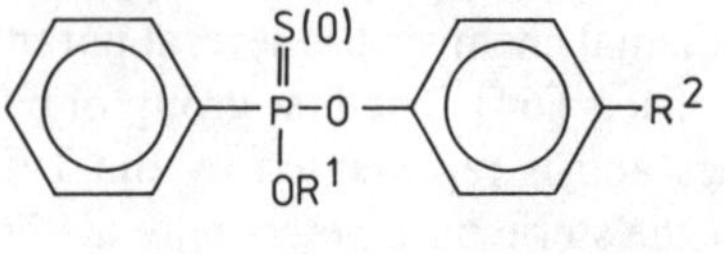

$R^1 = $ -CH$_3$; -CH$_2$CH$_3$; -CH$_2$CH$_2$CH$_3$

$R^2 = $ -H; -CH$_3$; -CH$_2$CH$_3$; -CH$_2$CH$_2$CH$_3$;
       -F; -Cl; -Br; -J; -OCH$_3$;
       -NO$_2$; -CN

Figure 7.5. Chemical structure of o-alkyl-O-arylphenylphosphonothioates. (Analyzed by W. Steurbaut, W. Dejonckheere, and R. H. Kips, *J. Chromatogr.*, **160**, 37, 1978.)

chromatography is carried out on the nonpolar stationary phase ($r = 0.700$). The significance of E_s decreases with the increasing polarity of the staionary phases in favor of electronic influences. Most probably, E_s again reflects dispersive rather than steric contributions to retention. Similar objections may be raised regarding the E_s applications in multiple-regression QSRR studies [6, 32, 47].

Recently Halkiewicz et al. [67] related GC retention on polar stationary phases to the corresponding retention on nonpolar phases using the length of the alkyl substituent, L, for a series of Cu(II) and Ni(II) complexes with dialkyldithiocarbamates. The L values for six alkyl substituents were taken from Verloop et al. [65], who obtained a relationship of the general form

$$t_P = k_1 t_{NP} - k_2 L + k_3 \qquad (7.7)$$

where t_P and t_{NP} are retention times of an individual solute determined on a polar and a nonpolar phase, respectively; L is the Verloop parameter for the alkyl substituent; and k_1, k_2, and k_3 are regression coefficients. Halkiewicz et al. [67] discussed Eq. (7.7) in terms of steric hindrance by the alkyl substituent affecting retention. The parameter L was assumed to be a measure of the steric hindrance.

This author finds the above interpretation delusive and resulting from the instinctive interpretation of the negative sign at the $k_2 L$ term. It must be noted that $k_2 L$ relates t_P to t_{NP}; if t_{NP} is the dependent variable and t_P and L are independent variables, the sign at the L term will be positive. On the other hand, the relationships reported by Halkiewicz et al. [67] [Eq. (7.7)] provide an additional proof for the validity of the general equation derived by Kaliszan and co-workers [10, 68] and relating GC retention data generated on polar stationary phases to those obtained on nonpolar phases by means of a dispersive molecular descriptor [Eq. (7.8)]:

$$I_P = k_1 I_{NP} - k_2 (MR) + k_3 \qquad (7.8)$$

where I_P and I_{NP} are the corresponding retention indices from polar and nonpolar phases; MR is the molar refractivity or the other parameter characterizing the ability of solutes to participate in dispersive interactions in a chromatographic system; and k_1, k_2, and k_3 are constants. Derivation of Eq. (7.8) is given in Chapter 9, on multiparameter QSRRs.

In view of the above discussion, the parameter L in Halkiewicz et al. [67] [Eq. (7.7)] is a solute dispersive descriptor rather than a steric parameter. Natural intercorrelation between L and MR, or other dispersive descriptors, for the alkyl moieties considered precludes any conclusive discussion of pure steric effect terms.

As mentioned, a unique group of solutes for the study of the importance of shape for chromatographic retention compose PAH. In their pioneering works, Janini et al. [69] and Zielinski and Janini [70] noted that in GC the retention of PAH on liquid crystal stationary phases depends on the shape of the solute molecule. In studies on structure–biological activity relationships of PAH , Kaliszan et al. [71] introduced the so-called shape parameter η. This parameter is defined as the ratio of the longer to the shorter side of a rectangle having a minimum area that can envelop a molecule (Fig. 7.6).

The parameter η was next found [72] to be of significance in two-parameter regression equations (along with the dispersive descriptor χ) for descriptions of GC retention indices determined on the nematic BMBT phase [N,N-bis(p-methoxybenzylidene)-α,α'-bi-p-toluidine]. For the data given in Table 7.1 the following equation was derived:

$$I_{nem} = 310\eta + 860\chi - 1710 \tag{7.9}$$

$$n = 15; \quad R = 0.9949; \quad s = 109.3; \quad F_{12, 2} = 584; \quad F_{12, 2, \alpha = 0.05} = 3.89$$

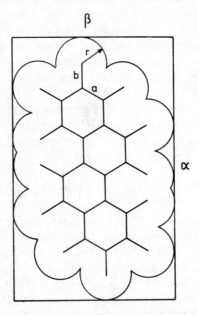

Figure 7.6. Determination of shape parameter $\eta = \alpha/\beta$ as exemplified by chrysene, where a and b denote approximate lengths of C–C and C–H bonds (1.4 and 1.1 Å, respectively); $r\,(= 1.2\ \text{Å})$ is the van der Walls radius of the hydrogen atom. (After A. Radecki, H. Lamparczyk, and R. Kaliszan, *Chromatographia*, **12**, 595, 1979. With permission.)

The analogous data determined on a normal isotropic phase were satisfactorily described by a one-parameter equation involving the dispersive descriptor (connectivity index χ [73]). Based on that observation, a relationship was proposed that would allow one to calculate the shape parameter from retention indices determined, for a given compound, on both the nematic phase, I_{nem}, and the isotropic phase, I_{iso}:

$$\eta = k_1 I_{nem} - k_2 I_{iso} + k_3 \tag{7.10}$$

where k_1, k_2, and k_3 are regression coefficients.

Correlation analysis done for 14 coplanar PAH (compounds in Table 7.1 except coronene) yielded a relationship in the form of Eq. (7.10), with correlation coefficient $R = 0.91$. In retrospect, however, those results should be reconsidered critically. The objection is that there is quite a high intercorrelation between the two retention indices applied for derivation of the regression Eq. (7.10). Probably, the relation holds true only as long as one remains inside the limited group of solutes used to derive the regression. Any other solute, even one similar in structure, may not fit the equation.

Table 7.1. GC Retention Indices Determined on Nematic BMBT Phase, I_{nem}, and Isotropic OV-101 Phase, I_{iso}, and Structural Descriptors for Polycyclic Aromatic Hydrocarbons[a]

| | Retention Index | | Shape | Connectivity |
Compound	I_{nem}	I_{iso}	Parameter	Index, $^1\chi$
C_{14}-Phenanthrene	2800	1836	1.459	4.815
Anthracene	2874	1846	1.559	4.809
C_{16}-Pyrene	3301	2139	1.119	5.559
C_{18}-Triphenylene	4017	2525	1.111	5.982
Benz[a]anthracene	4169	2516	1.572	6.220
Chrysene	4198	2526	1.683	6.226
C_{20}-Benzo[e]pyrene	4650	2858	1.149	6.975
Perylene	4739	2888	1.268	6.970
Benzo[a]pyrene	4834	2870	1.347	6.975
C_{22}-Dibenz[ac]anthracene	5099	3142	1.118	7.637
Benzo[ghi]perylene	5262	3185	1.119	7.720
Dibenz[ah]anthracene	5325	3137	1.782	7.631
Anthanthrene	5439	3215	1.347	7.714
Benzo[b]chrysene	5488	3159	1.938	7.637
C_{24}-Coronene	5821	3498	1.000	8.464

[a] According to Radecki et al. [72]. With permission.

In the case of nematic GC stationary phases and isomeric PAH as solutes, the meaning of molecular shape for retention differences seems to be unquestionable. The question arises whether the shape parameter η adequately reflects the structural differences of importance for retention. For closely related compounds, such as the 11 isomeric monomethylbenz[a]anthracenes analyzed by Lamparczyk [74], the shape parameter is significant in two-parameter regression equations linking relative retention on nematic phases to the corresponding data determined on an isotropic phase. In this case, the intercorrelation between the variables is acceptable.

Recently, Suprynowicz et al. [75] published the paper concerning the correlations between retention on liquid crystalline phases and structure of dimethylnaphthalene (DMN) isomers. Discussing the effect of molecular shape on solute retention, the authors [75] concluded that the shape parameter η had little, if any, relevance to chromatographic distribution. As has subsequently been proved [76], the above conclusion resulted from a misunderstanding of the shape parameter concept.

It should be emphasized that the shape parameter η must be defined unequivocally. It is expressed numerically by the ratio of the longer to the shorter side of the minimum rectangular envelope around the structure, drawn proportionally to the atomic dimensions (Fig. 7.7). The criterion of the smallest rectangle enveloping the structure considered is the decisive one for reproducibility. The criterion applied by Suprynowicz et al. [75] to get their shape parameter is not clear. Obviously, the molecular shape may not be the main factor governing the retention. One may expect to detect the significance of molecular shape for retention of a series of solutes when the prevailing bulky properties of the compounds studied are identical or similar. In the last instance, the shape parameter and a dispersive parameter must be considered together.

Keeping the above conditions in mind, Lamparczyk et al. [76], reconsidered the experimental data discussed by Suprynowicz et al. [75]. Exemplary data are given in Table 7.2. For the data tabulated, the following equation was derived (to facilitate the calculations, the numerical values of the retention index were divided by 1000 and from the $^1\chi$ values a constant number 4.200 was subtracted):

$$I_{obsd} \times 10^{-3} = 1.465 + 0.116\eta + 3.514(^1\chi - 4.200) \tag{7.11}$$

$$n = 8; \quad R = 0.9408; \quad s = 0.0097; \quad p = 0.02$$

In fact, for 2,7-DMN the experimental retention index given by Suprynowicz et al. [75] deviates significantly from the regression equation [Eq. (7.7)] and has not been included in the regression.

There would be nothing unexpected if one solute of a group of 9 does not

Figure 7.7. Smallest rectangular envelopes for DMN series. For different bond types the following interatomic distances were used: 1.4 Å (C–C aromatic); 1.5 Å (C–C aromatic to aliphatic); 1.1 Å (C–H single bond); and 1.2 Å (van der Walls radius of hydrogen atom). (After H. Lamparczyk, R. Kaliszan, and A. Radecki, *J. Chromatogr.*, **361**, 442, 1986. With permission.)

Table 7.2. Structural Parameters and Retention Indices for
Dimethylnaphtalene (DMN) Isomers[a]

DMN Isomer (Positions of Substitution)	Shape Parameter,[b] η	Connectivity Index,[c] $^1\chi$	Retention Index (129.4°C)	
			Observed[c]	Calculated[b]
1,7	1.00	4.232	1692	1694
1,3	1.03	4.232	1704	1697
1,4	1.03	4.238	1715	1718
2,7	1.60	4.226	1716	—
1,6	1.32	4.232	1723	1731
1,5	1.20	4.238	1728	1738
2,6	1.58[d]	4.226	1738	1740
2,3	1.42	4.232	1748	1742
1,2	1.26	4.238	1756	1745

[a] From Lamparczyk et al. [76]. With permission.

[b] Determined according to Fig. 7.7 and calculated using Eq. (7.11).

[c] From Suprynowicz et al. [75].

[d] Two smallest envelopes may be drawn of equal surface area. The η for the more compact (less elongated) envelope would be 1.42 (see Fig. 7.7).

follow a statistical relationship. What is peculiar with this compound, however, is that this single solute, as opposed to the rest of the group studied by Suprynowicz et al. [75], showed strikingly irregular behavior on the plots of retention index versus temperature given in the original publication. As evident from Table 7.2, the elution order calculated by means of Eq. (7.11) for the remaining eight DMNs is exactly following experimental findings.

As the separation of PAH has long since been of special analytical interest, the shape parameter has also been employed in QSRR studies of retention data generated by liquid chromatography. In their extensive publication, Wise et al. [77] used a slightly modified shape parameter, that is, length-to-breadth ratio (L/B). They considered the rectangle that envelops the molecule and at the same time maximizes the L/B. The retention data were determined on reversed-phase C-18 columns and were presented as the logarithms of retention indices calculated analogously to the GC Kováts indices. In addition to the shape parameter, the connectivity index was used as a measure of the size of the molecules. It was found that both the L/B and the shape parameter η were useful in predicting differences in a liquid chromatographic elution of isomeric PAH, but they could not be used for comparison of PAH of different molecular weights. The data for methyl-substituted derivatives of three PAH isomers are given in Table 7.3. The correlation between a retention parameter

Table 7.3. Structural and Retention Parameters of Methyl-substituted Isomers of Benzo[c]phenanthrene, Benz[a]anthracene, and Chrysene[a]

Solute	Connectivity Index $^1\chi$	Shape Parameters η^b	L/B	LC Retention (C-18) log I^c
Benzo[c]phenanthrene				
4-methyl	6.643	1.11	1.36	4.04
3-methyl	6.637	1.36	1.36	4.01
5-methyl	6.643	1.22	1.22	3.97
1-methyl	6.643	1.21	1.21	3.62
2-methyl	6.637	1.13	1.17	3.83
6-methyl	6.643	1.10	1.12	3.94
Benzo[a]anthracene				
3-methyl	6.631	1.71	1.71	4.39
9-methyl	6.631	1.71	1.71	4.39
4-methyl	6.637	1.64	1.64	4.33
10-methyl	6.637	1.59	1.59	4.24
8-methyl	6.637	1.58	1.57	4.19
12-methyl	6.643	1.35	1.51	4.10
7-methyl	6.643	1.50	1.50	4.14
2-methyl	6.631	1.49	1.50	4.09
1-methyl	6.637	1.34	1.47	4.14
11-methyl	6.637	1.35	1.45	4.13
5-methyl	6.637	1.35	1.43	4.28
6-methyl	6.637	1.22	1.38	4.10
Chrysene				
2-methyl	6.643	1.85	1.85	4.52
1-methyl	6.643	1.71	1.71	4.43
3-methyl	6.643	1.63	1.63	4.29
4-methyl	6.637	1.35	1.51	4.18
5-methyl	6.637	1.34	1.48	4.14
6-methyl	6.643	1.47	1.48	4.14

[a] Data from Wise et al. [77]. Reproduced from the *J. chromatogr. Sci.*, by permission of Preston Publications, A Division of Preston Industries, Inc.

[b] From Kaliszan et al. [71].

[c] Logarithm of liquid chromatographic retention index determined on a polymeric octadecylsilica column.

and L/B for the isomeric solutes considered is illustrated in Fig. 7.8. For the set of solutes from Table 7.3 (except 1-methylbenzo[c]phenanthrene) the correlation coefficient of the linear relationship illustrated in Fig. 7.8 is $r = 0.936$. For the subgroup of six isomeric methylchrysenes the corresponding correlation is excellent, $r = 0.991$.

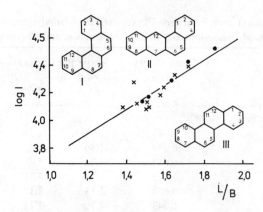

Figure 7.8. Linear correlation of length-to-breadth ratio (L/B) versus liquid chromatographic retention parameter for methyl-substituted isomers of benzo [c] phenanthrene (I) (■), benz [a] anthracene (II) (×), and chrysene (III) (●). (After S. A. Wise, W. J. Bonnett, F. R. Guenther, and W. E. May, *J. Chromatogr. Sci.*, **19**, 457, 1981. By permission of Preston Publications, A Division of Preston Industries, Inc.)

In a more recent paper Wise and Sander [78] analyzed the effect of L/B on reversed-phase HPLC retention of PAH isomers of molecular weight 278. These authors concluded that polymeric C-18 chemically bonded stationary phases exhibited a greater selectivity for molecular shape than monomeric phases and that shape selectivity increased with increasing loading for polymeric phases.

Wise and Sandler [78] explained the observation that planar and linear PAH interact more strongly with bonded phases than the nonplanar and nonlinear PAH in terms of the "slot model". According to this model, the bonded phase consists of a number of narrow "slots" into which the solute molecules can penetrate. Planar molecules would be able to fit more easily into these narrow slots and interact strongly with the C-18 stationary phase, whereas the nonplanar molecules would not penetrate as far into the slots, and so would interact less strongly with the stationary phase.

Jinno and Kawasaki [79] considered logarithms of the capacity factor, log k', determined for 26 PAH on various chemically bonded reversed-phase HPLC columns (phenyl, ethyl, octyl, and octadecyl columns). The log k' values determined on the individual chemically bonded phases were highly inter-correlated. For each phase considered, the authors obtained a high correlation ($r > 0.99$) between log k' and the so-called correlation factor F. The size factor F for a given solute was calculated as $F =$ (number of double bonds) + (number of primary and secondary carbon atoms) − (0.5 for a nonaromatic ring). For the whole set of 26 nonisomeric PAH, the correlations between log k' and L/B as the shape parameter were low. When the two structural descriptors F and L/B were used together in a multiregression analysis, Jinno and Kawasaki [79] found some improvement of correlation in relation to F alone, but only in the case of short-chain phases (i.e., phenyl and ethyl phases). The conclusion [79] that the shape of a solute is a more dominant factor for retention on bonded phases with shorter chains than on phases with longer chains is in opposition to the already discussed results by Wise and co-workers [77, 78].

The results of Wise et al. [77, 78] seem more convincing as derived from experiments on large series of well-selected PAH. In the Jinno and Kawasaki [79] work the size parameter F predominates, and thus, the shape contribution to retention becomes relatively insignificant. In any case, even for the relatively diverse set of PAH considered by Jinno and Kawasaki [79], the one-parameter equation relating retention to L/B is best (although statistically meaningless) in the case of the octadecyl stationary phase ($r = 0.41$), as compared to the short-chain phases (octyl $r = 0.29$, phenyl $r = 0.32$).

In another QSRR study involving the structural descriptors F and L/B and HPLC capacity factors of PAH, Jinno and Okamoto [80] analyzed data determined on the following chemically bonded stationary phases: phenyl-, diphenyl-, triphenyl-, and benzyl-bonded silicas. Again, the contribution of L/B to retention is relatively low, but among the phases studied, triphenylsilica has the highest ability to molecular shape recognition [80].

Jinno and Kawasaki [81] also analyzed 12 PAH by a normal-phase HPLC mode on pure silica and aminosilica phases with n-hexane as the mobile phase. The retention data (log k') thus determined, along with the literature data obtained on dimethylsilica, phenylsilica, octylsilica, and octadecylsilica in the reversed-phase HPLC mode, were subjected to QSRR analyses. The correlations between the capacity factors measured on various phases were high [the lowest was the intercorrelation between log k' values as determined on the pure silica phase and those obtained in an octadecylsilica reversed-phase system ($r = 0.9474$)]. In such a situation, the statistics of the two-parameter equations relating log k' to the size factor F and the shape parameter L/B were also similar. The significance of the shape parameter in regression equations was not given.

In conclusion, it should be stressed that at the present stage of development of QSRR studies, extension of the existing shape parameters beyond the coplanar positional isomers is rather difficult to realize. Conversely, the same holds true in the case of chromatographic evaluation of the molecular shape of solutes. Since molecular shape is not as important for retention as solute polarity and its dispersive properties, the latter two characteristics must be precisely determined at first.

References

1. R. P. W. Scott, The role of molecular interactions in chromatography, *J. Chromatogr.*, **122**, 35, 1976.

2. B. L. Karger, L. R. Snyder, and C. Eon, An expanded solubility parameter treatment for classification and use of chromatographic solvents and adsorbents. Parameters for dispersion, dipole and hydrogen bonding interactions, *J. Chromatogr.*, **125**, 71, 1976.

3. B. Abegaz and S. Gerba, Use of retention time plots as a means of identifying isomeric methyl ketononanoates, *J. Chromatogr.*, **324**, 440, 1985.

4. A. T. James, Gas–liquid chromatography. Separation and microestimation of volatile aromatic amines, *Anal. Chem.*, **28**, 1564, 1956.

5. A. A. Ainshtein and T. I. Shulyatieva, Indeksy uderzhivaniya alkil i arilkhlorisilanov, *Zh. Anal. Khim.*, **27**, 816, 1972.

6. V. D. Shatz, V. A. Belikov, G. I. Zelchan, I. I. Solomennikova, and E. Lukevics, Chromatography of organometallic and organometalloidal derivatives of amino alcohols, *J. Chromatogr.*, **174**, 83, 1979.

7. R. I. Siborov, L. C. Romanenko, and S. A. Reznikov, Vliyanie dipolnogo momenta nitroaromaticheskikh uglevodorov na ikh uderzhivanie v gazozhidkostnoi khromatographii, *Zh. Phiz. Khim.*, **51**, 2913, 1977.

8. B. I. Anvaer, A. A. Zhukhovitskii, I. I. Litovtseva, V. M. Sakharov, and N. M. Turkel'taub, O svyazi mezhdu obemom uderzhivaniya v gazozhidkostnoi khromatographii i dielektricheskoi postoyannoi nepodvizhnoi phazy, *Zh. Anal. Khim.*, **19**, 178, 1964.

9. M. Gassiot-Matas and G. Firpo-Pamies, Relationships between gas chromatographic retention index and molecular structure, *J. Chromatogr.*, **187**, 1, 1980.

10. R. Kaliszan and H.-D. Höltje, Gas chromatographic determination of molecular polarity and quantum chemical calculation of dipole moments in a group of substituted phenols, *J. Chromatogr.*, **234**, 303, 1982.

11. J. Bermejo and M. D. Guillen, A study of Kováts retention indices of aliphatic saturated esters and their relation to the polarity of the stationary phase, *J. Chromatogr.*, **318**, 187, 1985.

12. C. Hansch and A. Leo, *Substituent Constants for Correlation Analysis in Chemistry and Biology*, Wiley, New York, 1979.

13. J. K. Seydel and K.-J. Schaper, *Chemische Struktur und biologische Aktivität von Wirkstoffen. Methoden der Quantitativen Struktur—Wirkung—Analyse*, Verlag Chemie, Weinheim, New York, 1979.

14. R. L. Stern, B. L. Karger, W. J. Keane, and H. C. Rose, Studies on the gas–liquid chromatographic separation of diastereoisomeric esters, *J. Chromatogr.*, **39**, 17, 1969.

15. S. H. Langer, C. Zahn, and J. Pantazoplos, Selective gas–liquid chromatographic separation of aromatic compounds with tetrahalophthalate esters, *J. Chromatogr.*, **3**, 154, 1960.

16. D. E. Martire and P. Riedl, A thermodynamic study of hydrogen bonding by means of gas–liquid chromatography, *J. Phys. Chem.*, **72**, 3478, 1968.

17. A. A. Gaile and Ya. I. Leitman, Zavisimost selektivnosti ot struktury rostvoritelei i razdelyaemykh komponentov. I. O primenimosti uravneniya Hammetta dlya otsenki selektivnosti aromaticheskikh rastvoritelei, *Zh. Phiz. Khim.*, **41**, 2886, 1967.

18. A. Ono, Relation between substituent constants of alkylbenzenes and the separation of them by gas liquid chromatography, *Nippon Kagaku Zasshi*, **92**, 429, 1971; *Chem. Astr.* **76**, 24516t, 1972.

19. W. L. Jones and R. Kieselbach, Units of measurements in gas chromatography, *Anal. Chem.*, **30**, 1590, 1958.

20. B. L. Karger, I. Elmehrik, and R. L. Stern, Examination of the Hammett equation for gas–liquid chromatography, *Anal. Chem.*, **40**, 1227, 1968.

21. I. V. Parizheva, A. A. Gaile, and V. A. Proskuryakov, Zavisimost selektivnosti ot struktury rastvoritelei i razdelyaemykh komponentov. XIV. Selektivnost ephirov aliphaticheskikh dikarbonovykh kislot, *Zh. Phiz. Khim.*, **48**, 2533, 1974.

22. A. P. Zakharov, A. A. Gaile, and V. A. Proskuryakov, Zavisimost selektivnosti ot struktury rastworitelei i razdelyaemykh komponentov. XXVI. Sravnenie izbiratelnosti *N*-proizvodnykh nasyshchennykh i aromaticheskikh aminov po otnosheniyu k sisteme geksan-benzol, *Zh. Phiz. Khim.*, **51**, 2108, 1977.

23. G. H. E. Nieuwdorp, C. L. de Ligny, and N. G. van der Veen, Description of substituent effects in gas–liquid chromatography, *J. Chromatogr.*, **154**, 133, 1978.

24. A. Ono, Studies on the separation of alkylbenzenes by gas–liquid chromatography. VII. A linear free-energy relationship in the gas–liquid chromatographic separation of alkylbenzenes, *J. Chromatogr.*, **110**, 233, 1975.

25. V. V. Keiko, B. V. Prokop'ev, L. P. Kuzmenko, N. A. Kalinina, and V. B. Modonov, Ispolzovanie additivnoi skhemy rascheta indeksov uderzhivaniya v gazo-zhidkostnoi khromatographii. Soabshch. 3. Nekotorye zakonomernosti proyavleniya induktsionnogo ephpekta, *Izv. AN SSSR, Ser. Khim.* 2697, 1972.

26. B. Weinstein, *Application of the rule of six to the separation of diastereoisomeric esters by gas*–liquid chromatography. An exchange of comments, *Anal. Chem.*, **38**, 1238, 1966.

27. S. A. Giller, A. V. Eremeev, and V. G. Andrianov, Izuchenie khromatographicheskogo povedeniya nekotorykh alkilgidrazinov, *Zh. Phiz. Khim.*, **52**, 2333, 1978.

28. E. Lukevics, V. D. Shatz, and R. Ya. Moskovich, Azotosoderzhashchie kremniiorganicheskie soedineniya. XLIII. Khromatographicheskoe issledovanie 1-zemeshchennykh piperidinov i pergidroazepinov, *Zh. Obshch. Khim.*, **44**, 1051, 1974.

29. R. Fellous, L. Lizzani-Cuvelier, R. Luft, and J.-P. Rabine, Comportement des éthers. III. Corrélations de données chromatographiques des éthers avec celles d'autres populations chimiques, *J. Chromatogr.*, **110**, 7, 1975.

30. R. Fellous, L. Lizzani-Cuvelier, R. Luft, and J.-P. Rabine, Comportement des éthers. IV. Etablissement de relations linéaires extrathermodynamiques entre paramétres structuraux et grandeurs de rétention, *J. Chromatogr.*, **110**, 13, 1975.

31. J. K. Haken, D. K. M. Ho, and C. E. Vaughan, Gas chromatography of

homologous esters. VII. The retention behaviour of puryvate esters and related carbonyl and carboxyl compounds, *J. Chromatogr.*, **106**, 317, 1975.

32. J. K. Haken, A. Nguyen, and M. S. Wainwright, Gas chromatography of esters. XII. Linear extrathermodynamic relationships, *J. Chromatogr.*, **178**, 471, 1979.

33. A. Radecki, J. Grzybowski, H. Lamparczyk, and A. Nasal, Relationship between retention indices and substituent constants of phenols on polar stationary phases, *J. High Resolut. Chromatogr. Chromatogr. Commun.*, **2**, 581, 1979.

34. W. Steurbaut, W. Dejonckhere, and R. H. Kips, Relationships between chromatographic properties, partition data and chemical structure of O-alkyl-O-arylphenyl-phosphonothioates, *J. Chromatogr.*, **160**, 37, 1978.

35. L. Buydens and D. L. Massart, Prediction of gas chromatographic retention indices from linear free energy and topological parameters, *Anal. Chem.*, **53**, 1990, 1981.

36. L. Buydens, D. L. Massart, and P. Geerlings, Prediction of gas chromatographic retention indexes with topological, physicochemical, and quantum chemical parameters, *Anal. Chem.*, **55**, 738, 1983.

37. L. Buyders, D. Coomans, M. Vanbelle, D. L. Massart, and R. Vanden Driessche, Comparative study of topological and linear free energy-related parameters for the prediction of GC retention indices, *J. Pharm. Sci.*, **72**, 1327, 1983.

38. J. L. Mokrosz and L. Ekiert, Linear free energy relationship in thin layer chromatography of 5-arylidenebarbiturates, *Chromatographia*, **18**, 401, 1984.

39. L. Ekiert, J. Bojarski, and J. Mokrosz, Some aspects of structure–chromatographic behaviour relationships in TLC of barbiturates, in H. Kalász, and L. S. Ettre, (Eds.), *Chromatography, the State of the Art*, Akadémiai Kiadó, Budapest, 1985, p. 403.

40. K. Jinno and K. Kawasaki, Retention prediction of substituted benzenes in reversed-phase HPLC, *Chromatographia*, **18**, 90, 1984.

41. K. Jinno and K. Kawasaki, Computer-assisted retention prediction system for reversed-phase micro high-performance liquid chromatography, *J. Chromatogr.*, **316**, 1, 1984.

42. K. Jinno and K. Kawasaki, Automated optimization of reversed-phase liquid chromatographic separations using a computer-assisted retention prediction, *J. Chromatogr.*, **298**, 326, 1984.

43. B. Pilarski, K. Ośmiałowski, and R. Kaliszan, The relationship between electron densities and the electronic substituent constants for substituted pyridines, *Int. J. Quant. Chem.*, **28**, 233, 1985.

44. B. Hetnarski and R. D. O'Brien, The charge-transfer constant. A new substituent constant for structure–activity relationships, *J. Med. Chem.*, **18**, 29, 1975.

45. D. Sharples and J. R. Brown, Correlation of the base specificity of DNA—Intercalating ligands with their physico-chemical properties, *FEBS Lett.*, **69**, 37, 1976.

46. C. A. Streuli, W. H. Müller, and M. Orloff, Computer applications in gas–liquid chromatography. I. Prediction of component resolution from limited input data, *J. Chromatogr.*, **101**, 17, 1974.

47. C. A. Streuli and M. Orloff, Computer applications in gas–liquid chromatography. II. Prediction of retention volume for hydrocarbons and heteroatomic organic molecules using molecular parameters and multiple regression analysis, *J. Chromatogr.*, **107**, 23, 1974.

48. M. Gassiot, E. Fernandez, G. Firpo, R. Garbó, and M. Martin, Empirical quantum chemical approach to structure–gas chromatographic retention index relationships. I. Sterol acetates, *J. Chromatogr.*, **108**, 337, 1975.

49. G. Firpo, M. Gassiot, M. Martin, R. Carbó, X. Guardino, and J. Albaiges, Empirical quantum chemical approach to structure–gas chromatographic retention index relationships. II. Cyclohexane derivatives, *J. Chromatogr.*, **117**, 105, 1976.

50. G. Firpo, M. Gassiot, M. Martin, and R. Carbó, Prediction of gas-chromatographic retention indexes through a correlation method, with the structural data of quantum-chemistry (in Spanish), *Affinidad*, **32**, 333, 1975; *Chem. Abstr.*, **83**, 188004v, 1975.

51. J. J. Lander and W. J. Svirbely, The dipole moments of catechol, resorcinol and hydroquinone, *J. Am. Chem. Soc.*, **67**, 322, 1945.

52. A. Garcia-Raso, F. Saura-Calixto, and M. A. Raso, Study of gas chromatographic behaviour of alkenes based on molecular orbital calculations, *J. Chromatogr.*, **302**, 107, 1984.

53. J.-E. Dubois, J. R. Chretien, L. Soják, and J. A. Rijks, Topological analysis of the behaviour of linear alkanes up to tetradecenes in gas–liquid chromatography on squalane, *J. Chromatogr.*, **194**, 121, 1980.

54. F. Saura-Calixto, A. Garcia-Raso, and M. A. Raso, Study of applications of magnitudes of energy and charge of molecular orbitals to GC retention—Esters, *J. Chromatogr. Sci.*, **22**, 1984.

55. K. Ośmiałowski, J. Halkiewicz, A. Radecki, and R. Kaliszan, Quantum chemical parameters in correlation analysis of gas–liquid chromatographic retention indices of amines, *J. Chromatogr.*, **346**, 53, 1985.

56. R. Kaliszan, K. Ośmiałowski, S. A. Tomellini, S.-H. Hsu, S. D. Fazio, and R. A. Hartwick, Non-empirical descriptors of sub-molecular polarity and dispersive interactions in reversed-phase HPLC, *Chromatographia*, **20**, 705, 1985.

57. R. Kaliszan, K. Ośmiałowski, S. A. Tomellini, S.-H. Hsu, S. D. Fazio, and R. A. Hartwick, Quantitative retention relationships as a function of mobile and C_{18} stationary phase composition for non-cogeneric solutes, *J. Chromatogr.*, **352**, 141, 1986.

58. Ośmiałowski, J. Halkiewicz, and R. Kaliszan, Quantum chemical parameters in correlation analysis of gas–liquid chromatographic retention indices of amines. II. Topological electronic index. *J. Chromatogr.*, **361**, 63, 1986.

59. J. T. Ayres and C. K. Mann, Column chromatography with a polynitrostyrene resin stationary phase, *Anal. Chem.*, **36**, 2185, 1964.

60. J. J. Burger and E. Tomlinson, π–π charge transfer interactions in high-performance liquid–solid chromatography, *Anal. Proc.*, **19**, 126, 1982.

61. L. Nondek and R. Ponec, Chemically bonded electron acceptors as stationary

phases in high-performance liquid chromatography, *J. Chromatogr.*, **294**, 175, 1984.

62. L. Nondek, Picramidopropyl silica gel for "charge-transfer" HPLC, *J. High Resolut. Chromatogr. Chromatogr. Commun.*, **8**, 302, 1985.

63. L. Nondek and M. Minárik, Chromatographic selectivity in liquid chromatography of polycondensated aromatic hydrocarbons on 3-(2,4-dinitroanilino)propyl silica, *J. Chromatogr.*, **324**, 261, 1985.

64. M. Charton, Introduction, in M. Charton and I. Motoc (Eds.), *Steric Effects in Drug Design*, Akademie-Verlag, Berlin, 1983, p. 5.

65. A. Verloop, W. Hoogenstraaten, and J. Tipker, Development and application of new steric substituent parameters in drug design, in E. J. Ariëns (Ed.), *Drug Design*, Vol. 7, Academic Press, New York, 1976, p. 165.

66. In. I. Chumakov, M. S. Alyabyeva, and B. D. Kaboulov, Steric effects in liquid–solid chromatographic analysis of pyridine and its 2-alkylsubstituted derivatives, *Chromatographia*, **8**, 242, 1975.

67. J. Halkiewicz, H. Lamparczyk, and A. Radecki, Behaviour of copper (II) and nickel (II) dialkyldithiocarbamates on polar and non-polar stationary phases, *Z. Anal. Chem.*, **320**, 577, 1985.

68. R. Kaliszan, M. Pankowski, L. Szymula, H. Lamparczyk, A. Nasal, B. Tomaszewska, and J. Grzybowski, Structure–olfactory activity relationship in a group of substituted phenols, *Pharmazie*, **37**, 499, 1982.

69. G. M. Janini, K. Johnston, and W. L. Zielinski, Jr., Use of a nematic liquid crystal for gas–liquid chromatographic separation of polyaromatic hydrocarbons, *Anal. Chem.*, **47**, 670, 1975.

70. W. L. Zielinski and G. M. Janini, Utility of high-temperature thermotropic liquid crystals as stationary phases for novel gas–liquid chromatographic separations, *J. Chromatogr.*, **186**, 237, 1979.

71. R. Kaliszan, H. Lamparczyk, and A. Radecki, A relationship between repression of dimethylnitrosamine-demethylase by polycyclic aromatic hydrocarbons and their shape, *Biochem. Pharmacol.*, **28**, 123, 1979.

72. A. Radecki, H. Lamparczyk, and R. Kaliszan, A relationship between the retention indices on nematic and isotropic phases and the shape of polycyclic aromatic hydrocarbons, *Chromatographia*, **12**, 595, 1979.

73. L. B. Kier and L. H. Hall, *Molecular Connectivity in Chemistry and Drug Research*, Academic, New York, 1976.

74. H. Lamparczyk, Separation of twelve monomethylbenz[a]anthracenes on the liquid crystal and isotropic stationary phases in relation to the shape of their molecules, *J. High Resolut. Chromatogr. Chromatogr. Commun.*, **8**, 90, 1985.

75. Z. Suprynowicz, W. M. Buda, M. Mardarowicz, and A. Patrykiejew, Correlations between retention on liquid crystalline stationary phases and chemical structure. I. Dimethylnaphthalenes, *J. Chromatogr.*, **333**, 11, 1985.

76. H. Lamparczyk, R. Kaliszan and A. Radecki, Correlation between retention on liquid crystalline phases and chemical structure, *J. Chromatogr.*, **361**, 442, 1986.

77. S. A. Wise, W. J. Bonnett, F. R. Guenther, and W. E. May, A relationship between reversed-phase C_{18} liquid chromatographic retention and the shape of polycyclic aromatic hydrocarbons, *J. Chromatogr. Sci.*, **19**, 457, 1981.

78. S. A. Wise and L. C. Sander, Factors affecting the reversed-phase liquid chromatographic separation of polycyclic aromatic hydrocarbon isomers, *J. High Resolut. Chromatogr. Chromatogr. Commun.*, **8**, 248, 1985.

79. K. Jinno and K. Kawasaki, Correlation between the retention data of polycyclic aromatic hydrocarbons and several descriptors in reversed-phase HPLC, *Chromatographia*, **17**, 445, 1983.

80. K. Jinno and M. Okamoto, Molecular-shape recognition of polycyclic aromatic hydrocarbons in reversed phase liquid chromatography, *Chromatographia*, **18**, 495, 1984.

81. K. Jinno and K. Kawasaki, Correlation of the retention data of polyaromatic hydrocarbons obtained on various stationary phases used in normal- and reversed-phase liquid chromatography, *Chromatographia*, **18**, 44, 1984.

TOPOLOGICAL INDICES AS RETENTION DESCRIPTORS

From the point of view of QSRRs, efforts to translate molecular structures into unique characteristic structural descriptors expressed as numerical indices [1–3] are particularly interesting. This can be attempted by means of the chemical graph theory, where a chemical structural formula is expressed as a mathematical graph. The formula shows how bonds connect different atoms in a molecule. The mathematical graph describes abstract vertices joined by edges. Each molecular graph may be represented either by a matrix, a polynomial, a sequence of numbers, or a numerical index (topological index).

Most of the proposed topological indices are derived from a vertex adjacency relationship in the graph, or they are related to topological distances in the graph. Consider the two illustrative graphs G_1 and G_2 depicted in Fig. 8.1. The corresponding adjacency matrices $A(G_1)$ and $A(G_2)$ and the distance matrices $D(G_1)$ and $D(G_2)$ are shown in page 139.

The adjacency matrix entries, $a_{i,j}$, have value 1 when the vertices i and j are adjacent and zero otherwise. The distance matrix entries, $d_{i,j}$, are equal to the number of edges connecting the vertices i and j on the shortest path between them.

$$A(G_1)=\begin{array}{c|cccccccc} & \begin{array}{cccccccc} & & & j & & & & \end{array} \\ i & 1\ 2\ 3\ 4\ 5\ 6\ 7\ 8 \\ \hline 1 & 0\ 1\ 0\ 0\ 0\ 0\ 0\ 0 \\ 2 & 1\ 0\ 1\ 0\ 0\ 1\ 0\ 0 \\ 3 & 0\ 1\ 0\ 1\ 0\ 0\ 1\ 0 \\ 4 & 0\ 0\ 1\ 0\ 1\ 0\ 0\ 1 \\ 5 & 0\ 0\ 0\ 1\ 0\ 0\ 0\ 0 \\ 6 & 0\ 1\ 0\ 0\ 0\ 0\ 0\ 0 \\ 7 & 0\ 0\ 1\ 0\ 0\ 0\ 0\ 0 \\ 8 & 0\ 0\ 0\ 1\ 0\ 0\ 0\ 0 \end{array}$$

$$A(G_2)=\begin{array}{c|cccccccccc} & \begin{array}{cccccccccc} & & & & j & & & & & \end{array} \\ i & 1\ 2\ 3\ 4\ 5\ 6\ 7\ 8\ 9\ 10 \\ \hline 1 & 0\ 1\ 0\ 0\ 0\ 1\ 0\ 1\ 0\ 0 \\ 2 & 1\ 0\ 1\ 0\ 0\ 0\ 0\ 0\ 0\ 0 \\ 3 & 0\ 1\ 0\ 1\ 0\ 0\ 0\ 0\ 0\ 0 \\ 4 & 0\ 0\ 1\ 0\ 1\ 0\ 0\ 0\ 0\ 0 \\ 5 & 0\ 0\ 0\ 1\ 0\ 1\ 0\ 0\ 0\ 0 \\ 6 & 1\ 0\ 0\ 0\ 1\ 0\ 0\ 0\ 0\ 0 \\ 7 & 1\ 0\ 0\ 0\ 0\ 0\ 0\ 1\ 0\ 1 \\ 8 & 0\ 0\ 0\ 0\ 0\ 0\ 1\ 0\ 1\ 0 \\ 9 & 0\ 0\ 0\ 0\ 0\ 0\ 0\ 1\ 0\ 0 \\ 10 & 0\ 0\ 0\ 0\ 0\ 1\ 0\ 0\ 0 \end{array} \qquad (8.1)$$

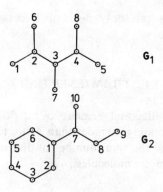

Figure 8.1. Hydrogen-suppressed graphs of 2,3,4-trimethylpentane (G_1) and *sec*-butylcyclohexane (G_2). Atom vertices are numbered consecutively.

$$
D(G_1)=\quad
\begin{array}{c|cccccccc}
 & \multicolumn{8}{c}{j} \\
i & 1 & 2 & 3 & 4 & 5 & 6 & 7 & 8 \\
\hline
1 & 0 & 1 & 2 & 3 & 4 & 2 & 3 & 4 \\
2 & 1 & 0 & 1 & 2 & 3 & 1 & 2 & 3 \\
3 & 2 & 1 & 0 & 1 & 2 & 2 & 1 & 2 \\
4 & 3 & 2 & 1 & 0 & 1 & 3 & 2 & 1 \\
5 & 4 & 3 & 2 & 1 & 0 & 4 & 3 & 2 \\
6 & 2 & 1 & 2 & 3 & 4 & 0 & 3 & 4 \\
7 & 3 & 2 & 1 & 2 & 3 & 3 & 0 & 3 \\
8 & 4 & 3 & 2 & 1 & 2 & 4 & 3 & 0
\end{array}
$$

$$
D(G_2)=\quad
\begin{array}{c|cccccccccc}
 & \multicolumn{10}{c}{j} \\
i & 1 & 2 & 3 & 4 & 5 & 6 & 7 & 8 & 9 & 10 \\
\hline
1 & 0 & 1 & 2 & 3 & 2 & 1 & 1 & 2 & 3 & 2 \\
2 & 1 & 0 & 1 & 2 & 3 & 2 & 2 & 3 & 4 & 3 \\
3 & 2 & 1 & 0 & 1 & 2 & 3 & 3 & 4 & 5 & 4 \\
4 & 3 & 2 & 1 & 0 & 1 & 2 & 4 & 5 & 6 & 5 \\
5 & 2 & 3 & 2 & 1 & 0 & 1 & 3 & 4 & 5 & 4 \\
6 & 1 & 2 & 3 & 2 & 1 & 0 & 2 & 3 & 4 & 3 \\
7 & 1 & 2 & 3 & 4 & 3 & 2 & 0 & 1 & 2 & 1 \\
8 & 2 & 3 & 4 & 5 & 4 & 3 & 1 & 0 & 1 & 2 \\
9 & 3 & 4 & 5 & 6 & 5 & 4 & 2 & 1 & 0 & 3 \\
10 & 2 & 3 & 4 & 5 & 4 & 3 & 1 & 2 & 3 & 0
\end{array}
\tag{8.2}
$$

The simplest topological index derived from the adjacency matrix $A(G)$ is the sum of all matrix elements, A':

$$A' = \sum_{i,j=1}^{N} a_{i,j} \tag{8.3}$$

where N is the total number of atoms in the molecule. Since the adjacency matrix is symmetrical, the index can be reduced to the sum A of the upper triangular adjacency submatrix, the so-called total adjacency of a molecule:

$$A = \frac{1}{2} \sum_{i,j=1}^{N} a_{i,j} \tag{8.4}$$

Since $a_{i,j} = 1$, if there is a bond connecting atoms i and j, index A equals the number of bonds in a molecule. Thus, the total adjacency index A is of low discriminative value. It can only distinguish between molecules having different numbers of cycles.

8.1. THE WIENER INDEX

The half-sum of the off-diagonal elements of the distance matrix $D(\mathbf{G})$ is a much more important topological index than A. It is termed the Wiener index w, and it corresponds to the total number of distances between all pair of vertices in acyclic and cyclic molecules [4]:

$$w = \frac{1}{2} \sum_{i,j} d_{i,j} \tag{8.5}$$

The smaller the Wiener index, the greater the compactness of the molecule. Another topological parameter resulting from Wiener's pioneering work is the so-called polarity number $p(W)$ of an alkane, which equals the number of pairs of vertices separated by three edges. In other words, $p(W)$ corresponds to half of the number of distances of length 3 in the distance matrix $D(\mathbf{G})$:

$$p(W) = \frac{1}{2} \sum_i d_{3,i} \tag{8.6}$$

8.2. THE HOSOYA INDEX

Hosoya [5] introduced an index Z defined as

$$Z = \sum_{k=0}^{[N/2]} p(\mathbf{G}, k) \tag{8.7}$$

In Eq. (8.7) $p(\mathbf{G}, k)$ is the number of ways in which k edges are chosen from the graph G in such a way that no two of them are adjacent: $[N/2]$ is the maximum k-number of graphs. By definition, $p(\mathbf{G}, 0) = 1$, and $p(\mathbf{G}_1, 1)$ equals the number of edges in the graph. Thus for G_1 in Fig. 8.1 the value $p(\mathbf{G}, 1) = 7$. The determination of values $p(\mathbf{G}_1, 2)$ and $p(\mathbf{G}_1, 3)$ is illustrated in Fig. 8.2 for 2,3,4-trimethylpentane [2]. The values for $p(\mathbf{G}_1, 4)$ and higher are zero for 2,3,4-trimethylpentane. Thus, the Hosoya index for the compound is

$$Z = 1 + 7 + 12 + 4 = 24 \tag{8.8}$$

$p(\mathbf{G}_1, 2) = 12$

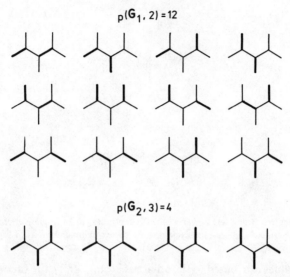

$p(\mathbf{G}_2, 3) = 4$

Figure 8.2. Illustration of determinations of number of ways in which $K = 2$ and $K = 3$ edges are chosen from graph G_1 of 2,3,4-trimethylpentane in such a way that no two of them are adjacent. Chosen edges are marked. (After A. T. Balaban, I. Motoc, D. Bonchev, and O. Mekenyan, *Steric Effects in Drug Design*, Akademie Verlag, Berlin, 1983, p. 21.)

8.3. MOLECULAR CONNECTIVITY INDEX

Molecular connectivity index χ, at present the most popular topological index, was introduced by Randić [6] for characterization of molecular branching. The Randić connectivity index may be considered as a modification of the so-called Zagreb group indices [7]. Both the Randić and Zagreb approaches are based on the concept of the degree D_i of the vertex i in the hydrogen-suppressed molecular graph. The D_i equals the number of bonds (edges) in the graph ramifying on the atom (vertex) i, and thus, instead of the term *degree of vertex i*, the term *valency of vertex* is often used. The degrees of vertices for 2,3,4-trimethylpentane and *sec*-butylcyclohexane, numbered sequentially as in Fig. 8.1, are given in Fig. 8.3 in parentheses.

The index originally introduced by Randić [6] has been calculated by means of the equation

$$^1\chi = \sum_{s=1}^{t} (D_i D_j)_s^{-1/2} \qquad i \neq j \tag{8.9}$$

where s refers to an edge in a graph; t is the total number of edges; D_i and D_j

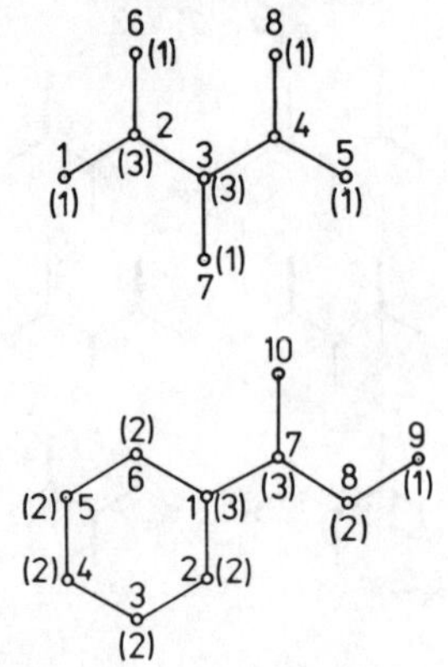

Figure 8.3. Degrees (valencies) of vertices (in parentheses) in hydrogen-suppressed graphs of 2,3,4-trimethylpentane and *sec*-butylcyclohexane. Vertex nummeration as in Fig. 8.1.

represent values attributed to the adjacent atoms i and j; the superscript 1 on χ denotes the so-called first-order connectivity index.

Due to the works by Kier and co-workers [1], a generalized connectivity index concept has been elaborated. According to that concept, index $^1\chi$ is the specific case obtained by considering the edge i–j of length $h=1$. The summation, like that applied for generating $^1\chi$, may be extended over all possible paths of length h:

$$^h\chi = \sum_{s=1}^{t} (D_i D_j \cdots D_{h+1})^{-1/2} \tag{8.10}$$

where s refers to a single path of length h, while t is the total number of paths of length h in a graph.

Apart from $^1\chi$, or the first-order connectivity index, the occasionally used indices are the zero- and second-order connectivity indices:

$$^0\chi = \sum_{s=1}^{N} (D_i)_N^{-1/2} \tag{8.11}$$

where N is the total number of vertices and s is just a vertex. The second-order index is

$$^2\chi = \sum_{s=1}^{t} (D_i D_j D_k)_s^{-1/2} \tag{8.12}$$

where s is a single path of length 2 and t is the number of paths of length 2 in the graph.

For the sake of illustration, the calculation of simple connectivity indices for an exemplary molecule is given in Fig. 8.4.

Hall and Kier [8] further extended the idea of a connectivity index in order to account for the nature of the atoms and the unsaturation of bonds. Instead of the valency of a vertex i, D_i, they proposed to use the atom connectivity, Δ_i^v, which they defined as follows:

$$\Delta_i^v = Z_i^v - H_i \tag{8.13}$$

where Z_i^v represents the number of valence electrons of atom i, and H_i is the number of hydrogen atoms attached to this atom. Thus, the general formula

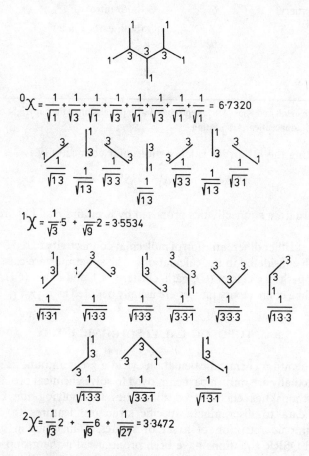

Figure 8.4. Illustration of calculation of zero- ($^0\chi$), first- ($^1\chi$), and second-order ($^2\chi$) connectivity indices for 2,3,4-trimethylpentane.

Table 8.1. Heteroatom Connectivities or Delta Values[a]

Group	Δ^v	Group	Δ^v
NH_4^+	1	–S–	0.944
NH_3	2		
–NH_2	3	=$\overset{\|}{S}$=	3.58
–NH–	4	H_3O^+	3
=NH	4	H_2O	4
–$\overset{\|}{N}$–	5	–OH	5
≡N	5	–O–	6
=N– (pyridine)	5	=O	6
–$\overset{\|}{N}$– (quaternary)	6	O (both nitro)	6
		O (both carboxylate)	6
–$\overset{\|}{N}$= (nitro)	6	–F	(–)20
		–Cl	0.690
		–Br	0.254
		–I	0.085

[a] Proposed by Hall and Kier [1, 8]. Reproduced with permission of the copyright owner, the American Pharmaceutical Association.

for calculating the hth-order valence molecular connectivity is

$$^h\chi^v = \sum_{\text{paths}} (\Delta_i^v \Delta_j^v \cdots \Delta_{h+1}^v)^{-1/2} \tag{8.14}$$

The heteroatom connectivities proposed by Kier and Hall [1] are given in Table 8.1.

There is a further differentiation of molecular connectivity based on the type of subgraph considered in the calculations. The subgraphs are classified into four types: path (P), cluster (C), path–cluster (PC), and chain (CH) [1]. The corresponding connectivity indices are usually denoted as χ_P, χ_C, χ_{PC}, and χ_{CH}.

8.4. TOPOLOGICAL ELECTRONIC INDEX

Standard quantum chemical calculations yield a great amount of numerical data that actually are only in part employed to solve chemical problems. This author's groups has encountered difficulties in applying the CNDO/2-generated data to discriminate specific structural features to determine chromatographic retention of a set of solutes. Difficulties in generating meaningful QSRR equations have been pronounced if the group of solutes studied is characterized by a marked structural diversity, and the chromatographic data are obtained employing polar phases.

Analyzing the electronic charge distribution in solute molecules, our group came to the conclusion that information could be extracted from the charge that would reflect in a numerical form the specific polar properties of the solutes [9]. The available topological indices comprise some information related to interatomic distances and to individual atoms connections. They do not utilize information concerning the electronic structure of the solutes. The CNDO/2 molecular orbital method of calculation is assumed to generate reliable data related to the electron charge distribution on individual atoms in a molecule. At present, when quantum chemical calculations can be done routinely and are inexpensive, the topological indices based on such calculations may become readily accessible. Most probably, the quantum chemical data may be transformed into various topological indices of value for specific discriminative purposes. The topological electronic index T^E proposed recently [9] has been calculated as follows. First, calculations of the electronic charge distribution in solute molecule are done by the CNDO/2 molecular orbital method. Next, for individual atoms of the solute, Cartesian coordinates are determined assuming the same geometry as applied for the CNDO calculations. To each individual vertex atom a number q is assigned equal to its electronic excess charge. Then, the distances $r_{i,j}$ between each pair of vertices are calculated (here, the distances are understood to be the shortest lines joining the vertices i and j in three-dimensional space). For every pair of vertices the absolute value of the excess-charge difference is divided by the square of the respective interatomic distance. The resulting numbers are summed for all the possible atomic pairs in the molecule:

$$T^E = \sum_{i,j} \frac{|q_i - q_j|}{r_{i,j}^2} \qquad i \neq j \tag{8.15}$$

The procedure for determination of T^E is illustrated in Fig. 8.5. All the calculations of T^E from the electronic charge distribution data are easily accomplished by means of a small personal computer. The calculation program, written in BASIC language by Dr K. Ośmiałowski, follows (courtesy of the author):

```
 5 REM Program yields topological electronic index
10 PRINT "Compound name"
15 INPUT a$
20 PRINT "how many atoms?"
25 INPUT w: PRINT w
30 PRINT "give coordinates x, y, z of atoms and corresponding excess charges q"
35 PRINT TAB 1; "i"; TAB3; "x"; TAB10; "y"; TAB18; "z"; TAB25; "q"
38 DIMx(w): DIMy(w): DIMz(w): DIMq(w)
40 FOR i=1 To w
```

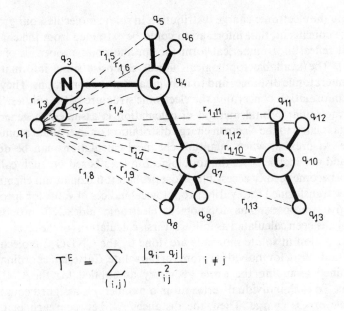

$$T^E = \sum_{(i,j)} \frac{|q_i - q_j|}{r_{ij}^2} \quad i \neq j$$

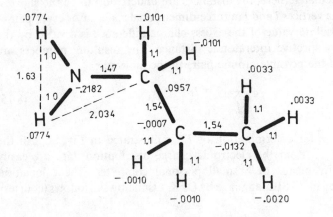

$$T^E = \frac{[.0774 - .0774]}{1.63^2} + \frac{[.0774 - (-.2182)]}{1.0^2} + \frac{[.0774 - .0957]}{2.034^2} + \cdots = 1.532$$

Figure 8.5. Illustration of procedure for determination of topological electronic index T^E. Solute: *n*-propylamine; upper part illustrates distances between hydrogen atom with excess charge q_1 and remaining atoms in molecule; lower part, nummerical data given for excess charges on individual atoms and for several exemplary interatomic distances. (After K. Ośmialowski, Halkiewicz, and R. Kaliszan, *J. Chromatogr.*, **361**, 63, 1986. With permission.)

```
45  INPUT x(i); "␣"; y(i); "␣"; z(i); "␣"; q(i)
50  PRINT TAB1; i; TAB3; x(i); TAB10; y(i); TAB18; z(i); TAB25; q(i)
55  NEXT i
60  LET s=0
65  DIM a(w,w): DIMr(w,w)
70  FOR i=z TO w
75  FOR j=1 TO i-1
80  LET    r(i,j)=SQR    ((x(i)-x(j)*(x(i)-x(j))+(y(i)-y(j))*(y(i)-y(j))+(z(i)
           -z(j))*(z(i)-z(j)))
85  LET a(i,j)=ABS(q(i)-q(j))/(r(i,j)*r(i,j))
90  LET s=s+a (ij)
95  NEXTj:NEXTi
100 PRINT "SIGMA ␣=␣"; s
```

8.5. LINEAR RELATIONSHIPS BETWEEN RETENTION DATA AND TOPOLOGICAL INDICES

The prevailing majority of QSRR reports employing topological indices concern the Randić–Kier connectivity index. On the other hand, it is difficult to prove the advantage of a single specific topological index over the remaining known indices because, for the sets of solutes analyzed, high intercorrelations among the existing topological indices are observed. From the data reported by Motoc and co-workers [2] and determined for a large series of various hydrocarbons, a correlation matrix [Eq. (8.16)] may be composed that comprises the most commonly applied topological indices:

$$
\begin{array}{c|ccc}
 & {}^1\chi & z & w \\
\hline
{}^1\chi & 1 & 0.95 & 0.98 \\
Z & & 1 & 0.96 \\
w & & & 1 \\
\end{array}
\tag{8.16}
$$

Certainly, for hydrocarbons the topological indices are also intercorrelated with the number of carbon atoms in a molecule. Accordingly, these indices can be viewed as "bulk" parameters, with a "shape" component of rather secondary importance.

Despite the limitations, molecular connectivity indices have gained extraordinary popularity among QSRR workers. Randić [6] reported a correlation of his ${}^1\chi$ index and an empirical branching index of Kováts for several C_5–C_8 alkane isomers. Karger et al. [10] found the molecular connectivity useful for prediction of reversed-phase TLC and HPLC retention data of phenols and alcohols containing various normal, branched, and cyclic hydrocarbon moieties.

QSRR studies based on regression analysis and employing the first-order connectivity index were published in 1977 by Kaliszan and Foks [11]. For the set of 20 reversed-phase TLC R_M data extrapolated to a primary aqueous system, correlation with a $^1\chi$ index is characterized by the coefficient $r = 0.95$. It must be emphasized that the solutes studied contained various alkoxy, aryloxy, aminoalkoxy, and alkylamino substituents.

Analyzing GC retention data for cyclic alcohols and esters (Fig. 8.6), Kaliszan [12] found a significant correlation of the retention index determined on a nonpolar stationary phase and the $^1\chi$ index. Retention data for QSRR analysis were from Heintz et al. [13]. The Kováts indices determined on three different phases are given in Table 8.2 along with the $^1\chi$ values.

Statistical evaluation of the quality of regression analysis of the relationship

$$I = a\,^1\chi + b \tag{8.17}$$

is given in Table 8.3.

As opposed to the correlation obtained for the retention data determined on the nonpolar phase SE-30, the analogous equations derived for the polar stationary phases were of little statistical value. Thus, the results presented in Table 8.3 prove that in the case of a structurally diverse group of solutes interacting with polar stationary phases, the connectivity index does not suffice to discriminate individual compounds. Whereas on nonpolar phases the nonspecific dispersive interactions determine retention, the more specific dipole orientation and dipole induction interactions become important, or even predominate, in the case of polar phases and polar solutes. It can be argued that it may also happen that χ correlates well with retention of a particular set of solutes on polar phases. One can imagine a group of solutes of equal polarity but differing in "bulk" properties as is the case with some homologous series. Generally, however, the connectivity index should be viewed as a dispersive molecular descriptor. As such, the index is a valuable structural parameter for QSRR studies.

Since 1977 many papers have been published reporting QSRRs generated by means of connectivity indices. The QSRR equations derived hold true for limited, closely congeneric series of solutes, usually nonpolar in character, for which retention.data are determined in nonpolar chromatographic phase systems. The above observations have been confirmed for many times. The overestimation of the discriminative power of the χ index leads often to (or results from) the statistically invalid relationships, which are notoriously published in the literature.

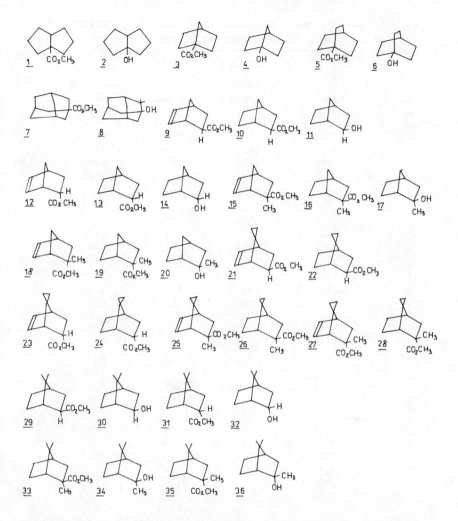

Figure 8.6. Group of methyl esters and alcohols considered in GC retention index versus connectivity index correlation analysis. (After R. Kaliszan, *Chromatographia*, **10**, 529, 1977. With permission.)

Connectivity indices as the GC retention predictors were successfully applied in QSRR studies of various isomeric, aliphatic, aromatic, and polycyclic hydrocarbons [14–22]. Hydrocarbons, as nonpolar solutes, form an especially grateful group for QSRRs employing the χ index. As expected, the correlations are good, at least as long as the solutes of the same degree and position of unsaturation are considered.

Table 8.2. Retention Indices for Compounds Given in Fig. 8.6 Measured at 140°C and Their Molecular Connectivities

Compound Number[a]	Retention Index[b]			Connectivity Index,[c] $^1\chi$
	Methyl Silicone	Polyethylene Glycole Carbowax 20M	Polyester EGSSX	
1	1200	1544	1720	5.227
2	1310	1660	—	4.312
3	1120	1487	1677	4.950
4	942	1495	1669	3.785
5	1247	1631	1804	5.450
6	1061	1858	1775	4.285
7	1431	1831	2031	6.399
8	1253	1831	2031	6.012
9	1098	1493	1693	4.669
10	1135	1512	1702	4.986
11	—	1544	1704	3.860
12	1103	1527	1735	4.744
13	1131	1507	1694	4.386
14	—	1558	1711	3.860
15	1130	1480	1650	5.106
16	1156	1488	1649	5.348
17	998	1471	1635	4.183
18	1125	1493	1676	5.106
19	1174	1511	1676	5.348
20	999	1496	1636	4.183
21	1230	1643	1839	5.637
22	1262	1639	1822	5.545
23	1245	1685	1896	5.637
24	1268	1643	1820	5.545
25	1254	1614	1789	5.999
26	1291	1623	1790	6.407
27	1270	1658	1846	5.999
28	1295	1631	1800	4.621
29	1268	1635	1809	5.539
30	—	1667	1800	4.621
31	1265	1620	1801	5.539
32	—	1687	1830	4.621
33	1302	1624	1783	6.109
34	1133	1574	1719	4.944
35	1303	1624	1784	6.109
36	1146	1600	1755	4.944

[a] Refer to numbers in Fig. 8.6.

[b] From Heintz et al. [13].

[c] From Kaliszan [12]. With permission.

Table 8.3. Regression Coefficients, Standard Deviations, and Correlation Coefficients of Equations the Form of Eq. (8.17)[a]

Statistics of Eq. (8.17)	Column		
	SE-30[b]	Carbowax 20M	EGSSX
Slope a	148.5	73.47	77.72
Intercept b	397.1	1208	1360
Standard deviation s	32	74	73
Correlation coefficient r	0.9518	0.5866	0.6227

[a] Derived from data given in Table 8.2 by Kaliszan [12]. With permission.
[b] Compound 2 omitted.

Other congeneric groups of solutes were also studied. Kier and Hall [23] related retention indices of separate groups of aliphatic alcohols, ketones, ethers, and esters to various connectivity indices. For each group considered separately, good correlations were obtained between retention indices determined on a nonpolar phase, squalene, and the first-order connectivity numbers. However, when the data determined on a polar stationary phase were considered, the $^1\chi$ gave no satisfactory description of retention. To get higher correlation coefficients, different connectivity indices were used in the two-parameter equations. Such different-order normal molecular connectivity indices as well as valence molecular connectivity indices, used in various combinations, allowed Millership and Woolfson [24, 25] to derive multiparameter regression equations characterized by high correlation coefficients. The relationships reported concerned diverse groups of solutes comprising aldehydes, ketones, esters, and alcohols as well as a group of drugs including amphetamines, local anaesthetics, and morphinelike compounds. The same type of multiple regression was recently done by Sabljić [26] in the case of chlorinated benzenes. However, the two connectivity indices used together in one regression equation are highly intercorrelated, which is especially evident for the $^1\chi$ and $^4\chi_{PC}$ parameters. On the other hand, the observation by Sabljić [26] that $^1\chi$ is a better structural descriptor for QSRR studies of GC retention data generated for chlorobenzenes on a nonpolar phase SE-30 than are such topological indices like the Wiener number w or the Balaban index J [2] seems valuable.

An exceptionally large number of multiparameter equations involving up to five different-order connectivity indices ($^0\chi-^4\chi$) along with such parameters as molecular mass and number of aromatic rings were calculated by Bulycheva et al. [27]. Unfortunately, for the closely congeneric polycyclic aromatic hydrocarbons studied, all the structural parameters considered are strongly intercorrelated, and thus, the GC retention indices cannot be correlated with more than one of them. Similar reservations concern the multiparameter regression equations in which the connectivity indices are applied along with a collinear parameter like molar volume [28] or other such topological indices as the Wiener number w or the Hosoya index Z [29].

Molecular connectivity indices correlate well with retention data of related solutes. However, if the solutes differ in specific polar properties, the χ index fails to account for such differences, even if the solutes considered are apparently similar. An illustration may be the relationship found [30] between the first-order molecular connectivity and the gas-chromagraphically-derived free energies of solution, ΔG, in the stationary phase for a group of 39 methyl esters of saturated mono-, di-, tri-, tetra-, penta-, and hexaunsaturated fatty acids. The ΔG values were taken from the literature [31]. As given in Table 8.4 and illustrated in Fig. 8.7, the ΔG values are precisely correlated to $^1\chi$ for each subgroup of solutes, but the regression lines for individual subgroups are shifted in parallel.

In Fig. 8.7, the discriminative power of $^1\chi$ toward isomers may be clearly seen. However, the connectivity index does not adequately reflect the retention differences caused by varying unsaturation. Certainly, different unsaturation gives different electrical properties of the solutes; dipole orientation and/or dipole induction interactions with the stationary phases may become different. Molecular connectivity does not account for such differences in solutes properties. Of course, the less pronounced changes of structure among the solutes of

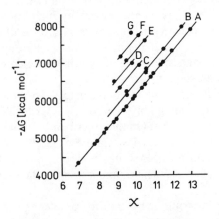

Figure 8.7. Partial molar free energies of solution in SILAR 5CP as function of connectivity indices for methyl esters of saturated (A), monounsaturated (B), diunsaturated (C), triunsaturated (D), tetraunsaturated (E), pentaunsaturated (F), and hexaunsaturated (G) fatty acids. (After R. Kaliszan *Chromatographia*, **12**, 171, 1979. With permission.)

Table 8.4. Connectivity Index $^1\chi$ and Partial Molar Free Energies of Solution, ΔG, for Methyl Esters of Higher Fatty Acids in SE-30 and SILAR 5CP Phases[a]

Acid	$^1\chi$	$-\Delta G$ (cal/mol) SE-30	SILAR 5CP
n-Dodecanoic	6.975	4320	4320
n-Tetradecanoic	7.975	4990	4920
n-Pentadecanoic	8.475	5320	5220
n-Hexadecanoic	8.975	5660	5520
n-Heptadecanoic	9.475	5990	5820
n-Octadecanoic	9.975	6330	6120
n-Nonadecanoic	10.475	6660	6420
n-Eicosanoic	10.975	7000	6720
n-Heneicosanoic	11.475	7340	7010
n-Docosanoic	11.975	7670	7310
n-Tetracosanoic	12.975	8340	7910
12-Methyltridecanoic	7.831	4870	4800
14-Methylpentadecanoic	8.831	5540	5400
16-Methylheptadecanoic	9.831	6210	6000
18-Methylnonadecanoic	10.831	6880	6600
12-Methyltetradecanoic	8.369	5230	5140
14-Methylhexadecanoic	9.369	5900	5740
16-Methyloctadecanoic	10.369	6570	6330
18-Methyleicosanoic	11.369	7240	6930
9-Hexadecenoic	8.625	5570	5620
11-Octadecenoic	9.625	6250	6210
9-Octadecenoic	9.625	6220	6200
6-Octadecenoic	9.625	6220	6180
11-Eicosenoic	10.625	6880	6780
5-Eicosenoic	10.625	6890	6750
13-Docosenoic	11.625	7560	7370
9-Docosenoic	11.625	7550	7360
15-Tetracosenoic	12.625	8230	7960
9,12-Octadecadienoic	9.275	6210	6330
9,12,15-Octadecatrienoic	8.925	6210	6480
6,9,12-Octadecatrienoic	8.925	6130	6410
11,14-Eicosadienoic	10.275	6880	6910
11,14,17-Eicosatrienoic	9.925	6880	7070
8,11,14-Eicosatrienoic	9.925	6780	6980
5,8,11,14-Eicosatetraenoic	9.574	6720	7020
5,8,11,14,17-Eicosapentaenoic	9.224	6730	7180
7,10,13,16-Docosatetraenoic	10.574	7380	7620
7,10,13,16,19-Docosapentaenoic	10.224	7380	7780
4,7,10,13,16,19-Docosahexaenoic	9.874	7320	7820

[a] Data compiled from reports by Golovnya and Kuzmenko [31] and Kaliszan [30]. With permission.

the same class are not reflected very precisely by χ. For example, Jinno and Kawasaki [28] found a rather moderate correlation of reversed-phase HPLC capacity factors of a series of alkylbenzenes with $^1\chi$. Among the retention data determined on three different reversed-phase columns, namely C-2, C-8, and C-18, the octadecylsilica-derived data are the least collinear with $^1\chi$. In the case of the C-18 phase the correlation coefficient is only $r=0.84$; for the remaining two columns the corresponding values are about $r=0.94$.

The same authors, in a separate article [32], reported the lack of correlation ($r=0.176$) between log k' from reversed-phase HPLC and the molecular connectivity for a set of benzene compounds containing common substituents (e.g., NH_2, NO_2, CN, $COOCH_3$, and Cl). The correlation coefficient of GC retention indices with molecular connectivity for a mixture of alkanes, alkenes, alcohols, ether, esters, aldehydes, and ketones chromatographed on a polar phase diethylene glycol succinate (DEGS) has been found to be low ($r=0.17$) [29]. For the same diverse set of solutes, the respective correlation for the data determined on a nonpolar phase squalane has been much higher ($r=0.90$). Rather moderate, at best, have been the correlations between logarithms of the capacity factor from reversed-phase HPLC and molecular connectivity indices of a set of 12 p-benzoquinones and p-hydroquinones reported by Horváth and co-workers [33]. Capacity factors determined on three alkyl-silica phases with three different mobile phases yielded three-parameter correlation equations with $^0\chi^v$, $^1\chi^v$, and $^2\chi^v$ as the independent variables characterized by an R coefficient ranging from 0.566 to 0.877. On the other hand, Szasz et al. [34] reported satisfactory description of GC, HPLC, and TLC retention data of pyrimidine and quinazoline derivatives by two-parameter regression equations involving various molecular connectivity indices.

To accommodate different families of solutes by a single common regression equation, some corrections to the molecular connectivity data have occasionally been introduced [35–37]. Shatz and co-workers [36, 37] proposed a topological index of hydrophobicity χ_H:

$$\chi_H = \chi + \sum_i (\delta\chi_i) \qquad (8.18)$$

where the corrections $\delta\chi_i$ to molecular connectivity χ (which are related to the specific structural fragments) have been determined experimentally. The corrections $\delta\chi_i$ are $\delta\chi_i=0$ for alkanes, $\delta\chi_i=-1.597$ for alkylbenzenes; and $\delta\chi_i=-3.076$ for ketones.

The popularity of the molecular connectivity indices among QSRR researchers worldwide is increasing [38]. Generally, however, it may be observed that the molecular-connectivity-versus-retention correlations are significant when congeneric solutes are analyzed. The term *congeneric* is used rather intuitively. Some solutes that are apparently similar in structure are not congeneric in their chromatographic behavior. On the other hand, there are the solutes of diverse structure that behave regularly during chromatographic separation. A series of ketones for which retention indices were determined using linear temperature-programmed capillary gas chromatography are congeneric [39]. Raymer et al. [39] related the chromatographic data obtained on a nonpolar phase SE-30 to $^0\chi^v$, $^1\chi^v$, and $^2\chi^v$. The best correlation, $r = 0.969$, was found with the $^1\chi^v$ index. Correlation is quite high, bearing in mind differences in hydrocarbon fragments of the solutes studied. These authors [39] also report a multiple correlation, $r = 0.976$, employing all three indices. The question arises, however, whether the three indices applied are not intercorrelated.

On the other hand, a series of mononitrated PAH and a series of dinitrated PAH appeared noncongeneric in QSRR studies of GC retention indices determined on the methyl phenyl silicone (SE-52) stationary phase [40]. Moreover, none of the two series was congeneric with a series of unsubstituted PAH. The QSRR analysis by Doherty et al. [40] employing various types of molecular connectivities revealed that no multivariant regression analysis gave statistically significantly better correlations than single-variant analyses. The data given by Doherty et al. [40] prove a high degree of intercorrelation among the connectivity indices differing in order and subgraph type for mono- and dinitro-PAH. Specifically, in the case of mononitrated PAH the best has been the correlation of retention index I with the $^1\chi^v$ index ($r = 0.970$). For dinitrated PAH the I-versus-$^4\chi^v_{PC}$ relationship is characterized by $r = 0.982$ and is of the same value as the I-versus-$^3\chi^v_P$ relationship ($r = 0.981$).

In liquid chromatography, meaningful QSRRs that employ molecular connectivity data concern the retention data determined in partitioning systems, usually by the reversed-phase technique. The regular behavior of homologues in reversed-phase LC and the relationship between retention data and the topological parameter, like molecular connectivity, are not surprising [41]. The evidence for that type of retention data (log k') versus connectivity index relationship in adsorption chromatography has been presented by Markowski et al. [42]. Correlations ranging from $r = 0.887$ to $r = 0.989$, depending on the chromatographic stationary–mobile phase system, were derived for a limited set of PAH.

The advantages and limitations of molecular connectivity as a structural descriptor for QSRRs can be well observed for barbiturates. Barbiturates form a group of solutes differing in the structure of their hydrocarbon substituents in the pyrimidine ring. As such, the compounds appeared suitable for QSRR studies. First correlations between retention data and the connectivity index for limited sets of barbiturates were published in 1978. Millership and Woolfson [16] analyzed GC data and Bonjean and Luu Duc [43] analyzed

R_M values obtained by the reversed-phase TLC technique. Later, a number of papers were published in this area [44–51].

Individual authors claimed that either the variously modified connectivity index or a combination of several type of connectivities allowed precise prediction of retention. It must be noted here that intercorrelations among various molecular connectivities are apparently lower in the case of barbiturates than has been observed for PAH. Yet, it is not clear from the published material whether the connectivity indices applied simultaneously in an individual multiparameter regression analyses are orthogonal enough for meaningful results.

Probably, the most conclusive results for QSRR studies of barbiturates are those from the extensive analyses of Bojarski and co-workers [46, 50, 51]. According to these studies, the different modifications of the first-order connectivity index have no unambiguous advantages over the classical $^1\chi$. On the other hand, it is more difficult to assign any defined physical meaning to some of those modifications and transformations. The exception may be the first-order valence connectivity $^1\chi^v$ applied by Stead et al. [47] to GC retention data.

Correlations with connectivity indices are good for a closely related subseries of barbiturates (e.g., when simple saturated aliphatic substituents are present). The more diverse the structure of the solutes considered, the lower the correlation coefficients independent of the chromatographic technique used for generating the retention data. If, for example, the relationship log k' versus the first-order valence index is considered [46], when a subgroup of aliphatic-substituted derivatives have a correlation coefficient that is $r = 0.9920$, r drops to 0.9655 after inclusion of alkenyl-substituted barbiturates and to 0.9143 when N-methyl-C-5-disubstituted derivatives are additionally considered. For the available derivatives, the respective correlation is $r = 0.7426$.

The last correlation coefficient realistically reflects the potency of the connectivity index as a molecular descriptor for QSRR studies of different solutes of a given class. Although the potency of molecular connectivity as a parameter quantitatively reflecting molecular branching is unquestionable, its ability to differentiate a solute participation in complex chromatographic interactions should not be overestimated. In fact, Clark and co-workers [45] obtained high ($r > 0.99$) multiple-correlation coefficients for three-parameter equations relating the reversed-phase HPLC capacity ratios (log k') to $(^1\chi)^2$, $(^3\chi_P)^2$, and $^3\chi_P^v$. However, there is no information concerning the intercorrelation of the variables used. The squared terms, χ^2, are rather unexpected. Different three connectivities are applied in the regression equation describing the logarithm of the octanol–water partition coefficient, log P and log k', whereas log P and log k' data are strongly intercorrelated [45].

In conclusion, the molecular connectivity is most probably a useful descriptor of nonspecific, dispersive increments to retention. This observation is supported by the results of multiple-regression studies by Massart and co-workers [29, 52, 53]. These authors analyzed GC retention indices determined on phases of different polarity for a large set of nonheterocyclic solutes. Multiparameter QSRR equations were obtained in which various connectivity indices were present along with other topological indices, substituent electronic and hydrophobic constants, and quantum chemically calculated electronic indices. In the case of nonpolar GC stationary phases, when the dispersion forces predominate in the intermolecular interactions, the chromatographic behavior of solutes can be explained, at least in part, by means of topological parameters. However, in the case of solutes of varying polarity, the significance of topological indices for prediction of retention diminishes due to differences in inductive interactions with the stationary phase not accounted for by molecular connectivity. With the polar phases and diverse sets of solutes, the connectivity parameters become of secondary importance for retention as the specific polar interactions become decisive.

For a number of topological indices in use in chemistry [1–3], some applications in QSRR studies found also the Wiener number w and the Hosoya index Z. Molecular connectivity indices are generally highly inter-correlated with both w and Z.

Bonchev et al. [19] published a QSRR equation employing the Wiener number. The relationship describes GC retention index I of a series of monoalkylbenzenes and *ortho*-dialkylbenzenes as follows:

$$I = (244 \pm 4)\, w^{(0.297 \pm 0.003)} \tag{8.19}$$

Trinajstić and co-workers [54] reported two-parameter regression equations relating both GC and HPLC retention data of a set of polycyclic aromatic hydrocarbons to their Wiener number w and its square, w^2. The Wiener number has also been applied in QSRR studies of a series of chlorobenzenes [26] along with the $^1\chi$ index and in the studies of a set of various nonheterocyclic structures [29] along with $^1\chi$ and Z. There is no convincing evidence proving the advantage of one of the topological indices over another. Whereas Bonchev et al. [19] advocate the superiority of the Wiener numbers w over the Randić index χ, Buydens and Massart [29] and Sabljić [26] report that the sequence would be χ as the best, then w, and next Z.

References

1. L. B. Kier and L. H. Hall, *Molecular Connectivity in Chemistry and Drug Research*, Academic, New York, 1976.
2. A. T. Balaban, I. Motoc, D. Bonchev, and O. Mekenyan, Topological indices for

structure–activity correlations, in M. Charton and I. Motoc, (Eds.), *Steric Effects in Drug Design*, Akademie-Verlag, Berlin, 1983, p. 21.

3. A. Sabljić and N. Trinajstić, Quantitative structure–activity relationships: The role of topological indices, *Acta Pharm. Jugosl.*, **31**, 189, 1981.

4. H. Wiener, Correlation of heats of isomerization, and differences in heats of vaporization of isomers, among the paraffin hydrocarbons, *J. Am. Chem. Soc.* **69**, 2636, 1947.

5. H. Hosoya, Topological index. A newly proposed quantity characterizing the topological nature of structural isomers of saturated hydrocarbons, *Bull. Chem. Soc. Japan*, **44**, 2332, 1971.

6. M. Randić, On characterization of molecular branching, *J. Am. Chem. Soc.*, **97**, 6609, 1975.

7. I. Gutman and N. Trinajstić, Graph theory and molecular orbitals. Total π-electron energy of alternate hydrocarbons, *Chem. Phys. Lett.*, **17**, 535, 1972.

8. L. H. Hall and L. B. Kier, Structure–activity studies using valence molecular connectivity, *J. Pharm. Sci.*, **66**, 642, 1977.

9. K. Ośmiałowski, J. Halkiewicz, and R. Kaliszan, Quantum chemical parameters in correlation analysis of gas–liquid chromatographic retention indices of amines. II. Topological electronic index, *J. Chromatogr.*, **361**, 63, 1986.

10. B. L. Karger, J. R. Gant, A. Harthopf, and P. H. Weiner, Hydrophobic effects in reversed-phase liquid chromatography, *J. Chromatogr.*, **128**, 65, 1976.

11. R. Kaliszan and H. Foks, The relationship between the R_M values and the connectivity indices for pyrazine carbothioamide derivatives, *Chromatographia*, **10**, 346, 1977.

12. R. Kaliszan, Correlation between the retention indices and the connectivity indices of alcohols and methyl esters with complex cyclic structure, *Chromatographia*, **10**, 529, 1977.

13. M. Heintz, M. Gruselle, A. Druilhe, and D. Lefort, Relations entre structure chimique et donnees de retention en chromatographie en phase gazeuse. VI. Alcohols et esters methyliques des structures cycliques, *Chromatographia*, **9**, 367, 1976.

14. Y. Michotte and D. L. Massart, Molecular connectivity and retention indexes, *J. Pharm. Sci.*, **66**, 1630, 1977.

15. R. Kaliszan and H. Lamparczyk, A relationship between the connectivity indices and retention indices of polycyclic aromatic hydrocarbons, *J. Chromatogr. Sci.*, **16**, 246, 1978.

16. J. S. Millership and A. D. Woolfson, The relation between molecular connectivity and gas chromatographic retention data, *J. Pharm. Pharmacol.*, **30**, 483, 1978.

17. M. Randić, The structural origin of chromatographic retention data, *J. Chromatogr.*, **161**, 1, 1978.

18. T. R. McGregor, Connectivity parameters as predictors of retention in gas chromatography, *J. Chromatogr. Sci.*, **17**, 314, 1979.

19. D. Bonchev, Ov. Mekenjan, G. Protić, and N. Trinajstić, Application of

topological indices to gas chromatographic data: Calculation of the retention indices of isomeric alkylbenzenes, *J. Chromatogr.*, **176**, 149, 1979.

20. E. K. Whalen-Pedersen and P. C. Jurs, Calculation of linear temperature programmed capillary gas chromatographic retention indices of polycyclic aromatic compounds, *Anal. Chem.*, **53**, 2184, 1981.

21. K. D. Bartle, M. L. Lee, and S. A. Wise, Factors affecting the retention of polycyclic aromatic hydrocarbons in gas chromatography, *Chromatographia*, **14**, 69, 1981.

22. D. L. Vassilaros, R. C. Kong, D. W. Later, and M. L. Lee, Linear retention index system for polycyclic aromatic compounds. Critical evaluation and additional indices, *J. Chromatogr.*, **252**, 1, 1982.

23. L. B. Kier and L. H. Hall, Molecular connectivity analyses of structure influencing chromatographic retention indexes, *J. Pharm. Sci.*, **68**, 120, 1979.

24. J. S. Millership and A. D. Woolfson, A study of the relationship between gas chromatographic retention parameters and molecular connectivity, *J. Pharm. Pharmacol.*, **31**, 444P, 1979.

25. J. S. Millership and A. D. Woolfson, Molecular connectivity and gas chromatographic retention parameters, *J. Pharm. Pharmacol.*, **32**, 610, 1980.

26. A. Sabljić, Calculation of retention indices by molecular topology. Chlorinated benzenes, *J. Chromatogr.*, **319**, 1, 1985.

27. Z. Yu. Bulycheva, B. A. Rudenko, L. V. Dylevskaya, and V. Ph. Kutenev, O korrelyatsii gazokhromatographicheskikh indeksov uderzhivaniya i indeksov svyazannosti vysokikh poryadkov dlya politsiklicheskikh aromaticheskikh uglevodorodov, *Zh. Anal. Khim.*, **40**, 330, 1985.

28. K. Jinno and K. Kawasaki, Correlations between retention data of isomeric alkylbenzenes and physical parameters in reversed-phase micro high-performance liquid chromatography, *Chromatographia*, **17**, 337, 1983.

29. L. Buydens and D. L. Massart, Prediction of gas chromatographic retention indices from linear free energy and topological parameters, *Anal. Chem.*, **53**, 1990, 1981.

30. R. Kaliszan, The relationship between the connectivity indices and the thermodynamic parameters describing the interaction of fatty acid methyl esters with polar and nonpolar stationary phases, *Chromatographia*, **12**, 171, 1979.

31. R. V. Golovnya and T. E. Kuzmenko, Thermodynamic evaluation of the interaction of fatty acid methyl esters with polar and non-polar stationary phases, based on their retention indices, *Chromatographia*, **10**, 545, 1977.

32. K. Jinno and K. Kawasaki, Retention prediction of substituted benzenes in reversed-phase HPLC, *Chromatographia*, **18**, 90, 1984.

33. J.-X. Huang, E. S. P. Bouvier, J. D. Stuart, W. R. Melander, and Cs. Horváth, High-performance liquid chromatography of substituted *p*-benzoquinones and *p*-hydroquinones. II. Retention behaviour, quantitative structure–retention relationships and octanol–water partition coefficients, *J. Chromatogr.*, **330**, 181, 1985.

34. Gy. Szasz, O. Papp, J. Vámos, K. Hankó-Novák, and L. B. Kier, Relationships

between molecular connectivity indices, partition coefficients and chromatographic parameters, *J. Chromatogr.*, **269**, 91, 1983.

35. Y. Sasaki, T. Takagi, H. Kawaki, and A. Iwata, Novel substituent entropy constant σ_{s0} represents the molecular connectivity χ and its related indices, *Chem. Pharm. Bull.*, **31**, 330, 1980.

36. V. D. Shatz, O. V. Sakhartova, L. A. Brivkalne, and V. A. Belikov, Vybor uslovii elyuirovaniya v obrashchenno-phazovoi khromatographii. Indeks svyazyvaemosti i uderzhivanie uglevodorodov i prosteishikh kislorodsoderzhashchikh soedinenii, *Zh. Anal. Khim.*, **39**, 894, 1984.

37. O. V. Sakhartova and V. D. Shatz, Vybor uslovii elyuirovaniya v obrashchenno-phazovoi khromatographii. Priblizhennaya apriornaya otsenka uderzhivaniya poliphunktsionalnykh kislorodsoderzhashchikh soedinenii, *Zh. Anal. Khim.*, **39**, 1496, 1984.

38. S. Zhang, G. Zhao, and S. Wan, Relation between thermodynamic properties of adsorption, topological parameters and retention volumes ln V_g^T for some alkanes and their oxygenated products on Porapak-Q (in Chinese), *Sepu*, **2**, 112, 1985; *C. A.* **103**, 93346a, 1985.

39. J. Raymer, D. Wiesler, and M. Novotny, Structure–retention studies of model ketones by capillary gas chromatography, *J. Chromatogr.*, **325**, 13, 1985.

40. P. J. Doherty, R. M. Hoes, A. Jr. Robbat, and C. M. White, Relationship between gas chromatographic retention indices and molecular connectivities of nitrated polycyclic aromatic hydrocarbons, *Anal. Chem.*, **56**, 2697, 1984.

41. H. Colin and G. Guiochon, Selectivity for homologous series in reversed phase liquid chromatography. I. Theory, *J. Chromatogr. Sci.*, **18**, 54, 1980.

42. W. Markowki, T. Dzido, and T. Wawrzynowicz, Correlation between chromatographic parameters and connectivity index in liquid–solid chromatography, *Pol. J. Chem.*, **52**, 2063, 1978.

43. M.-C. Bonjean and C. Luu Duc, Connectivite moleculaire: Relation dans une serie de barbituriques, *Eur. J. Med. Chem.*, **13**, 73, 1978.

44. L. Ekiert, Z. Grodzińska-Zachwieja, and J. Bojarski, Quasi-column chromatography of barbituric acid derivatives. II. Separation of non-polar adsorbents using binary water–organic solvent mixtures as the mobile phase, *Chromatographia*, **13**, 472, 1980.

45. M. J. M. Wells, C. R. Clark, and R. M. Patterson, Correlation of reversed-phase capacity factors for barbiturates with biological activities, partition coefficients, and molecular connectivity indices, *J. Chromatogr. Sci.*, **19**, 573, 1981.

46. J. Bojarski and L. Ekiert, Relationship between molecular connectivity indices of barbiturates and chromatographic parameters, *Chromatographia*, **15**, 172, 1982.

47. A. H. Stead, R. Gill, A. T. Evans, and A. C. Moffat, Predictions of gas chromatographic retention characteristics of barbiturates from molecular structure, *J. Chromatogr.*, **234**, 277, 1982.

48. M. J. M. Wells, C. R. Clark, and R. M. Patterson, Investigation of N-alkylbenzamides by reversed-phase liquid chromatography. III. Correlation of

chromatographic parameters with molecular connectivity indices for C_1 to C_5 N-alkylbenzamides, *J. Chromatogr.*, **235**, 61, 1982.

49. K. Bojarski and L. Ekiert, The evaluation of molecular connectivity indices and van der Waals volumes for correlation of chromatographic parameters, in H. Kalasz (Ed.), *Proceedings of the Symposium on Advanced Liquid Chromatography*, Akademiai Kiado, Budapest, 1982, p. 35.

50. J. Bojarski and L. Ekiert, Evaluation of modified valence molecular connectivity index for correlation of chromatographic parameters, *J. Liq. Chromatogr.*, **6**, 73, 1983.

51. L. Ekiert, J. Bojarski, and J. Mokrosz, Some aspects of structure–chromatographic behaviour relationships in TLC of barbiturates, in H. Kalasz and L. S. Ettre, (Eds.), *Chromatography, the State of the Art*, Akademiai Kiado, Budapest, 1985, p. 403.

52. L. Buydens, D. L. Massart, and P. Geerlings, Prediction of gas chromatographic retention indexes with topological, physicochemical, and quantum chemical parameters, *Anal. Chem.*, **55**, 738, 1983.

53. L. Buydens, D. Coomans, M. Vanbelle, D. L. Massart, and R. Vanden Driessche, Comparative study of topological and linear free energy-related parameters for the prediction of gas chromatographic retention indices, *J. Pharm. Sci.*, **72**, 1327, 1983.

54. N. Adler, D. Babić, and N. Trinajstić, On the calculation of the HPLC parameters for polycyclic aromatic hydrocarbons, *Fresenius Z. Anal. Chem.*, **322**, 426, 1985.

MULTIPARAMETER
STRUCTURE–CHROMATOGRAPHIC
RETENTION RELATIONSHIPS

Structure–retention relationships have been studied by numerous chromatographers with predictive purposes in view. To get quantitative or semiquantitative relationships, allowing the prediction of the retention behavior of an individual solute of a given class, usually empirically derived physicochemical data have been applied. Thus, the vapor pressure or boiling point of a solute have been the physicochemical parameters of importance for GC retention evalution. Chromatographically determined empirical parameters related to a functional group or a molecular structural fragment contribution to retention have often been described in all chromatographic techniques.

The predictive relationships derived with such data are often highly reliable. The fundamental disadvantage of this approach to QSRRs is that the relationships derived hold true only for the unique chromatographic system used for deriving the QSRRs. Moreover, the interpretation of the empirical retention parameters in terms of other empirical parameters is not very informative from the theoretical point of view. In addition, the availability of reliable empirical parameters is limited.

Here, the emphasis is put on the nonempirical structural descriptors and on the QSRRs, allowing the prediction of retention from the information encoded in structural formula of the solute exclusively. Nonetheless, some representative QSRR approaches employing the empirically derived retention descriptors are presented here for instructive purposes. These approaches may be applied by analogy to derive predictive relationships for the particular solute group–stationary phase–mobile phase chromatographic system.

9.1. STRUCTURAL FRAGMENT CONTRIBUTIONS TO RETENTION

Since the early works by Martin [1] followed by numerous publications by Green et al. [2], a great many reports have appeared in which substituent increments to retention were determined. The contributions of the substituents to retention have often been found constant and additive for a given

162

chromatographic system. Based on this observation, reliable predictions of retention behavior have frequently been reported, especially in the case of reversed–phase LC or of GC on nonpolar stationary phases.

As pointed out in Chapter 1, for the linear free-energy relationships (LFERs) to be found between real and model systems, changes either in entropy or enthalpy must be constant, or the enthalpy changes must be linearly related to entropy changes. The enthalpy–entropy compensation found for different chromatographic system substantiates the comparison of the chromatographically derived physicochemical data for compounds significantly differing in properties. From that point of view, the LFER-based approach to chromatographic distribution is of special interest, assuming that the free energy of the process comprises the independent contributions of the component functional group.

Tomlinson et al. [3] determined functional group contributions to reversed-phase HPLC retention for sets of methyl to decyl benzoates, both unsubstituted and o, m, p-OCH$_3$, $-$NO$_2$, and –Cl substituted. These contributions were determined at conditions where enthalpy–entropy compensation was found. Tomlinson et al. [3] defined group contribution τ as

$$\tau = \log r_{ij} = \log(k'_i k'_j) = \log[(t_{R_j} - t_{R_0})(t_{R_i} - t_{R_0})] \tag{9.1}$$

where k' and t_R are the capacity ratio and retention times of solutes j and i, which differ by a functional group; r is the chromatographic selectivity coefficient; and t_{R_0} is the retention time of an unretarded tracer. The τ values for different groups at different positions of substitution were found to depend linearly on the mobile phase composition as expressed by the mobile phase surface tension (Fig. 9.1).

The relationship observed in Fig. 9.1 is in agreement with the Horváth theory [4] of solvophobic chromatography. Thus, the hypothetical τ values at the 100% water mobile phase could be derived. These values were fairly well correlated ($r = 0.915$) to the standard [5] n-octanol–water hydrophobic substituent constants obtained for a phenoxyacetic acid series. Tomlinson et al. [3] were also able to calculate the functional group contributions in free-energy, enthalpy, and entropy terms.

When the enthalpy–entropy compensation is observed, the extrapolation of retention data in a simple binary system to a pure aqueous eluent offers the possibility of obtaining consistent, comparable data concerning capacity factors and group contributions. As proposed by Hafkenscheid and Tomlinson [6], such theoretical capacity factors, together with the function of solute melting points and entropies of fusion, can be used to give estimates of aqueous solubilities according to the formula resembling the Yalkowsky–

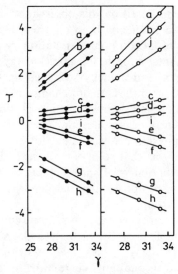

Figure 9.1. Relationships between reversed-phase HPLC functional group contribution τ and mobile phase surface tension γ (mNm), using octyl (●) and octadecyl (○) stationary phases. Solutes are eight homologous series of ring-substituted alkylbenzoates: (*a*) 3-Cl; (*b*) 4-Cl; (*c*) 3-OCH$_3$; (*d*) 4-OCH$_3$; (*e*) 3-NO$_2$; (*f*) 4-NO$_2$; (*g*) 2-OCH$_3$; (*h*) 2-NO$_2$; (*i*) 2-Cl; (*j*) 2-CH$_2$. (After E. Tomlinson, H. Poppe, and C. Kraak, *Int. J. Pharm.*, **7**, 225, 1981. With permission.)

Valvani equation [7]:

$$-\log X_\mathrm{w} = ak'_\mathrm{w} + b\,\Delta S_\mathrm{f}(T_\mathrm{m}-20) + c \qquad (9.2)$$

where X_w is the mole fraction aqueous solubility; ΔS_f is the entropy of fusion; and T_m is the melting point of the solute.

There are not many reports published in which enthalpy–entropy compensation have been analyzed. Normally, the LFERs are assumed to exist between the systems studied. One of the more interesting approaches based on such an assumption is that presented by Chen and Horváth [8]. These authors evaluated substituent contributions to chromatographic retention by using a multiple-regression analyses and indicator variables (theoretical basis of the approach have already been discussed in Chapter 5). Retention data, κ_i, (log k' from HPLC or R_M from PC) of n congeners containing m possible substituents were expressed by a set of linear equations involving substituent parameters and indicator variables. The substituent parameter measures the change in chromatographic retention upon replacing a hydrogen in a reference substance by a given substituent. The indicator variable is set to unity when a given substituent is present in an individual derivative and to zero otherwise.

Chen and Horváth [8] analyzed the PC retention data obtained by Kuchař et al. [9] for aromatic–aliphatic acids. The R_M values predicted on the basis of the statistically evaluated ΔR_M values have been consistent with observed data. Similar analyses were done with a set of aromatic amines, acids, and amino acids chromatographed in reversed-phase HPLC systems. In this instance, the number of retention increments was greater than that of the substituents, as a given substituent in different positions may provide different contributions to retention, depending on the particular molecular environment. Chen and Horváth [8] have also found that the substituent parameters calculated from the corresponding retention data obtained with three slightly different systems (octadecyl silica stationary phases: Partisil, Spherisorb, and LiChromosorb; temperature 296 and 343 K) show satisfactory correlation. The conclusion has been drawn that the prediction of retention data from structural parameters is possible for different chromatographic conditions. Recently, Horváth and co-workers [10] derived QSRRs based on functional group contributions to the reversed-phase HPLC capacity factors. This has been done for a set of 12 substituted p-benzoquinones and p-hydroquinones chromatographed on three alkylsilica phases with three different mobile phases.

The structural fragment contribution to retention may be employed for structural analysis of complex molecules. As an example, the results by DiBussolo and Nes [11] are presented. These authors assigned the contributions that individual molecular features made to the HPLC retention relative to the cholesterol of several dozen sterols. The features examined included the presence and absence of various methyl and ethyl groups, double bonds, and hydroxyl groups in the nucleus and/or side chain, the chirality at the C_3 and C_{24} atoms, the configuration about certain double bonds, and the length of the side chain and its branching. Comparison of observed retention data for a given sterol with those calculated by means of structural contributions can be used to gain structural information on a sterol from its chromatographic behavior.

GC retention index increments for substituent groups of multiple-substituted pyrazine derivatives have been determined by Mihara and Enomoto [12]. The increments were assigned separately for the Carbowax 20M and the OV-101 phases.

In extensive studies on amidines, Oszczapowicz and co-workers attempted to predict GC retention indices of solutes on an SE-30 nonpolar column [e.g., 13]. These authors reported the prediction of Kováts indices on SE-30 based on regression equations involving the corresponding retention indices of two structural elements constituting the solute molecule.

Among the numerous reports concerning fragmental contribution to GC retention indices, the majority deal with hydrocarbons as a group of solutes.

Polychlorinated biphenyls studied by Sissons and Welti [14] resemble hydrocarbons with respect to their chromatographic behavior. As most of the procedures developed for the calculation of retention indices apply only to the stationary phase used in the calculation (usually squalane), they will not be discussed here.

A method described by Tejedor [15] deserves mention as it allows the calculation of GC retention indices of mono and dialkylaromatic compounds (at 100°C) on stationary phases whose polarity lies between that of squalane and that of Ucon 50 HB 280X (and that corresponds to Rohrschneider x constants of between 0 and 1.71). Tejedor [15] calculated the contributions to retention of normal and branched alkyl groups as well as the dependence of these contributions on the polarity of the stationary phase. However, attempts to apply the regularities found for prediction of retention of the hydrocarbons studied on more polar stationary phases (Apiezon L, Carbowax 20M) were unsuccessful.

A research group led by Takács worked for many years on a method for a priori calculations of GC retention indices, allowing the assessment of the contributions of atoms and bonds and interactions with stationary phases. The results of numerous publications by the group have been summarized in the literature [16]. These authors divided the retention index into three components:

$$I^{\text{st ph}}_{\text{substance}}(T) = I_a + I_b + I^{\text{st ph}}_{\text{int}}(T) \tag{9.3}$$

where I is the retention index (in retention index units) under isothermal conditions; T is the column temperature (°C or K); I_a is the atomic index contribution; I_b is the bond index contribution; and I_{int} is the interaction index contribution.

The bond index increments are calculated in retention index units (i.u.) from the bond energy by the factor:

$$f_c = 0.018 \, \text{i.u./kJ} \tag{9.4}$$

The energy of the bonds in the ethane molecule is as in Fig. 9.2. Thus, the individual bond index increments are

$$I_b(\text{C–H}) = 0.018 \times 412.8 = 7.43 \, \text{i.u.} \tag{9.5}$$

$$I_b(\text{C–C}) = 0.018 \times 349.1 = 6.28 \, \text{i.u.} \tag{9.6}$$

The index I_b for ethane is

$$I_b(\text{ethane}) = I_b(\text{C–C}) + 6I_b(\text{C–H}) = 50.86 \, \text{i.u.} \tag{9.7}$$

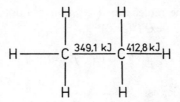

Figure 9.2. Energy of bonds in ethane molecule (according to Budahegyi et al. [16]).

The atomic index increment I_a is

$$I_a = \sum_{g=1}^{h} \left(\frac{\text{atomic weight}}{10}\right)_g = \frac{\text{molecular weight}}{10} \tag{9.8}$$

where g is a serial number and h is the number of atoms in the molecule. Thus, for ethane the atomic index contribution is

$$I_a(\text{ethane}) = \frac{6 \times 1 + 2 \times 12}{10} = 3.00 \text{ i.u.} \tag{9.9}$$

In the case of ethane the molecular index contribution to retention, I_M, is

$$I_M(\text{ethane}) = I_a + I_b = 3.00 + 50.86 = 53.86 \text{ i.u.} \tag{9.10}$$

When the energy of a given bond is not known, it may be evaluated based on the bond length. Budahegyi et al. [16] give as an example the procedure of the calculation of the C_1-C_2 bond energy in oestradiol from the interatomic distance:

$$E_{C_1-C_2} = \frac{71.463}{0.1390} = 514.1 \text{ kJ} \tag{9.11}$$

where 71.463 kJ·nm is the quantum chemical conversion factor for C–C aromatic bonds and 0.1390 nm is the C_1-C_2 bond length.

To employ Eq. (9.3), one needs also the interaction index increment. This interaction index contribution is composed of two parts:

$$I_{\text{int}}^{\text{st ph}}(T) = I_{i,g}^{\text{st ph}}(T) + I_{i,i}^{\text{st ph}}(T) \tag{9.12}$$

where $I_{i,g}^{\text{st ph}}(T)$ is the general interaction index contribution, and $I_{i,i}^{\text{st ph}}(T)$ is the individual interaction index contribution.

The general interaction index contribution is calculated at a given column temperature T from the molecular index contribution, $I_M = I_a + I_b$, and the solute vapor pressure index, $I^0(s, T)$:

$$I_{i, g}(s, T) = I^0(s, T) - I_a - I_b \qquad (9.13)$$

The vapor pressure index is

$$I^0(s, T) = 100 \left[\frac{\log p^0(s) - \log p^0(z)}{\log p^0(z+1) - \log p^0(z)} + z \right] \qquad (9.14)$$

where z and $z + 1$ mean carbon atom numbers in the reference homologues. The general interaction index contribution is obtained by the summation

$$I_{i, g}(s, T) = \sum_{g=1}^{q} I_{i, g}(g)_T \qquad (9.15)$$

where g is a serial number and q is the number of increments in the molecule. For a given solute s and the stationary phase st ph, the individual interaction index contribution is

$$I_{i, i}^{st\ ph}(s, T) = \sum_{g=1}^{q} I_{i, i}(g)_T^{st\ ph} \qquad (9.16)$$

or

$$I_{i, i}^{st\ ph}(s, T) = I^{st\ ph}(s, T) - I^0(s, T) \qquad (9.17)$$

An example of Budahegyi et al. is used to illustrate the calculation procedure. The solute considered is 2,3,4-trimethylpentane of molecular weight 114.0. From Eq. (9.8):

$$I_a = \frac{114.0}{10} = 11.40 \text{ i.u.} \qquad (9.18)$$

The bond energies of the compound are given in Fig. 9.3. The bond index increment values are calculated as follows:

$$
\begin{aligned}
I_b(\text{C1–H1}) &= 0.018 \times 429.68 = 7.73 \text{ i.u.} \\
I_b(\text{C1–C2}) &= 0.018 \times 303.04 = 5.45 \text{ i.u.} \\
I_b(\text{C2–H2}) &= 0.018 \times 419.05 = 7.54 \text{ i.u.} \\
I_b(\text{C3–H31}) &= 0.018 \times 419.35 = 7.55 \text{ i.u.} \\
I_b(\text{C3–C31}) &= 0.018 \times 303.20 = 5.46 \text{ i.u.} \\
I_b(\text{C31–H331}) &= 0.018 \times 438.49 = 7.89 \text{ i.u.} \\
I_b(\text{C3–C4}) &= 0.018 \times 298.24 = 5.37 \text{ i.u.}
\end{aligned}
\qquad (9.19)
$$

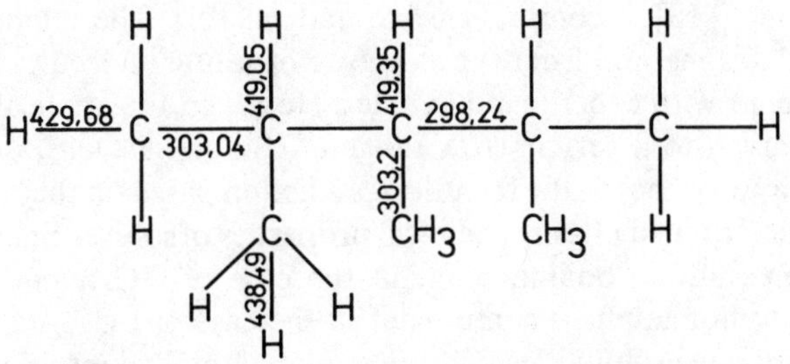

Figure 9.3. Energy (in kJ) of bonds in 2,3,4-trimethylpentane (according to Budahegyi et al. [16].)

The total bond index contribution for 2,3,4-trimethylpentane is

$$I_b = 12 \times 7.73 + 4 \times 5.45 + 2 \times 7.54 + 2 \times 5.37 + 7.55 + 5.46 + 3 \times 7.89$$
$$= 177.06 \text{ i.u.} \tag{9.20}$$

and the molecular index contribution is

$$I_M = I_a + I_b = 11.40 + 177.06 = 188.46 \text{ i.u.} \tag{9.21}$$

Next, Budahegyi et al. [16] calculate the interaction index contribution using Apolane-87 as the stationary phase at 70°C, where the retention index of 2,3,4-trimethylpentane is $I_{(CH_3)_3C_5H_9}^{Apo\,87}(70°C) = 755.5$ i.u. For the calculation, Eq. (9.3) is applied:

$$I_{int}^{Apo\,87}(70°C) = I_{(CH_3)_3C_5H_9}^{Apo\,87}(70°C) - I_M$$
$$= 755.5 - 188.46 = 567.04 \text{ i.u.} \tag{9.22}$$

The vapor pressure index for the solute at 70°C, $I^0[(CH_3)_3C_5H_9, 70°C]$ is 749.0 i.u.; then the general interaction index contribution can be calculated:

$$I_{i,g}[(CH_3)_3C_5H_9, 70°C] = I^0[(CH_3)_3C_5H_9, 70°C] - I_a - I_b$$
$$= 749 - 11.40 - 177.06 = 560.54 \text{ i.u.} \tag{9.23}$$

Now, from Eq. (9.12) the individual interaction index contribution may be obtained:

$$I_{i,i}^{Apo\,87}[(CH_3)_3C_5H_9, 70°C] = I_{int}^{Apo\,87}(70°C) - I_{i,g}[(CH_3)_3C_5H_9, 70°C]$$
$$= 567.04 - 560.54 = 6.50 \text{ i.u.} \tag{9.24}$$

The empirical approach to retention prediction elaborated by Budahegyi and co-workers [16] is comprehensible and reliable. The retention indices predicted by this method from the structure of solute molecules are often in good agreement with experimental values. However, the applicability of the method is limited to a very narrow range of compounds (e.g., isoalkanes).

There are approaches to the retention prediction based on the knowledge of empirically determined physicochemical properties of solutes. Such properties are vapor pressure or boiling point in the case of GC retention and the solubility in a mobile phase component in the case of LC.

Relative aqueous solubilities of a set of phenylthiohydantoin amino acids have been employed by Jinno [17] to derive QSRR equations describing log k' determined in various reversed-phase HPLC systems. Jinno has also been able to compensate for the changes in composition of a mobile phase by means of his experimentally determined multiparameter relationship.

Certainly, the numerous QSRR equations relating the logarithm of the n-octanol–water partition coefficients as well as the substituent or fragmental hydrophobic constants to the reversed-phase LC retention data are also based on empirical structural parameters. These will be reviewed separately in Chapter 11 because of the exceptional importance of reversed-phase HPLC and TLC for quantitation of hydrophobic properties of chemical substances. Separately mentioned here should be the contributions by Haken and co-workers [18] in the case of GC retention indices and by Jinno and Kawasaki [19–21] in the case of reversed-phase HPLC capacity factors. Both groups of researchers studied various combinations of individual electronic, steric, and hydrophobic substituent constants, reporting meaningful correlations for a particular regression equation. For example, Jinno and Kawasaki [19] used Eq. (9.25) to describe the logarithm of the capacity factor, log k', of a set of substituted phenols chromatographed on a C-18 reversed-phase material, with acetonitrile–water (65:35) as the mobile phase, in terms of the Hammet σ and Hansch π substituent constants:

$$\log k' = 0.177\,\pi + 0.182\,\pi\sigma(1-\pi) - 0.367 \qquad (9.25)$$

Equation (9.25) is characterized by the correlation coefficient $R = 0.954$. It is not easy, however, to assign physical meaning to the product of π and σ.

9.2. EMPIRICAL PHYSICOCHEMICAL DATA AS INDEPENDENT VARIABLES IN QSRR EQUATIONS

In extensive studies of the relationships between physicochemical properties and GC retention indices, Dimov and co-workers [22, 23] used information concerning the vapor pressure and molecular volume of the solute to calculate the so-called physicochemical index (PCI). According to the QSRR approach proposed, the theoretical retention index is predicted as the sum of PCI and the empirically determined correction-structural number (STN). The STN index has been calculated for several hydrocarbon classes by regression analysis of the influence of different structural elements of the solutes on GC retention on a nonpolar phase. Dimov and Papazova [23] present the following equation to calculate the STN correction factor to the PCI index, which is calculated on the basis of the vapor pressure and molecular volume of alkanes:

$$STN = 4.28 - 0.79n_R + 0.27n_i - 0.05n_d + 2.46n_{CH_3} + 1.61n_B$$
$$- 0.11n_L - 2.21n_O + 0.57n_q \qquad (9.26)$$

where n is the number of the element in a molecule described by the respective subscript. Subscript R denotes substituents; L denotes the carbon atoms in the straight chain of the isoparaffin; d is a CH_2 group between the substituents R at different carbon atoms; CH_3 is a methyl group; B are butane chains; and so on. Equation (9.26), involving eight independent variables, has been derived from the retention data of a set of only 17 isoparaffins. The structural descriptors in Eq. (9.26) are difficult to identify unequivocally. Intercorrelation among the descriptors has not been studied. For the other, even closely related classes of solutes (e.g., cycloparaffins), the STN equation derived is quite different from Eq. (9.26). This makes the empirical approach by Dimov and co-workers of limited practical value for retention prediction and not promising for further progress in QSRR theory.

The boiling temperature or the data derived from it have been used in QSRR studies of GC data directly as independent variables in multiple regression. Calculated from the boiling point, the standard retention index was applied in studies of GC retention of phenols by Radecki et al. [24] along with substituent constants.

Correlation between gas chromatographic specific retention volumes, V_g, and boiling point at 760 Torr, T_b, was studied by Castello and D'Amato [25] for a set of straight and branched-chain bromoalkanes. The curvilinear relationship is apparent on the graphs presented in Castello and D'Amato [25], although the data points are scattered independent of the stationary phase on which the V_g data has been determined. In contradistinction to the $V_g = f(T_b)$ relationship, the plots of ln V_g versus the number of carbon atoms in the 1-bromoalkanes studied are precisely linear.

However, the results recently published by Bermejo and co-workers [26–28] are highly intriguing. These authors derived regression equations of high predictive value using solute boiling point and a molecular volume parameter as the independent variables determining Kováts indices on phases of varying polarity. Pulling aside QSRRs of esters [26], where the statistics may cause some objections, one finds interesting the equations derived for chlorinated 1,4-dimethylbenzenes chromatographed on squalane, SE-54, Apiezon M, Ucon LB 55X, OV-210, and 2,4-trixylenyl phosphate (TXP) [27, 28]. The equations obtained are of the general form

$$I = a T_b \pm b V_W + c \qquad (9.27)$$

where I is the retention index, T_b is the solute boiling point, and V_W is its van der Waals volume. It must be noted here that the two-parameter regression equations have been derived for only four reference compounds. At first sight such regression equations may be treated as fortuitous and devoid of any predictive value. The data provided by Bermejo et al. [28] are convincing, however, and suggest that the equations found statistically result from some fundamental physicochemical relationships. Chromatographic indices and the calculated molecular properties for a set of deuterated and protiated solutes studied by Bermejo et al. [28] are given in Table 9.1.

The empirical equations of the form of Eq. (9.27) were derived for T_b and V_W for solutes 2, 4, 6, and 16 in Table 9.1, for which the corresponding data pairs were 186°C and 80.14 mL/mol; 200°C and 78.84 mL/mol; 222°C and 89.62 mL/mol; and 254°C and 87.02 mL/mol, respectively. For the four phases of increasing polarity (quantified by McReynolds number P_M), the two-parameter equations derived are:

squalene: $I = 2.86 T_b + 7.06 V_W - 48.53$ (9.28)
$\quad\quad\quad P_M = 0 \quad n = 16; \quad R = 1.0000; \quad s = 0.7$

SE-54: $I = 4.56 T_b + 0.46 V_W + 189.20$ (9.29)
$\quad\quad\quad P_M = 337 \quad n = 16; \quad R = 0.9999; \quad s = 1.8$

UCON LB 550X: $I = 6.73 T_b - 6.56 V_W + 449.77$ (9.30)
$\quad\quad\quad P_M = 996 \quad n = 16; \quad R = 0.9999; \quad s = 2.4$

TXP: $I = 7.85 T_b - 9.05 V_W + 513.66$ (9.31)
$\quad\quad\quad P_M = 1420 \quad n = 16; \quad R = 0.9999; \quad s = 2.7$

As evident from Eqs. (9.28)–(9.31), the contribution of V_W is smallest in the case of SE-54 phase. For this phase, there is practically a linear one-parameter relationship between retention index and T_b, from which the boiling points given in Table 9.1 are calculated. These T_b values and retention indices on squalane allowed Bermejo et al. [28] to calculate V_W by means of Eq. (9.28).

Table 9.1. Kováts Indices Determined on Four Stationary Phases, Boiling Point T_b,[a] and van der Waals volume V_W[a] of Deuterated Protiated 1,4-Dimethylbenzenes[b]

| Compound | Experimental Kováts Indices | | | | T_b (°C) | V_W (mL/mol) |
	Squalane	SE-54	UCON LB 550X	TXP		
[²H₁₀]-1,4-Dimethylbenzene	858	880	964	1013	—	—
1,4-Dimethylbenzene	865	886	970	1020	—	—
[²H₉]-2-Chloro-1,4-dimethylbenzene	1044	1071	1174	1246	185.0	79.80
2-Chloro-1,4-dimethylbenzene	1051	1077	1180	1253	186.4	80.23
[²H₉]-1-(Chloromethyl)-4-methylbenzene	1074	1130	1272	1361	197.9	78.69
1-(Chloromethyl)-4-methylbenzene	1080	1136	1277	1367	199.2	79.15
[²H₈]-2,5-Dichloro-1,4-dimethylbenzene	1213	1236	1350	1437	220.9	89.20
2,5-Dichloro-1,4-dimethylbenzene	1220	1242	1356	1444	222.2	89.66
[²H₈]-2,3-Dichloro-1,4-dimethylbenzene	1231	1267	1387	1487	227.6	89.03
2,3-Dichloro-1,4-dimethylbenzene	1237	1273	1393	1493	228.8	89.40
[²H₈]-2-Chloro-1-(chloromethyl)-4-methylbenzene	1240	1297	1451	1565	234.1	87.68
2-Chloro-1-(chloromethyl)-4-methylbenzene	1247	1303	1456	1571	235.3	88.18
[²H₈]-3-Chloro-1-(chloromethyl)-4-methylbenzene	1249	1310	1471	1584	236.8	87.86
3-Chloro-1-(chloromethyl)-4-methylbenzene	1256	1315	1476	1590	237.9	88.26
[²H₈]-1,4-Bis(chloromethyl)benzene	1288	1381	1587	1714	252.2	87.14
1,4-Bis(chloromethyl)benzene	1294	1389	1591	1722	254.0	87.26
[²H₇]-2,3,5-Trichloro-1,4-dimethylbenzene	1390	1416	1534	1643	259.8	98.51
2,3,5-Trichloro-1,4-dimethylbenzene	1406	1422	1540	1649	261.0	100.29

[a] Calculated by Eq. (9.28).

[b] As reported by Bermejo et al. [28]. With permission.

173

The V_W values calculated are consequently lower for the deuterated analog in each pair of solutes. The same holds true for T_b and the Kováts index. Unfortunately, the experimental boiling points and molar volumes for the deuterated analogs are unknown.

Bermejo et al. [28] have not discussed the systematic changes in contribution of V_W to I with increasing polarity of the stationary phase used. They interpreted the changes in sign of the regression coefficient at V_W as if V_W opposed the boiling point effect on more polar phases.

This author considers both boiling point and chromatographic retention as resulting from the abilities of solute molecules to participate in composite intermolecular interactions, whereas the van der Waals volume is a parameter quantifying the abilities of a solute to participate in dispersive interactions. From that point of view the different regression coefficients in Eqs. (9.28)–(9.31) result from differences in magnitude of solute–solute interactions in liquid state as compared to the solute–chromatographic phase interactions. The explanation will be described in the next section.

Roth and Novak [29] attempted to predict GC retention data based on the relationship between the specific retention volume and the activity coefficient of the solute in a given chromatographic system:

$$V_g^0 = \frac{273.15\,R}{\gamma_D P_D^0 x_1 M_1} \tag{9.32}$$

where V_g^0, R, γ_D, P_D^0, x_1, and M_1 are the specific retention volume of the solute, molar gas constant, Raoult-law activity coefficient of the solute, saturation vapor pressure of the solute, molar fraction of the stationary phase (solvent) in the solute–stationary phase mixture within the gas chromatographic zone, and the molar mass of the solvent, respectively. Normally, x_1 approaches unity, and M_1 of the stationary phase is known. Also, the vapor pressure data for a large number of compounds may be found in literature. Thus, to use Eq. (9.32), one needs to know the value of the activity coefficient. To evaluate activity coefficients, the UNIFAC model of Prausnitz and co-workers [30] may be applied. The model is a combination of the universal quasi-chemical equation (UNIQUAC) with the solution-of-groups concept. According to the model, the activity coefficients in multicomponent liquid mixtures are calculated by virtue of structural parameters derived from van der Waals volumes. Roth and Novak [29] used literature [30] UNIFAC interaction parameters. The calculated and experimental relative retention volumes for a set of ketones and alcohols are moderately intercorrelated as illustrated in Fig. 9.4. As reported [29], the differences between the calculated and measured V_g^0 values reach 10% of the value measured. Thus, the original UNIFAC

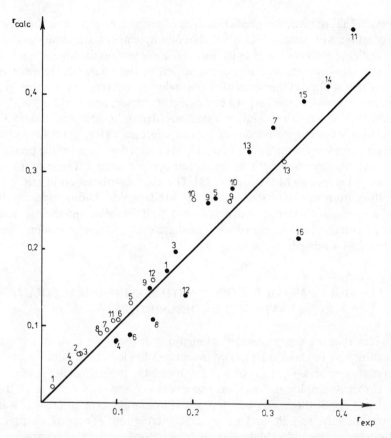

Figure 9.4. Comparison of calculated and experimental relative retention volumes. (○) Ketones–dioctyl phthalate (relative to 2-nonanone); 1 = acetone; 2 = 2-pentanone; 3 = 3-pentanone; 4 = 3-methyl-2-butanone; 5 = 2-hexanone; 6 = 3-hexanone; 7 = 3-methyl-2-pentanone; 8 = 4-methyl-2-pentanone; 9 = 2-heptanone; 10 = 4-heptanone; 11 = 2,4-dimethyl-3-pentanone; 12 = cyclopentanone; 13 = cyclohexanone. (●) Alcohols–diglycerol (relative to 1-nonanol); 1 = methanol; 2 = ethanol; 3 = 1-propanol; 4 = 2-propanol; 5 = 1-butanol; 6 = 2-butanol; 7 = 1-pentanol; 8 = 2-pentanol; 9 = 2-methyl-1-butanol; 10 = 3-methyl-1-butanol; 11 = 1-hexanol; 12 = 2-hexanol; 13 = 2-methyl-1-pentanol; 14 = 3-methyl-1-pentanol; 15 = 4-methyl-1-pentanol; 16 = 2-octanol. (After M. After Roth and J. Novák, *J. Chromatogr.*, **258**, 23, 1983. With permission.)

model can be used only for a rough estimation of GC retention data. On the other hand, chromatography may certainly be exploited to refine the UNIFAC model.

Recently, Petrović et al. [31] applied the functional group contribution concept to the analysis of liquid chromatographic retention data.

When dealing with the empirically determined chromatographic retention descriptors, one should not forget about the fundamental Rohrschneider apppach [32] to retention prediction. The essence of the concept is that under the conditions of classical GC, where the column temperature, carrier gas inlet, and outlet pressures as well as its flow rate are constant during the analyses, the individual interaction of any substance with a given phase can be described by means of the interactions of a few selected reference substances (e.g., benzene, ethanol, methyl ethyl ketone, nitromethane, and pyridine) and the specific constants of the examined substance (Rohrschneider's constants). The difference of the retention index of a given solute on a given stationary phase and on the reference squalane phase, ΔI_s, is obtained as a sum of the products of the stationary-phase-specific and solute-specific factors. The concept was further developed by McReynolds [33]. The approach to retention prediction resulting from works of Rohrschneider, McReynolds, Ladon, and Sandler [34] and many others may be of practical value if the solute and the stationary phase constants are known, which is the case with the more commonly used phases and a number of solutes.

9.3. MULTIPARAMETER QSRRs WITH NONEMPIRICAL SOLUTE STRUCTURE DESCRIPTORS

The idea that the intermolecular interactions determining chromatographic retention may be classified into two main types has long been popular among chromatographers. On this concept are based the theoretical approaches to the chromatographic distribution process (see Chapter 3) or at least their empirical verification. The two types of chromatographic interactions are, most generally, specific and nonspecific interactions. There is no precise definition of the individual class of interactions, and their separation is intuitive. However, chemists try to identify, and then to quantify, the factors determining a particular class of intermolecular interactions. Thus, the specific interactions are associated with electric charge distributions within the molecules, with their shape, and with the presence of particular molecular fragments, groups, or elements. These features determine what was once called *molecular polarity*. On the other hand, the interactions classified as nonspecific are usually ascribed to molecular bulk properties and are joined with such additive scalar quantities as molar volume or refractivity. Their parameterization may easily be accomplished. Certainly, solute polarity as a vector quantity is especially difficult to parameterize by a single number or a set of numbers. There is no precise universal measure of molecular polarity but for the solution of particular chemical problems the polarity of a series of compounds may often be compared by means of dipole moments, electronic substituent constants, electric charge distributions, and so on.

Once the relationships between chromatographic retention data and molecular descriptors are established, it will be possible to derive quantitative data concerning the specific as well as the nonspecific properties of chemicals, which next may be used for prediction of the other physicochemical or biological behavior. Especially valuable for predictive reasons, and the resulting rational design of the required structures, seem to be the nonempirically derived molecular descriptors. With such structural descriptors and the quantitative structure–property relationships established, chemists can to elegant chemistry with pencil and paper (or rather computer) instead of the dirty job with tubes and flasks. Let us see how far we are from that enticing goal.

Based on the results of their earlier works [35, 36] on QSRRs involving some quantum chemical parameters, Gassiot-Matas and Firpo-Pamies [37] separated GC retention index into two components: one depending on the bulk properties of the solute and the other resulting from the polar interactions of the solute with the stationary phase. To parameterize bulk or steric properties of solutes, these authors applied the first-order molecular connectivity $^1\chi$. To characterize the polar term in correlation equations, the square dipole moment of the molecule, μ^2, was used. The general equation analyzed statistically has the form

$$I = a\,{}^1\chi + b\mu^2 + c \tag{9.33}$$

where I is the Kováts index, and a, b, and c are regression coefficients. Using the experimental (or extrapolated and interpolated) dipole moments along with the calculated connectivity indices, Gassiot-Matas and Firpo-Pamies [37] were able to describe Kováts indices of related solutes on different stationary phases.

In fact, it is not clear from their data [37] what is the statistical significance of the μ^2 term in the regression equations presented. For example, one set of solutes form eight homologous hydrocarbons plus nine aliphatic alcohols. For such a group, dipole moment μ assumes practically only two values, zero for hydrocarbons and about $1.7D$ for alcohols. In other words, the parameter μ may play the role of indicator variable accomodating two groups of solutes by one regression equation. Retention indices of each of the two groups separately would be well described by the single-parameter equation $I = f(^1\chi)$. Using any other indicator variable D (e.g., $D = 1$ for hydrocarbons and $D = 0$ for alcohols), one would probably get the regression equation $I = f(^1\chi, D)$ of the same quality as $I = f(^1\chi, \mu^2)$.

Independent of whether μ^2 is or is not a good descriptor of molecular polarity, the approach to QSRR studies based on the separation of nonspecific and polar contribution to retention proposed by Gassiot-Matas and Firpo-Pamies [37] has been highly valuable for development of this research area.

Analyzing GC retention indices of a large group of substituted phenols, Kaliszan and Höltje [38] used solute molecular refractivity as a bulk (dispersive) descriptor. The data analyzed are given in Table 9.2. The retention indices considered were taken from the literature [39] and were determined on a nonpolar phase dimethylpolysiloxane (SE-30) and two polar phases: 3-cyanopropylmethylpolysiloxane (OV-225) and polyneopentyl glycol adipate (NGA). Also, the experimental dipole moments μ_{obs} were taken from the literature (detailed sources are given in ref. 38) as the values determined at 25°C in benzene solution. MR data were calculated according to Hansch et al. [40] and μ_{calc} values were derived by Kaliszan and Höltje [38] employing the CNDO/2 method of calculation.

Regression analysis was done to the general relationship

$$I = a(MR) + b\mu^2 + c \qquad (9.34)$$

Dipole moments determined in benzene at room conditions did not fit Eq. (9.34) for four compounds: 4-OH$^-$, 3-OH$^-$, and 3-NH$_2$-substituted phenols, for which the experimental dipole moments are significantly lower than expected from Eq. (9.34), and 2-nitrophenol, the experimental dipole moment of which is too high to fit Eq. (9.34). For the reduced set of 16 derivatives, the following relationships were found:

$$I(\text{SE-30}) = 66.67\,(\pm 14.42)\mu_{obs}^2 + 25.34\,(\pm 5.22)(MR) + 57.53$$
$$n = 16; \quad R = 0.9679; \quad s = 31 \qquad (9.35)$$

$$I(\text{OV-225}) = 164.72\,(\pm 36.61)\mu_{obs}^2 + 24.48\,(\pm 12.90)(MR) + 408.33$$
$$n = 16; \quad R = 0.9472; \quad s = 77 \qquad (9.36)$$

$$I(\text{NGA}) = 155.69\,(\pm 33.63)\mu_{obs}^2 + 23.99\,(\pm 12.19)(MR) + 599.81$$
$$n = 16; \quad R = 0.9477; \quad s = 73 \qquad (9.37)$$

The correlations obtained in Eqs. (9.35)–(9.37) are highly significant statistically, although the predictive power of the equations is probably too low for analytical purposes. It should be noted that the increment of the dispersive interactions on all three phases employed is nearly constant, but the μ_{obs}^0 term depends strongly on the characteristics of the individual stationary phase.

Most probably, the μ_{obs}^2 term in Eqs. (9.35)–(9.37) is only a crude approximation of the polar interaction contribution to the GC retention index. Thus, if the separation of chromatographic interactions into two distinctive classes (i.e., nonpolar and polar interactions) is justified, further

Table 9.2. Retention Indices on Three Phases[a] and Molar Refractivities and Dipole Moments for Substituted Phenols[b]

Phenol	Kováts Indices			Molar Refractivity	Dipole Moment	
	SE-30	OV-225	NGA		μ_{obsd}	μ_{calc}
$2\text{-}CH_3$	1035	1587	1742	32.83	1.45	1.72
$4\text{-}CH_3$	1059	1654	1813	32.83	1.61	1.87
$3\text{-}CH_3$	1065	1648	1782	32.83	1.58	1.79
$2,6\text{-}(CH_3)_2$	1098	1593	1716	37.45	1.38	1.69
$2,4\text{-}(CH_3)_2$	1134	1660	1825	37.45	1.39	1.77
$3\text{-}C_2H_5$	1160	1742	1898	37.48	—	1.60
$4\text{-}C_2H_5$	1162	1746	1890	37.48	—	1.85
$3,5\text{-}(CH_3)_2$	1163	1706	1877	37.45	1.55	1.82
$2,3\text{-}(CH_3)_2$	1169	1693	1857	37.45	1.25	1.74
$2,4\text{-}Cl_2$	1183	1708	1877	38.21	1.59	2.12
$4\text{-}Cl$	1192	1922	2058	33.21	2.19	2.36
$3\text{-}Cl$	1194	1911	2061	33.21	2.14	2.03
$2,4,6\text{-}(CH_3)_3$	1204	1621	1778	42.07	1.40	1.78
$2,6\text{-}Cl_2$	1206	1727	1871	38.21	—	3.20
$4\text{-}OCH_3$	1210	1930	2050	35.05	1.92	2.21
$3\text{-}OCH_3$	1211	1940	2083	35.05	—	2.30
$2,3,5\text{-}(CH_3)_3$	1260	1823	1960	42.07	—	1.78
$4\text{-}Br$	1274	2054	2191	36.06	2.19	2.70
$3\text{-}CH_3\text{-}4\text{-}Cl$	1283	2025	2135	37.84	—	2.02
$4\text{-}NH_2$	1314	2154	2277	32.60	—	2.50
$4\text{-}OH$	1334	2330	2515	28.03	1.40	1.61 (3.21)
$3\text{-}NH_2$	1335	2219	2352	32.60	1.83	2.53 (2.61)
$2,4,6\text{-}Cl_3$	1349	1928	2067	43.21	1.62	1.67
$2,4,5\text{-}Cl_3$	1362	2039	2158	43.21	—	1.99
$3\text{-}OH$	1368	2371	2576	28.03	2.07	3.49
$3,5\text{-}Cl_2$	1391	2217	2343	38.21	2.18	2.17
$4\text{-}I$	1398	2230	2348	41.02	2.13	—
$4\text{-}CO_2CH_3$	1500	2376	2461	40.05	—	2.71
$4\text{-}COCH_3$	1578	2478	2529	38.36	—	3.15
$2\text{-}NH_2$	1242	2039	2196	32.60	—	2.85
$3\text{-}Br$	1270	2069	2214	36.06	—	2.34
$2\text{-}iso\text{-}C_3H_7\text{-}5\text{-}CH_3$	1271	1776	1932	45.75	—	1.82
$2,6\text{-}(tert\text{-}C_4H_3)\text{-}4\text{-}CH_3$	1494	1782	1830	70.01	—	1.75
$2\text{-}OCH_3$	1095	1544	1627	35.05	—	2.12
$2\text{-}NO_2$	1149	1556	1703	34.54	3.10	3.82
$2,6\text{-}(OCH_3)_2$	1347	1936	2014	41.89	—	3.00
$2\text{-}OCH_3\text{-}4\text{-}C_3H_7$	1392	1810	1884	48.98	—	2.28
$2\text{-}OCH_3\text{-}4\text{-}CHO$	1447	2199	2235	40.90	—	3.39
$2,6\text{-}(OCH_3)_2\text{-}4\text{-}CH_3$	1473	2076	2106	46.51	—	3.07
$2\text{-}OCH_3\text{-}4\text{-}COCH_3$	1531	2283	2326	38.36	—	2.89
$2,6\text{-}(OCH_3)_2\text{-}4\text{-}C_3H_7$	1624	2254	2256	55.82	—	3.20
$2,6\text{-}(OCH_3)_2\text{-}4\text{-}COCH_3$	1849	2685	2683	52.04	—	4.49
$2\text{-}OCH_3\text{-}4\text{-}CH_2CH=CH_2$	1367	1848	1923	48.51	—	2.18

[a] From Grzybowski et al. [39].
[b] From Kaliszan and Höltje [38]. With permission.

179

efforts should be directed more adequate parameterization of solute polarity. To prove the general validity of the relationship

$$I = aP + bD + c \tag{9.38}$$

where I is the retention index of a solute, P and D are its polar and dispersive properties descriptors, respectively, and a, b, and c are constants depending on the properties of the individual stationary phase, the following argument was proposed [38, 41]. If one has retention indices for a given set of compounds on a polar phase, I_P, and on a nonpolar phase, I_{NP}, then according to Eq. (9.38), one can write

$$I_P = a_P P + b_P D + c_P \tag{9.39}$$

$$I_{NP} = a_{NP} P + b_{NP} D + c_{NP} \tag{9.40}$$

where the subscripts P and NP denote polar and nonpolar stationary phase, respectively. Taking the descriptor P from Eq. (9.40),

$$P = (I_{NP} - b_{NP} D - c_{NP})/a_{NP} \tag{9.41}$$

Equation (9.39) is rewritten as

$$I_P = a_P (I_{NP} - b_{NP} D - c_{NP})/a_{NP} + b_P D + c_P \tag{9.42}$$

After rearrangement, one gets

$$I_P = \frac{a_P}{a_{NP}} I_{NP} - \left(\frac{a_P}{a_{NP}} b_{NP} - b_P\right) D + \text{const} \tag{9.43}$$

or

$$I_P = k_1 I_{NP} - k_2 D + k_3 \tag{9.44}$$

where k_1, k_2, and k_3 are constants.

The constant k_2 is defined as

$$k_2 = \frac{a_P}{a_{NP}} b_{NP} - b_P \tag{9.45}$$

where a and b are the descriptors of polar and dispersive properties, respectively, of the stationary phases. In Eq. (9.45) a_P is larger than a_{NP} [the same holds true for polar phases OV-225 and NGA as compared to the nonpolar phase SE-30 in Eqs. (9.35)–(9.37)], whereas b_{NP} and b_P are of comparable magnitude [see Eqs. (9.35)–(9.37) for comparison of the coefficients at MR variable]. Thus, k_2 is positive and the term $-k_2 D$ in Eq. (9.44) is negative. If the general relationship Eq. (9.38) is valid, then for the same sets of solutes, retention indices determined on polar phases should be related to the respective data determined on nonpolar phases by an equation of the form of Eq. (9.44) involving a negative term proportional to solute dispersive properties. The greater the differences in polarities of the stationary phases, the larger the absolute value of the $k_2 D$ term (in Eq. (9.44). There is a change of the sign at the $k_2 D$ term) if the retention index on a nonpolar phase is described in terms of the index obtained on a polar phase and a dispersive descriptor. A phase pair can be imagined for which the term $k_2 D$ in Eq. (9.44) becomes insignificant.

Now, experimental evidence would be required to prove the validity of Eq. (9.44). Such evidence may be derived from the data given in Table 9.2 and found in the existing literature. Kováts indices of a set of phenols determined on the polar phases OV-225 and NGA are related to the indices determined on the nonpolar phase SE-30 by equations of the form of Eq. (9.44), with MR as a nonspecific properties descriptor:

$$I(\text{OV-225}) = 1.95\,(\pm 0.16)\,I(\text{SE-30}) - 22.08\,(\pm 3.56)\,(\text{MR}) + 285.76$$
$$n = 43; \quad R = 0.9691; \quad s = 71 \tag{9.46}$$

$$I(\text{NGA}) = 1.81\,(\pm 0.20)\,I(\text{SE-30}) - 24.50\,(\pm 4.36)\,(\text{MR}) + 682.59$$
$$n = 43; \quad R = 0.9472; \quad s = 87 \tag{9.47}$$

The same type of relationship between GC retention indices generated on the phases of different polarity was reported by Grzybowski et al. [39]. Although these authors could not rationalize their relationships and made minor mistakes in calculations, they provide valuable observations, which certainly facilitated the above derived rationalization of the statistically detected regularities. Grzybowski et al. [39] showed impressive intuition deriving two-parameter statistical equations of the type of Eq. (9.44). Both regressors give poor correlation with the dependent variables in single-parameter equations. Correlation coefficients of the linear relationships

$$I(\text{OV-225}) = a\,I(\text{SE-30}) + b \tag{9.48}$$

or

$$I(\text{NGA}) = a'\,I(\text{SE-30}) + b' \tag{9.49}$$

as well as the analogous relationships with MR as the independent variable are low. In the case of Eq. (9.48) the respective value is $r = 0.77$, and for Eq. (9.49) it is $r = 0.71$. Correlations with MR are less than $r = 0.3$. In such a situation the combination of the two independent variables in an equation of the form of Eq. (9.44) is really lucky.

Lamparczyk and co-workers [42] provide evidences for the validity of Eq. (9.44). These authors related the GC retention time t on polar (NGA and OV-225) and nonpolar (OV-101) stationary phases for a series of copper and nickel complexes of dialkyldithiocarbomates by means of the parameter L, reflecting the length of the alkyl substituent. An example equation [42] is given below for the Cu(II) complexes:

$$t(\text{OV-225}) = 5.10(\pm 0.60)t(\text{OV-101}) - 4.53(\pm 1.49)L + 17.42(\pm 5.06)$$
$$n = 15; \quad R = 0.9844; \quad s = 2.38 \tag{9.50}$$

In fact, Lamparczyk et al. [42] regard the length of the alkyl substituent, L, as a measure of steric hindrance. For the solutes considered, however, the L parameter reflects as well the bulk (dispersive) properties of the individual compounds.

This author believes that the argumentation applied for deriving Eq. (9.44) may be applied in discussing the relationships between GC retention indices and solutes boiling temperatures. The QSRR equations, employing boiling point and a bulk nonspecific parameter as the independent variables used in combination, have been published for several classes of solutes. Radecki et al. [24] reported such equations using the so-called standard retention index calculated from the boiling point and molecular refractivity as the independent variables. In a series of papers Bermejo and co-workers [26–28] obtained relationships describing Kováts indices determined on different stationary phases in terms of boiling points and van der Waals volumes of the solutes. Example Bermejo equations have been discussed in other chapters. Comparing Eq. (9.44) and the previously discussed Eqs. (9.28)–(9.31), one can conclude that the relationships considered are similar. Most probably, the parameter V_W in Eqs. (9.28)–(9.31) corresponds to the dispersive descriptor D in Eq. (9.44). The question is whether the T_b parameter can be used instead of retention index I_{NP} in Eq. (9.44). However, if one realizes that the intermolecular interactions among the identical molecules of the solute in a liquid state are qualitatively of the same nature as other intermolecular interactions (e.g., solute–stationary phase interactions), one agrees that their separation into specific and nonspecific interactions is also valid. In other words, for solute–solute interactions in the liquid state, Eq. (9.38) holds true. Then, the boiling point may be considered as a measure of these complex interactions, similar to the retention index that measures solute–stationary phase interactions. The boiling point may be treated as a parameter describing the retention of a solute on a hypothetical phase composed of itself. If both polar and nonpolar contributions are in the same proportion regarding both the chromatographic retention index and boiling point, then for a set of solutes the two quantities should be linearly related. Very close to that is the situation in Eq. (9.29) where the V_W term is of relatively low significance (coefficient at V_W is 0.46). Normally, however, the polar and nonpolar contributions to the retention index are different as compared to the boiling point. Then, a correction term must be added into the equations relating I to T_b that is calculated by means of the bulk, nonspecific, dispersive solute descriptors. Analyzing the sign of the coefficient at the V_W parameter, it may be concluded that the chlorinated 1,4-dimethylbenzene solutes are more polar than the squalane phase [$+7.06 V_W$ term is present in Eq. (9.28)]; the solutes are of comparable polarity to the SE-54 phase [$+0.46 V_W$ in Eq. (9.29)]; and the solutes are less polar than the UCON LB 550X phase [$-6.56 V_W$ in Eq. (9.30)] and even more less polar than the TXP phase [$-9.05 V_W$ in Eq. (9.31)].

The argumentation given above provides a rational explanation to the relationships found by Bermejo and co-workers [26–28]. It explains both the change from the positive to the negative contribution of the van der Waals volume term in Eqs. (9.28)–(9.31) and the validity of the two-parameter regression equations, derived from only four data points, for all sets of derivatives studied.

Returning to Eqs. (9.39) and (9.40), one can relate the constants a and b to the polar and dispersive properties of stationary phases, respectively. Rearranging Eq. (9.39) or (9.40) yields

$$I_P - b_P D = a_P P + c_P \tag{9.51}$$

or

$$I_{NP} - b_{NP} D = a_{NP} P + c_{NP} \tag{9.52}$$

Thus, the term $I - bD$ is related to solute polarity.

Assuming D is reliably calculable (e.g., from MR) and having retention index I determined, one can get a quantity related to a solute chromatographic polarity. The quantity that must be known is the constant b. When $b_P = b_{NP}$ (i.e., when the two phases, polar and nonpolar, have equal dispersive properties),

$$b_P = b_{NP} = k_2/(k_1 - 1) \tag{9.53}$$

where k_1 and k_2 are the coefficients of the regression equation [Eq. (9.44)] as described numerically in Eqs. (9.46) and (9.47). For a first approximation, one can assume $b_P \approx b_{NP}$ for two phases of similar molecular weights.

If the dispersive characteristics of the phasess, b_P and b_{NP}, differ significantly, one of the two b values must be known. Then,

$$b_P = k_1 b_{NP} - k_2 \qquad (9.54)$$

or

$$b_{NP} = (k_2 + b_P)/k_1 \qquad (9.55)$$

A possible way to obtain data related to the dispersive properties of a chosen standard phase would be to find the relation between the dispersive descriptors and retention indices for a group of nonpolar compounds. The stationary phase chosen should be as nonpolar as possible. Alternatively, different values of b calculated from Eq. (9.55) can be compared with those obtained from Eq. (9.38). In order to get a statistically significant relationship for an equation of the type of Eq. (9.38), a certain number of retention indices, polarity descriptors, and dispersive descriptors are required.

To evaluate the chromatographic polarity measure, as expressed in Eqs. (9.54) and (9.55), for all the phenols studied [38], the electrostatistically determined total solute dipole moment was assumed as the polarity descriptor. Taking $b_{NP} = 25.34$ from Eq. (9.35) and $k_1 = 1.95$ and $k_2 = 22.08$ from Eq. (9.46), the $b_P = 27.3330$ can be calculated by Eq. (9.55). Next, the chromatographic polarity parameter is calculated from Eq. (9.51), where D is represented by MR.

The chromatographic polarity parameter, as calculated by Eq. (9.51) or (9.52), was correlated [38] with quantum chemically calculated dipole moments (Table 9.2) for all 43 phenols considered. The apparent correlation (Fig. 9.5) between the chromatographic polarity parameter, $I(NGA) - 27.3330$ (MR), and the quantum chemically calculated dipole moment, μ_{calc}, may be found, but only for 27 of 43 phenols considered.

Although the elimination of some data from the correlation studies seems to be justified, nonetheless the relationship illustrated in Fig. 9.5 is not precise. That observation provides additional evidence in support of the theoretical assumption that there is no identity of solute polarity and the total molecular dipole moment either electrostatistically determined or quantum chemically calculated. (See also Chapter 7 for a discussion.)

Nonempirical structural parameters were applied in multiparameter QSRR studies by Buydens et al. [43]. These authors analyzed GC retention indices for a group of solutes consisting of aliphatic ethers, esters, alcohols, ketones, and aldehydes. In the multiple-regression analysis they used two classes of nonempirical descriptors, molecular connectivities and quantum chemical indices. The following CNDO/2-calculated electronic parameters were used: the magnitude of the dipole moment, DM; the sum of the absolute values of the charges in a given molecule, QT; and the sum of the absolute values of the charges of the atoms constituting the functional group and the atoms in the

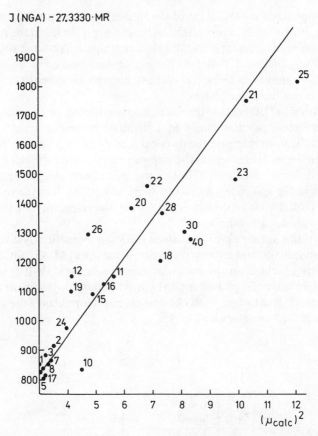

Figure 9.5. Correlation between chromatographic polarity parameter, $I - b(MR)$, and calculated dipole moments for phenols without space-consuming substituents. Solutes are numbered according to Table 9.2.

α position to this functional group, QA. As in the dipole–dipole interaction, the energy of the process is proportional to the square of dipole moments, and the square terms of the individual quantum chemical parameters were included in the regression analysis.

To get meaningful QSRRs, Buydens et al. [43] considered separately the subsets of mono- and bifunctional derivatives. Multiparameter regression equations involving quantum chemical indices, along with several topological indices and substituent constants, have shown the importance of electronic parameters for solute retention on polar stationary phases. Whereas for the nonpolar stationary phases the topological indices of monofunctional molecules can explain a great part of the total retention variance, for the polar stationary phases a combination of topological and electronic parameters is necessary. When the quantum chemical indices are included in the regression, they are preferred to the experimental Hammett sigma. The same authors [43] came to the conclusion that for the bifunctional molecules no parameter combination studied was really efficient enough to describe the specific interactions on the polar phases.

As reported [43], there is a significant intercorrelation among the structural parameters used simultaneously in individual regression equations. This makes it difficult to interpret the physical meaning of a particular parameter applied. However controversial the stepwise entry of the variables is, one can agree with Buydens et al. [43] that the parameters describing the local polarity, like the QA index, are much more important than the total dipole moment, DM. This is especially evident when the retention indices on polar stationary phases are considered.

Recently, this author's group studied QSRRs in a structurally diverse set of primary, secondary, and tertiary (heterocyclic) amines [44, 45]. GC retention indices were obtained on the nonpolar methyl silicone (OV-101) stationary phase and on two polar phases: methyl phenyl cyanopropyl silicone (OV-225) and neopentyl glycol adipate (NGA). The chemical formulas of the stationary phases employed are given in Fig. 9.6.

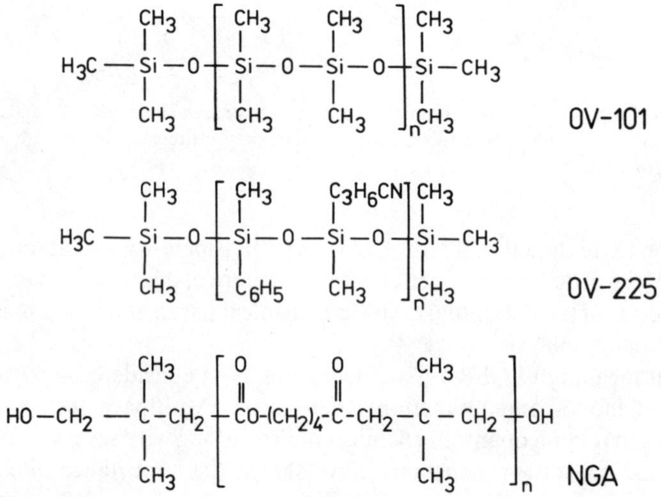

Figure 9.6. Chemical formulas of stationary phases employed for determination of retention indices of amines by Ośmiałowski et al. [44, 45].

The chromatographic data along with the structural parameters considered in QSRR analysis are given in Table 9.3.

Methods of determination of the individual structural parameters listed in Table 9.3 have been described earlier (see Chapter 6 for details).

Among the several CNDO/2-calculated electronic indices, the total energy E_T was found to greatly affect retention. In the case of the nonpolar phase OV-101, it describes 79% of the retention index variance among the solutes, but for the NGA phase the corresponding value is only 53%.

At first QSRRs were established [44] for the retention data generated on the nonpolar OV-101 phase. If the subseries of primary, secondary, and hetero-cyclic amines are analyzed separately, E_T is correlated to I(OV-101) at the same level as molar refractivity MR but only for the primary and secondary amines (Table 9.4). In the third case (i.e., for heterocyclic amines differing significantly in structure), E_T is much better correlated with retention than MR. Thus, the CNDO/2-calculated total energy seems to be the most reliable parameter to quantify the ability of a solute to participate in nonspecific (dispersive) interactions with stationary phase.

However, even on a nonpolar phase OV-101 about 20% of the variance in retention data has not been explained by the total energy. In such a situation attempts were undertaken to characterize a solute polarity increment to the retention. Initially, the CNDO/2-calculated dipole moment was used along with E_T in a two-parameter regression equation. Its significance level was only $p \approx 0.5$, however. Looking for a more reliable measure of chromatographic polarity, we considered electron excess charges on individual atoms in a molecule and proposed a submolecular polarity parameter Δ (see Chapter 6 for details). The regression equation relating the Kováts index on the OV-101 phase, I(OV-101), to total energy, E_T and the polarity parameter Δ is as follows:

$$I(\text{OV-101}) = -11.66\,(\pm 2.74)E_T - 1017\,(\pm 753)\Delta + 302\,(\pm 307)$$
$$n = 22; \quad R = 0.93; \quad s = 67 \tag{9.56}$$

Equation (9.56) is significant at the $p < 0.0001$ level; the terms E_T and Δ are significant at the $p < 0.0001$ and $p < 0.006$ levels, respectively.

Like the previously discussed local polarity indices proposed by Massart and co-workers [43], the parameter Δ differentiates solute polarities better than dipole moment. The amines described by Eq. (9.56) can hardly be considered as congeneric with respect to their polar properties. The lone electron pair of nitrogen atoms present in each compound makes comparable some feasible interactions with the stationary phase (e.g., EPD–EPA interactions). Having the charge transfer interactions in mind, we considered the

Table 9.3. Kováts GC Retention Indices Normalized to 130 °C and Structural Parameters of Amines[a]

Compound Number	Amine	Retention Indices[b]			Total Energy, E_T (a.u.)	Energy of HOMO, E_{HOMO} (a.u.)[c]	Dipole Moment, μ (D)	Molar Refractivity MR^d	Polarity Parameter, Δ (electrons)	Topological Electronic Index, T_E (electrons/A^2)
		OV-101	OV-225	NGA						
1	Allylamine	463	409	548	−38.100	−0.4964	1.7229	18.926	0.3101	1.5866
2	n-Butylamine	553	588	800	−48.597	−0.4843	1.8984	24.044	0.3110	1.6746
3	sec-Butylamine	471	575	729	−48.596	−0.4847	1.8975	24.044	0.3413	1.9182
4	tert-Butylamine	501	574	600	−48.593	−0.4895	1.8971	24.044	0.3600	2.1516
5	n-Pentylamine	635	805	877	−57.381	−0.4767	1.9067	28.688	0.3113	1.7722
6	n-Propylamine	466	457	521	−39.917	−0.4936	1.8930	19.400	0.3139	1.5321
7	Isopentylamine	615	715	883	−57.272	−0.4793	1.8896	28.688	0.3145	1.9085
8	Isopropylamine	469	494	521	−39.921	−0.5048	1.8317	19.400	0.3439	1.8751
9	Diallylamine	660	797	885	−62.519	−0.4686	1.7733	32.572	0.2833	2.1312
10	Di-n-propylamine	694	753	906	−65.964	−0.4655	1.8426	33.520	0.2917	2.1720
11	Diethylamine	527	467	600	−48.602	−0.4767	1.8613	24.323	0.2897	2.1247
12	Methyl-n-pentylamine	706	819	961	−65.963	−0.4626	1.8629	33.520	0.2786	1.9765
13	Methyl-n-hexylamine	871	875	1032	−74.583	−0.4590	1.9108	38.164	0.2786	2.0512
14	Methyl-n-butylamine	630	616	816	−57.279	−0.4670	1.8793	28.876	0.2786	1.8866
15	Di-n-butylamine	943	1020	1069	−83.242	−0.4595	1.8501	42.808	0.2875	2.4639
16	Pyrazine	696	940	1115	−54.621	−0.4564	0.0048	21.482	0.1723	1.0181
17	Pyridine	692	867	1086	−50.866	−0.4707	2.1019	23.890	0.2338	0.9799
18	β-Picoline	841	1059	1156	−59.554	−0.4652	2.1261	28.508	0.2198	0.9722
19	3-Chloropyridine	890	1134	1280	−66.356	−0.4757	2.1976	26.349	0.2570	1.1006
20	Chloropyrazine	895	1180	1365	−70.038	−0.4658	2.1565	28.757	0.3024	1.2350
21	2-Chloropyridine	870	1198	1455	−66.245	−0.4737	3.6424	26.349	0.3120	1.1321
22	4-Cyanopyridine	955	1640	1675	−68.643	−0.4809	0.9618	28.205	0.2471	1.3142

[a] From Ośmiałowski et al. [44, 45]. With permission.
[b] Retention indices extrapolated to 130 °C.
[c] a.u. = atomic units.
[d] Calculated from bond refractivities (see Table 6.6).

Table 9.4. Correlations between Retention Indices of Amines Determined on OV-101 Phase and Molar Refractivities MR and Total Energies E_T[a]

Class of Amines	Correlation Coefficient, r	
	With MR	With E_T
Primary (1–8)[b]	0.90	0.90
Secondary (9–15)[b]	0.98	0.98
Tertiary, heterocyclic (16–22)[b]	0.85	0.94

[a] From Ośmiałowski et al. [44]. With permission.
[b] Refers to numbers in Table 9.3

energy of the highest occupied molecular orbital. The resulting three-parameter regression equation is characterized by higher R and lower s coefficients as compared to Eq. (9.56):

$$I(\text{OV-101}) = -16.59\,(\pm 3.34)E_T - 1988\,(\pm 766)\Delta -$$

$$7098(\pm 3798)\,E_{\text{HOMO}} - 3076\,(\pm 1822)$$

$$n = 22; \quad R = 0.96; \quad s = 50 \tag{9.57}$$

However, the high intercorrelation between the E_T and E_{HOMO} data used in Eq. (9.57), characterized by $r = 0.77$, precludes the application of the two parameters in the same regression equation. By excluding compounds 1, 6, 8, and 15 (Table 9.3) from regression analysis the intercorrelation between E_T and E_{HOMO} may be decreased from 0.77 to 0.53. Then E_T and E_{HOMO} can be used together in regression analyses, which proves the significance of the E_{HOMO} term in the three-parameter equation at the $p < 0.004$ level. Such a result may be assumed as some evidence that the E_{HOMO} parameter is meaningful for a quantitative description of the retention of amines on the OV-101 phase.

Multiparameter regression analyses employing nonempirical parameters have also been applied by this author's group in QSRR studies of reversed-phase HPLC data [46, 47]. The coefficients a_{ij} and b_{ij} in the Soczewiński–Wachtmeister relationship [48] describing capacity factor k'_{ij} for a particular solute i determined on phase j as a linear function of the mole fraction X of one of the components of a binary solvent have been quantitatively related to solute structure:

$$\log k'_{ij} = a_{ij}X + b_{ij} \tag{9.58}$$

Thus, the coefficients a_i and b_i were determined on four stationary phases

$(j = 1, 2, 3, 4)$ of varying C-18 coverage with four compositions of the water–methanol mobile phase for 12 selected variously substituted benzene derivatives. Both a_i and b_i have satisfactorily been described by the two-parameter regression equation involving quantum chemically derived total energy E_T and the local polarity parameter Δ. For example, the coefficients a_i and b_i determined for the phase having 3.18×10^{-4} mol/g C-18 coverage are characterized by $R = 0.9616$ and $R = 0.9805$, respectively. The details of this approach to QSRRs, accounting for variations in solute, mobile phase, and stationary phase characteristics, will be discussed in the next section.

Returning to QSRR studies of amines, it must be observed that attempts to describe retenion indices on polar phases OV-225 and NGA by means of the two-parameter regression with E_T and Δ have been unsuccessful. On polar phases the specific interactions of the solutes have not been accounted for by the simple Δ parameter. Our attempts to find a structural descriptor of higher than Δ discriminative power resulted in introduction of the so-called topological electronic index (see Chapter 8 for details). We recalculated the QSRRs for retention data determined on the OV-101 phase, replacing the parameter Δ in Eq. (9.56) by the topological electronic index T^E. The resulting equation has the form

$$I(\text{OV-101}) = -13.39\,(\pm 1.38)E_T - 149.6\,(\pm 35.7)T^E + 161\,(\pm 96)$$

$$n = 22; \quad R = 0.9810; \quad s = 35; \quad p = 10^{-15} \tag{9.59}$$

Equation (9.59) is significant at the $p = 10^{-15}$ significance level, whereas the p values for the variables E_T and T^E are 0.0002 and 0.0024, respectively. Analogous equations in the case of the two polar phases studied are

$$I(\text{NGA}) = -20.85\,(\pm 4.17)E_T - 424.5\,(\pm 108.2)T^E + 456\,(\pm 290)$$

$$n = 22; \quad R = 0.9484; \quad s = 107; \quad p = 3.3 \times 10^{-9} \tag{9.60}$$

$$I(\text{OV-225}) = -20.19\,(\pm 5.15)E_T - 369.5\,(\pm 133.7)T^E + 270\,(\pm 359)$$

$$n = 22; \quad R = 0.9145; \quad s = 132; \quad p = 2.8 \times 10^{-7} \tag{9.61}$$

The statistical parameters for the respective one-parameter relationships with retention indices are given in Table 9.5.

Comparing the correlations in one-parameter equations (Table 9.5) to those observed in two-parameter equations [Eqs. (9.59)–(9.61)], one finds the topological electronic index a useful descriptor of structural differences among the solutes analyzed. Parameters E_T and T^E are nearly completely orthogonal (i.e. for their intercorrelation $r = 0.1031$). For that reason the two variables are ideally suitable for multiple-regression analyses.

Table 9.5. Correlation Coefficients r and Significance Levels p of Linear Relationships between Retention Indices I Determined on Three Stationary Phases for Amines Listed in Table 9.3 and Total Energy E_T and Topological Electronic Index T^E [a]

	E_T		T^E	
Phase	r	p	r	p
OV-101	0.8954	2×10^{-7}	0.3064	0.39
NGA	0.7295	0.0009	0.5338	0.04
OV-225	0.7295	0.0007	0.4733	0.09

[a] Based on data given by Ośmiałowski et al. [45].

The proposed topological electronic index T^E enables progress to be made in numerical differentiation of the polar or specific properties of the solutes. This index is calculated in such a way (see Chapter 8) that it reflects (at least to some extent) differences in solute size, shape, and constitution. Certainly, the QSRR equations involving nonempirical molecular descriptors are still lacking in precision, especially in the case of diverse noncongeneric solutes chromatographed on polar phases. Neither E_T nor T^E is an ideal structural descriptor, all the more so since the CNDO/2 molecular orbital method yields only approximate data, and the geometry assumed for the calculation may differ from that in the chromatographic system. Nonetheless, further search for more precise nonempirically derived molecular descriptors seems promising.

High intercorrelation between E_T and E_{HOMO} makes the observation questionable, but it should be noted here that the introduction of E_{HOMO} into Eq. (9.61) increases the correlation to $R = 0.9445$. The corresponding increases in R in the case of the OV-101 and NGA phases are less evident. This would be expected if one assumed E_{HMO} to be a parameter quantifying EPD–EPA interactions.

Reviewing the reported statistically significant, multiparameter QSRRs employing nonempirically derived structural parameters, this author noticed an equation by Kuchař et al. [49]. For GC Kováts indices I of a series of arylaliphatic acids determined on Apolane 87 these authors obtained the relationship

$$I = 28.77(MR)(ar) + 8.57(MR)(al) + 1009.3$$

$$n = 36; \quad R = 0.963; \quad s = 96.5; \quad F = 215.7 \tag{9.62}$$

where (MR)(ar) is the molecular refractivity of the aromatic substituents and (MR) (al) is a contribution of the connecting aliphatic chain to total solute molecular refractivity. If Eq. (9.62) is not fortuitous, it suggests that the dispersive contributions to retention by the individual parts of a solute need not be simply additive. Having such possibility in view, it would be interesting to analyze groups of solutes in which two or more distinctive and unequivocally identifiable types of component fragments are present. Perhaps different parts of a solute molecule contribute differently to retention.

The results obtained by Ecknig and co-workers [50, 51] in their systematic studies on GC retention prediction are interesting. These authors analyzed relationships of the type

$$\log V_g = A_0 + A_1 \varphi + A_2 D \tag{9.63}$$

where V_g is the specific retention volume; the parameters φ and D, characterizing polar and nonpolar interaction, respectively, are calculated on the basis of simplified geometric model; and A_0, A_1, and A_2 are empirical constants.

According to Ecknig and co-workers [50, 51], the parameter D is a measure of the energy of dispersive interactions between solute and stationary phase molecules. To calculate D, the following formula is employed:

$$D = \sum_{i=1}^{m} \sum_{j=1}^{n} D_{ij} K_i W_j \tag{9.64}$$

Thus, the interaction contributions D_{ij} of all of the interacting groups i and j of the solute and the stationary phase molecules, respectively, are summed after correction due to the shielding factor K_i and the contact point concentration W_j; the number of interacting groups i of the solute is m, and the corresponding number for the stationary phase molecule is n. The individual contributions, D_{ij}, are calculated from the interacting groups refractions R_i and R_j and the distance r_{ij} between the interacting groups (r_{ij} is the sum of the group van der Waals radii):

$$D_{ij} = \frac{R_i R_j}{r_{ij}^6} \tag{9.65}$$

The correction factors K_i and W_j introduced by Kleinert and Ecknig [50] require special explanation. Defined as the numbers of contact points of an atomic group of solute i and stationary phase moecule, j, K_i and K_j are the numbers of sites on which the molecule is able to interact with neighboring molecules. It is assumed that each atomic group can interact with the four nearest neighbors, and consequently, the value of K_i or K_j is 4 if interacting groups are not shielded and 0 if the groups are completely shielded. The factor W_j is the concentration of contact points belonging to the atomic group j in the molecule of stationary phase. The W_j value is calculated as the ratio of all interacting contact points of atomic group j (i.e., $\sum K_j$) to the total number of contact points of all different atomic groups contained in the stationary phase molecule. The contact points in the case of interacting molecules of ethanol (solute) and oxidipropionitrile (ODPN, stationary phase) are marked in Fig. 9.7 after Kleinert and Ecknig [50].

The parameter φ in Eq. (9.63) is considered by Kleinert and Ecknig [50] as a measure of the energy of intermolecular interactions caused by polar functional groups. Specifically, these authors considered dipole orientation φ_{or} and induction φ_{ind} interactions. To calculate φ (either φ_{or} or φ_{ind}), they employ the formula

$$\varphi = \sum_{u=1}^{m'} \sum_{v=1}^{n'} \varphi_{ij}^{uv} K_u W_v \qquad (9.66)$$

where u and v are the so-called contact segments belonging to the polar atomic groups in solute, i, and stationary phase, j, molecules. The contact segments are chosen based on the following rules: (a) each atom may have four nearest neighbors; (b) the orientations of the interacting groups, which provide minimum potential energy, are preferred; and (c) all of the contact segments of the solute interact with the stationary phase molecule if there is no steric hindrance to the interaction.

The contact segment is understood as that part of the molecular surface where the polar groups may contact during the intermolecular interaction. In Fig. 9.8 the contact segments of the ethanol–oxidipropionitrile interaction are given after Kleinert and Ecknig [50].

The correction factor K_u is the number of contact segments, u, characterized by equal interaction energy. The factor W_v is defined as the ratio of the sum of energetically equal contact segments v (i.e., $\sum v$) to the total sum of all contact segments of the stationary phase molecule that are geometrically able to interact with the corresponding contact segments u of the solute molecule.

When φ in Eq. (9.66) represents orientation forces, $\varphi_{ij}^{uv} = \varphi_{ij}^{or}$ is calculated from the classical Keesom formula:

$$\varphi_{ij}^{or} = -\frac{\mu_i \mu_j}{r_{ij}^3} \cdot [-2\cos\theta_i \cos\theta_j + \sin\theta_i \sin\theta_j + \cos(\phi_i - \phi_j)] \qquad (9.67)$$

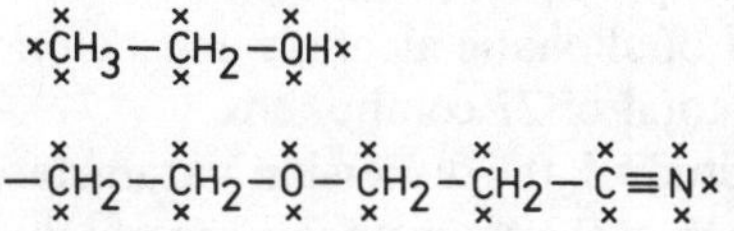

$$\overset{\times}{\underset{\times}{C}}H_3 - \overset{\times}{\underset{\times}{C}}H_2 - \overset{\times}{\underset{\times}{O}}H\times$$

$$\times\overset{\times}{\underset{\times}{N}} \equiv \overset{\times}{\underset{\times}{C}} - \overset{\times}{\underset{\times}{C}}H_2 - \overset{\times}{\underset{\times}{C}}H_2 - \overset{\times}{\underset{\times}{O}} - \overset{\times}{\underset{\times}{C}}H_2 - \overset{\times}{\underset{\times}{C}}H_2 - \overset{\times}{\underset{\times}{C}} \equiv \overset{\times}{\underset{\times}{N}}\times$$

Figure 9.7. Contact points of interacting molecules of ethanol and oxydipropionitrile as suggested by Kleinert and Ecknig [50].

Oxidipropionitrile

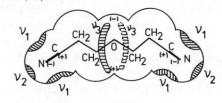

Ethanol

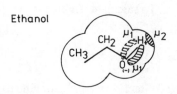

Figure 9.8. Most probable arrangement of contact segments (shaded areas) on molecular surfaces of oxydipropionitrile (stationary phase) and ethanol (solute) for polar interaction between these molecules. (After Th. Kleinert and W. Ecknig, *J. Chromatogr.*, **315**, 75, 1984. With permission.)

where μ_i an μ_j are the dipole moments of the groups i and j of solute and stationary phase molecules, respectively. The orientation angles and distance r_{ij} are depicted in Fig. 9.9.

The induction forces he been characterized by Kleinert and Ecknig [50] by means of the formula of Debye:

$$\varphi_{ij}^{ind} = \frac{\mu_i^2 \alpha_j (3\cos\theta_i + 1)}{2r_{ij}^6} \qquad (9.68)$$

where α_j is the group polarizability of the interacting group j.

Practical verification of the nonempirical approach by Kleinert and Ecknig [50] to the quantitation of dipersive and polar interactions of solutes with GC stationary phases has been reported in a subsequent paper by the group [51]. As the dependent variables in QSRR studies, the specific retention volumes determined on three phases of different polarities [i.e., OV-101, OV-17, and oxidipropionitrile (ODPN) were considered. The set of solutes analyzed consisted of aliphatic alcohols and ketones, halogenoalkanes, and nitroalkanes for a total of 27 compounds.

For the solutes studied, the dispersion parameter D was calculated for the interaction with each of the three phases separately. In the case of individual solutes the value of D was the lowest for the interaction with the OV-17 phase, higher for the solute–OV-101 phase interaction, and the highest in the case of

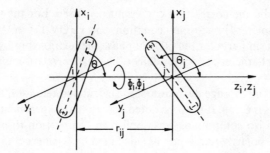

Figure 9.9. Keesom interactions between two dipoles i and j. θ_i, θ_j, ϕ_i, and ϕ_j are orientation angles and r_{ij} is distance between geometrical centers. (After Th. Kleinert and W. Ecknig, *J. Chromatogr.*, **315**, 71, 1984. With permission.)

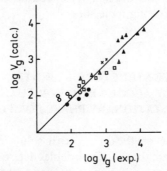

Figure 9.10. Correlation between experimental data and retention parameters on oxydipropionitrile stationary phase calculated by Eq. (9.63). Symbols: nitriles ($\times$), alcohols ($\square$), chloroalkanes ($\bigcirc$), bromoalkanes ($\bigcirc$), iodoalkanes ($\bullet$), ketones ($\triangle$), and nitroalkanes ($\triangle$). (After Th. Kleinert W. Ecknig, and J. Novák, *J. Chromatogr.*, **315**, 85, 1984. With permission.)

solute interaction with the ODPN phase. For example, for ethanol–stationary phase interactions the corresponding numerical values of D are 43.6×10^{-45}, 49.35×10^{-45}, and 52.51×10^{-45} mol^{-2}, respectively. As expected, for an individual phase the D paramter increased with the chain lengths of homologues.

The polar parameter φ was calculated by Kleinert and Ecknig [50] only for interactions with the polar phases, OV-17 and ODPN. In the case of the OV-17 phase, φ_{ij}^{ind} was calculated, whereas φ_{ij}^{or} was calculated for ODPN interactions with solutes. These polar parameters are the same for all species of a homologous series.

The QSRR equations obtained are of rather moderate quality. In the case of the nonpolar OV-101 phase linear relationship $\log V_g(\text{OV-101}) = f(D)$ is characterized by the correlation coefficient $r = 0.880$. For the OV-17 phase the two-parameter regression equation $\log V_g(\text{OV-17}) = f(D, \varphi^{\text{ind}})$ gives $R = 0.901$, and similarly for the ODPN phase the relationship $\log V_g(\text{ODPN}) = f(D, \varphi^{\text{or}})$ is characterized by $R = 0.899$. The last correlation is illustrated in Fig. 9.10.

In Fig. 9.10 the average deviation of the calculated retention data from the measured ones is 6.7%. It must be noted here that an experimental correction to φ values of iso compounds was applied in comparison to n compounds. Generally, the polarity parameter as calculated by Kleinert et al. [50, 51] has no real advantage over the quantum chemical polarity measures discussed earlier. Its calculation is relatively complex, and the resulting correlations with retention data are not high. On the other hand, the approach to QSRRs proposed by Ecknig and co-workers is a valuable attempt to consider mutual orientations of the interacting molecules.

9.4. QUANTITATIVE STRUCTURE–RETENTION RELATIONSHIPS SIMULTANEOUSLY ACCOUNTING FOR CHANGES IN MOBILE AND/OR STATIONARY PHASE COMPOSITION

Unlike GC separation, where the differences in solute–carrier gas interactions can be neglected, there are three main variables determining distribution of a solute between a mobile and a stationary LC zone. At a constant temperature of separation, the three variables are chemical structure of the solute, physicochemical properties of the mobile phase, and physicochemical properties of the stationary phase.

In HPLC, solute distribution is easily quantified by means of retention parameters, usually the capacity factor. If one gets numerical measures of properties of the solutes, of the mobile phases, and of the stationary phases, one can attempt to derive a general relationship linking the appropriate quantities and retention parameters together. If the quantitative relationships between the retention data as dependent variables and the numerically expressed properties of the solutes, the mobile phases, and the stationary phases as independent variables are precise enough, one may use them to predict the retention of a given solute at a given mobile phase composition on a given stationary phase.

The approach proposed by Jinno and Kawasaki [19–21] consisted in expressing the regression coefficients at individual structural descriptors by a function of the volume fraction of the organic modifier, X, in the binary solvent. Assume that a QSRR equation describes the capacity factor $\log k'_1$ of a set of solutes chromatographed on a given reversed-phase material with a mobile organic-aqueous solvent of composition X_1:

$$\text{For } X = X_1, \quad \log k'_1 = g_1 D_1 + h_1 D_2 + i_1 \tag{9.69}$$

and similarly for other compositions of the same solvents $X_2, X_3, \ldots, X_N$, the corresponding relationships hold true:

$$\text{For } X = X_2, \quad \log k'_2 = g_2 D_1 + h_2 D_2 + i_2 \tag{9.70}$$
$$\vdots \qquad\qquad \vdots$$
$$\text{For } X = X_N, \quad \log k'_N = g_N D_1 + h_N D_2 + i_N \tag{9.71}$$

where D_1 and D_2 are structural descriptors (their number may certainly be larger if there is enough data points for regression). For several classes of solutes chromatographed on four reversed-phase materials, with acetonitrile–water and methanol–water binary solvents of different composition, Jinno and Kawasaki [19–21] derived regression equations describing individual coefficients g, h, and i in Eqs. (9.69)–(9.71) in terms of volume fraction X of the organic modifier. For example, for a set of 31 substituted benzene derivatives $\log k'$ values determined on a C-18 column were described by the general equation

$$\log k' = g\pi + h(\text{HA} - \text{HD}) + i \tag{9.72}$$

The g, h, and i data are given in Table 9.6 for five compositions of two mobile phases. In Eq. (9.72) π is the Hansch hydrophobic substituent constant, HA is the number of electron acceptor groups in the solute molecule, and HD is the number of electron donor groups. The regression coefficients from Table 9.6 were next subjected to multiparameter regression analyses with X and its various trasformations (e.g., X^2, X^3, and X^4). In effect, g, h, and i were found to be complex functions of the volume fraction of the organic modifier in the solvent (Table 9.7).

To derive the two-parameter regression equations presented in Table 9.7, Jinno and Kawasaki [19–21] analyzed five data points only and tested various X^n terms (where $n = 0, 1, 2, 3, 4$). The functions derived are rather complex and difficult to rationalize. Nonetheless, according to these authors, there is a consistency between the observed and predicted reversed-phase HPLC retention parameters for several chromatographic systems involving three distinctive classes of solutes (i.e., alkylbenzenes + PAH, substituted benzenes, and phenols), four reversed-phase columns (i.e., C-18, C-8, C-2, bonded phenyl), and two binary solvents of varying composition (i.e., acetonitrile–water and methanol–water). Based on the above-discussed relationships, a computer-assisted retention prediction system for reversed-

Table 9.6. Regression Coefficients g, h, and i of Eq. (9.72) Derived from Retention and Structural Data of Standard Substituted Benzenes[a] for Individual Compositions of Binary Mobile Phases[b]

Volume Fraction of Organic Modifier X	Acetonitrile–water[c]				Methanol–water[d]			
	g	h	i	R	g	h	i	R
0.8	0.141	—	—	—	0.108	−0.055	−0.164	0.958
0.7	0.212	−0.074	−0.072	0.986	0.158	−0.073	0.067	0.968
0.6	0.212	−0.082	0.144	0.976	0.212	−0.088	0.325	0.949
0.5	0.303	−0.105	0.378	0.976	0.263	−0.096	0.607	0.955
0.4	0.427	−0.134	0.703	0.986	0.297	−0.105	0.911	0.932
0.3	0.563	−0.164	1.055	0.992	—	—	—	—

[a] See Table 6.3.

[b] According to Jinno and Kawasaki [19]. With permission. R is multiple correlation coefficient.

[c] Standards: aniline, nitrobenzene, dimethyl phthalate, o-nitrotoluene, p-chloroaniline, α-bromo-p-nitrotoluene.

[d] Standards: anisole, dimethyl phthalate, α-bromoacetophenone, p-chloroaniline, m-aminoacetophenone.

Table 9.7. Functions of Mobile Phase Composition of Organic Modifiers Determining Coefficients g, h, and i from Table 9.6[a]

Acetonitrile		Methanol	
Relationship	R[b]	Relationship	R[b]
$g = 0.769 X^3 - 1.661 X + 1.041$	1.000	$g = 0.129 X^4 - 0.505 X^2 + 0.377$	0.999
$h = -0.233 X^4 + 0.365 X - 0.273$	0.999	$h = 0.108 X^3 - 0.111$	0.999
$i = 2.593 X^2 - 5.405 X + 2.444$	1.000	$i = 0.678 X^3 - 3.444 X + 2.245$	1.000

[a] Numerical data according to Jinno and Kawasaki [19]. With permission.
[b] R is multiple-correlation coefficient.

phase HPLC was proposed [20, 21]. To use the system, one needs some retention data for standard substances (four or five according to the authors) on every column one wants to use at various mobile phase compositions, (e.g., methanol–water ranging from 80:20 to 30:70). The relationship between predicted and observed capacity factors, k', is illustrated in Fig. 9.11.

From the theories of the chromatographic distribution process (see Chapter 3), quadratic equations can be derived to describe the dependence of the logarithms of the capacity factors on the volume fraction of the organic solvent in reversed-phase HPLC using binary aqueous–organic mobile phases. As has been confirmed by numerous reports, simplified linear forms of this dependence are often adequate over a limited range of concentrations of organic solvents in the mobile phase.

For QSRRs the starting relationship is usually

$$\log k' = a - mX \qquad (9.73)$$

where X is the concentration of the organic modifier in a binary aqueous solvent, and a and m are constants. For practical reasons the linear correlation between the coefficients a and m is important. Such linear relationships have been found by Schoenmakers et al. [52] and analyzed from the point of view of QSRRs by Jandera [53–55] and Shatz and co-workers [56].

As pointed out by Snyder and co-workers [57], the coefficient a in Eq. (9.73) equals the logarithm of the capacity factor derived with pure water as the mobile phase, whereas m is a parameter connected with the properties of the organic component of the mobile phase.

To calibrate the retention scale, Jandera [53] measured the capacity factors, k', of a homologous series of solutes at several different concentrations of the organic solvent used in the binary mobile phase. From experimental data the constants a and m in Eq. (9.73) were evaluated by regression analysis. Next,

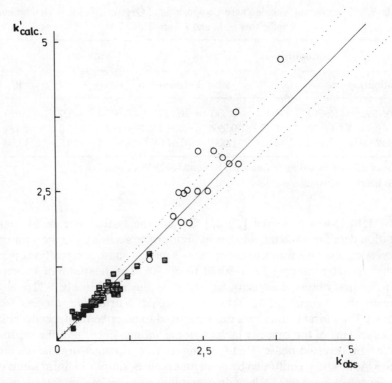

Figure 9.11. Relationship between calculated and observed reversed-phase HPLC capacity factors k' determined on C-18 stationary phase with mobile phase of methanol–water (75:25). Alkylbenzenes and PAH (○); substituted benzenes (■). (After K. Jinno and K. Kawasaki, *J. Chromatogr.*, **316**, 1, 1984. With permission.)

from the constants a and m, the other constants, a_0, a_1, q, and p, were calculated assuming the following relationships:

$$a = a_0 + a_1 n_c \tag{9.74}$$

$$m = q + pa \tag{9.75}$$

where n_c is the number of carbon atoms in the aliphatic saturated straight chain of calibration homologues. The validity of the assumptions expressed by Eqs. (9.74) and (9.75) has been proved in several publications by Jandera and by others [52–57]. The constants in Eqs. (9.74) and (9.75) were calculated by linear regression analysis using the a and m data for individual solutes of the calibration series.

If Eqs. (9.74) and (9.75) are combined with Eq. (9.73), the resulting relationship has the form

$$\log k' = (a_0 + a_1 n_c) - (q + pa)X$$
$$= (a_0 + a_1 n_c)(1 - pX) - qX \qquad (9.76)$$

In systems where the quadratic form of $\log k'$ versus X plots should be considered, Jandera [53] modified his Eq. (9.76) into the form:

$$\log k' = (a_0 + a_1 n_c)(1 - pX) - qX + (d_0 + d_1 n_c)X^2 \qquad (9.77)$$

where d_0 and d_1 are constants, that may also be determined from the capacity factors of calibration homologues.

According to Jandera [54], a_1 and p do not depend significantly on the character of the homologous series but on the organic solvent used as the less polar component of the mobile phase. On the other hand, a_0 and q depend also on the type of homologous series and on the stationary phase used. The constants a_0, a_1, and p determined from k' data of calibration homologues are considered to be generally valid also for all the other compounds. If this assumption is correct, then by Eq. (9.76) or (9.77), n_{ce} and q_i for an arbitrary compound may be calculated from its experimental k' measured at least at two different mobile phase compositions, X. Such determined constants n_{ce} and q_i are equivalent to the n_c and q of the calibration homologues and may be related to the nonspecific (Jandera [54] uses the word *lipophilic*) and polar contributions to retention, respectively.

Jandera [54, 55] has practically verified his theoretical approach to retention prediction considering selectivity α in various reversed-phase HPLC systems. Experiments were carried out using several C-8 and C-18 columns and binary solvents, methanol–water, acetonitrile–water, as well as methanol–acetonitrile–water ternary mobile phase. In order to allow a comparison of selectivity for various compounds under changing conditions, the retention of all compounds was related to a single standard compound with the capacity factor k'_s. Thus, for a given chromatographic system

$$\log \alpha = \log k'_i - \log k'_s \qquad (9.78)$$

Using Eq. (9.76) or (9.77) for capacity factors in Eq. (9.78), one gets

$$\log \alpha = a_1(1 - pX)\Delta n_c - X \Delta q \qquad (9.79)$$

or

$$\log \alpha = [a_1(1 - pX) + d_1 X^2]\Delta n_c - X \Delta q \qquad (9.80)$$

where Δ denotes the difference of the respective constants for a given solute and the standard compound. Jandera [54] assumes the first term in Eq. (9.79) or (9.80) to be the lipophilic contribution to selectivity, $\log \alpha_L$, which is controlled by the size of the nonpolar (hydrocarbon) part of the molecule and is proportional to Δn_c. The second term $(-X \Delta q)$ in Eqs. (9.79) and (9.80) is the polar contribution to selectivity, $\log \alpha^P$, which is proportional to the difference between the constants q_i of the solute and of the standard compound and to the concentration of the organic solvent in the mobile phase, X.

The Δn_c of a given sample compound is the equivalent to the number of carbon atoms in the alkyl chain of the calibration homologous series; $\log \alpha_L$ depends also on the character of both the column and the organic solvent used in the mobile phase. The value of $\log \alpha_L$ decreases with the increasing concentration of the organic solvent in the mobile phase, X.

The Δq characterizes the difference between the polarities of the functional groups in the molecules of the solute and of the standard. According to Jandera [54], the value of $\log \alpha_P$ can be attributed to the interactions of the chromatographed compounds with the organic solvent in the mobile phase.

Table 9.8 gives the constants Δn_c and Δq for several solutes chromatographed on a C-8 column with methanol–water mobile phase. The C_2–C_6 n-alkylbenzenes were used as the calibration homologous series. Toluene was the standard solute. Table 9.8 also gives the Hansch hydrophobic substituent constants [5] and the Snyder polarity indices [58].

Individual correlations Δn_c versus π and Δq versus P' are illustrated in Figs. 9.12 and 9.13.

The correlations given in Figs. 9.12 and 9.13 are convincing. Especially interesting is the relationship between the two polarity parameters. The outliers in Fig. 9.13 are benzyl alcohol and m-cresol, which possess a hydroxy group. This group is able to form hydrogen bonds with methanol and water. This in turn increases interactions with the mobile phase, giving higher Δq values than would be expected from Snyder indices.

The values of Δq obtained for the same solutes on a C-8 column are consistent for the two mobile phase systems, acetonitrile–water and methanol–water (Fig. 9.14). However, an analogous relationship derived from reversed-phase HPLC data determined on a C-18 column is characterized by a marked dispersion of the points for compounds belonging to a certain class with increasing values of Δq. Jandera [54] suggests the use of the relationship between Δq values determined in different chromatographic systems for identification of functional groups in organic compounds from reversed-phase HPLC data.

Using the Δn_c and Δq constants, Jandera [55] was able to predict the selectivity and capacity factors of solutes in binary and ternary mobile phases containing 50–90% of the organic solvent.

Table 9.8. Jandera Constants Δn_c and Δq and Hansch Lipophilic Constant π and Snyder P' Polarity Indices for Solutes Chromatographed on C-8 Column with Methanol–Water Eluent^a

Compound	Substituent (Benzene Ring)	π	Δn_c	Compound	P'	Δq
Toluene	CH$_3$	0.56	0	n-Decane	−0.3	−0.15
Ethylbenzene	C$_2$H$_5$	1.02	0.72	n-Hexane	0	−0.1
Bromobenzene	Br	0.86	0.55	Toluene	2.3	0
Chlorobenzene	Cl	0.71	0.25	Bromobenzene	2.7	0.06
Ethyl benzoate	COOC$_2$H$_5$	0.51	0.54	Chlorobenzene	2.7	0.06
Methyl benzoate	COOCH$_3$	0.01	−0.57	Phenetole	2.9	0.06
Anisole	OCH$_3$	−0.02	−0.70	Anisole	3.5	0.09
Nitrobenzene	NO$_2$	−0.28	−0.87	Acetophenone	4.4	0.18
Acetophenone	COCH$_3$	−0.55	−1.47	Nitrobenzene	4.5	0.19
Benzonitrile	CN	−0.57	−1.46	Benzonitrile	4.6	0.23
Phenyl acetate	OCOCH$_3$	−0.64	−1.59	Benzyl alcohol	5.5	0.41
Benzaldehyde	CHO	−0.65	−1.32	Aniline	6.2	0.29
Phenol	OH	−0.67	−1.43	m-Cresol	7.0	0.47
Acetanilide	NHCOCH$_3$	−0.97	−1.86			
Benzyl alcohol	CH$_2$OH	−1.03	−1.83			
Aniline	NH$_2$	−1.23	−1.88			

^{a} Data from Jandera [54]. With permission. Lipohilic constant from Hansch and Lee [5] and polarity indices from Snyder and Kirkland [58].

203

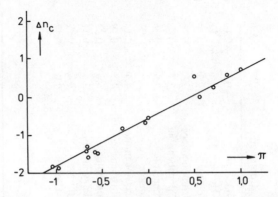

Figure 9.12. Correlation between carbon equivalents, Δn_c, and Hansch hydrophobic substituent constants π. Numerical data are given in Table 9.8. (After P. Jandera, *J. Chromatogr.*, **352**, 91, 1986. With permission.)

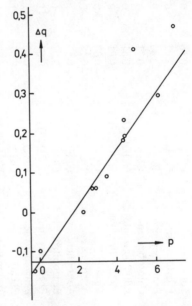

Figure 9.13. Correlation between constants Δ_q and Snyder polarity indices P'. Numerical data are given in Table (9.8). (After P. Jandera, *J. Chromatogr.*, **352**, 91, 1986. With permission.)

The experimentally derived constants Δn_c and Δq may be useful solute structural descriptors for retention prediction. As these constants quantify nonspecific and polar contributions to retention, their description in terms of nonempirical structural parameters would be desired. Perhaps, neither π nor P' are the proper structural descriptors to be related to Δn_c and Δq, respectively.

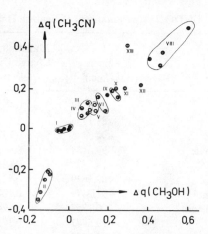

Figure 9.14. Dependence of Δ_q in acetonitrile–water mobile phases on Δ_q in methanol–water mobile phases for C-8 column. Classes of solutes: I = alkylbenzenes; II = alkanes; III = bromo- and chlorobenzenes; IV = alkyl aryl ethers; V = esters of aromatic carboxylic acids and aromatic alcohols; VI = alkyl aryl ketones and di-aryl ketones; VII = halogenated phenylureas; VIII = phenols and alkylphenols; IX = nitrobenzene; X = benzonitrile; XI = benzaldehyde; XII = n-butyl-N-phenylcarbamate; XIII = aniline. (After P. Jandera, *J. Chromatogr.*, **352**, 91, 1986. With permission.)

Recently, our group attempted to derive QSRR equations accounting for changes (limited) of all three chromatographic variables (i.e., solute structure, mobile phase composition, and stationary phase properties). Reversed-phase HPLC capacity factors have been the dependent quantities. A set of 12 substituted benzene derivatives with various functional groups were selected for QSRR studies (Table 9.9). Structural diversity among the solutes tested is evident, either from the point of view of nonspecific (also lipophilic) and polar properties. Yet the solutes selected form a related family of aromatic compounds. Changes in composition of the mobile phase methanol–water were also limited to the range of 35–65% v/v of methanol. The lower and the upper concentrations limited the linearity of the dependence of the logarithm of the capacity factor, $\log k'_{ij}$, for the ith solute determined on the jth stationary phase, on the mole fraction X of one of the components of the binary solvent (here it has been water):

$$\log k'_{ij} = a_{ij}X + b_{ij} \tag{9.81}$$

where a_{ij} and b_{ij} are constants for a given solute i chromatographed on an individual stationary phase j.

The third type of independent variable in our HPLC system was the

Table 9.9. CNDO/2MO Parameters of Benzene Derivatives[a]

Number	Compound	Total Energy (a.u.)[b]	Maximum Excess Charge Difference (electrons)	Dipole Moment (D)
1	Phenol	−65.5548	0.4328	1.7492
2	Acetophenone	−81.1756	0.5088	3.0417
3	Nitrobenzene	−94.8446	0.7774	5.0589
4	Methylbenzoate	−99.6184	0.6836	2.0376
5	p-Cresol	−74.2375	0.4265	1.7379
6	p-Ethylphenol	−82.6639	0.4275	2.3324
7	p-Propylphenol	−91.3479	0.4270	2.1472
8	4-sec-Butylphenol	−100.1758	0.4246	1.8457
9	Aniline	−59.5473	0.3737	1.5206
10	N-Methylaniline	−68.2305	0.3564	1.1504
11	4-Chloroacetophenone	−96.6241	0.5028	2.2894
12	3,4-Dichloroacetophenone	−112.3142	0.4922	1.2809

[a] Analysed by Kaliszan et al. [46, 47]. With permission.
[b] a.u. = atomic units.

dimethyloctadecylsilane (C-18) coverage of chemically bonded stationary phases. Three columns were used of varying C-18 coverage ranging from 1.54 $\times 10^{-4}$ to 4.96×10^{-4} mol/g. For the sake of simplicity, at that stage of research, the results obtained on the fourth column of the highest C-18 coverage were not used in QSRR studies because a marked deviation from the linearity of $\log k'$ versus C-18 coverage was observed for the coverage 6.6×10^{-4} mol/g (see Fig. 9.15).

Keeping the above limitations and approximations in mind, we attempted to derive statistical equations allowing prediction of the retention of a class of solutes on the octadecylsilica phases of varying C-18 coverage with a methanol–water solvent of varying composition based on nonempirical parameters, characterizing solute structure.

Based on the observed linearity ($r = 0.99$) of $\log k'$ versus X for five concentrations of the methanol–water mobile phase and for four C-18 phases of different hydrocarbon coverages, we determined the coefficients a_{ij} and b_{ij} of Eq. (9.81) by linear regression. The numerical values obtained are listed in Table 9.10.

Kaliszan et al. [46, 47] assumed that for a given stationary phase coverage, a_{ij} and b_{ij} are some functions of the structure of the solutes. Using multiparameter regression analyses, we have found that a_{ij} and b_{ij} are satisfactorily

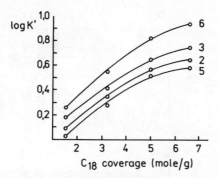

Figure 9.15. Plot of log k' versus C-18 coverage for solutes 2, 3, 5, and 6 (Table 9.9) with mobile phase of water–methanol (65:35) in reversed-phase HPLC system. (After R. Kaliszan, K. Ośmiałowski, S. A. Tomellini, S.-H. Hsu, S. D. Fazio, and R. A. Hartwick, *J. Chromatogr.*, **352**, 141, 1986. With permission.)

described by a two-parameter equation involving the quantum chemically calculated total energy of a solute, E_{Ti}, and its polarity parameter Δ_i, defined as the maximum excess electronic charge difference in a molecule (see Chapter 7). Thus, the a_{ij} and b_{ij} determined for an individual solute i on the stationary phase j are:

$$a_{ij} = \alpha_j E_{Ti} + \beta_j \Delta_i + \gamma_j \qquad (9.82)$$

$$b_{ij} = \alpha'_j E_{Ti} + \beta'_j \Delta_i + \gamma'_j \qquad (9.83)$$

where α_j, β_j, γ_j, α'_j, β'_j and γ'_j are regression coefficients derived by the conventional least-squares method. For example, the coefficients a_{12} and b_{12} determined for the phase having 3.18×10^{-4} mol/g C-18 coverage ($j = 2$) are

$$a_{12} = 0.534(\pm 0.0122)E_{Ti} + 3.3473(\pm 1.5866)\Delta_i - 0.4034(\pm 0.9334)$$

$$n = 12; \quad R = 0.9616; \quad s = 0.2481 \qquad (9.84)$$

$$b_{12} = -0.0415(\pm 0.0068)E_{Ti} - 2.3986(\pm 0.8663)\Delta_i - 0.9429(\pm 0.5083)$$

$$n = 12; \quad R = 0.9805; \quad s = 0.1351 \qquad (9.85)$$

Having the a_{ij} and b_{ij} data for i compounds determined on j phases from $i \times j$ regression equations of the form of Eq. (9.81), one can attempt to describe them in terms of E_{Ti} and Δ_i of the ith compound. If Eqs. (9.82) and (9.83) are

Table 9.10. Coefficients a and b of Eq. (9.81) for Individual Solutes Chromatographed on Phases of Different Hydrocarbonaceous (C-18) Coverage[a]

Number	Coverage, 1.54×10^{-4}		Coverage, 3.18×10^{-4}		Coverage, 4.96×10^{-4}		Coverage, 6.6×10^{-4}	
	a	b	a	b	a	b	a	b
1	−2.6959	0.4536	−2.5581	0.6639	−2.8606	0.9593	−2.6078	0.9413
2	−3.0864	0.9465	−3.0568	1.1786	−3.4881	1.5156	−2.5891	1.2605
3	−2.7960	0.9358	−2.7800	1.1596	−3.5946	1.6078	−2.9937	1.5324
4	−3.4745	1.3031	−3.5860	1.6051	−3.9130	1.9501	−3.6501	1.9751
5	−3.0970	0.8482	−2.9067	1.0646	−3.3485	1.4121	−3.0797	1.4014
6	−3.3854	1.1933	−3.5349	1.5152	−3.7593	1.8154	−3.5988	1.8782
7	−4.0343	1.6683	−4.1325	2.0059	−4.3633	2.3292	−4.2734	2.4389
8	−4.4055	1.9693	−4.6025	2.3567	−4.8630	2.7080	−4.7467	2.8292
9	−2.4448	0.6029	−2.1299	0.5849	−2.5553	0.8762	−2.1672	0.7902
10	−2.8347	0.9873	−2.6928	1.0879	−2.8933	1.3476	−2.8162	1.4047
11	−3.4547	1.3868	−3.8161	1.7990	−3.8413	2.0147	−3.8169	2.1259
12	−3.8477	1.8329	−4.3011	2.2945	−4.3579	2.5474	−4.3741	2.6970

[a] Compounds are numbered as in Table 9.9 [47]. With permission.

208

statistically significant, Eq. (9.81) for the jth phase may be rewritten as

$$\log k'_{ij} = (\alpha_j E_{Ti} + \beta_j \Delta_i + \gamma_j)X + (\alpha'_j E_{Ti} + \beta'_j \Delta_i + \gamma'_j) \tag{9.86}$$

The phases under study differed in their C-18 coverage. A linearity among $\log k'$ values determined for alkylbenzenes on hydrocarbonaceous stationary phases of different alkyl chain lengths (C-2, C-8, and C-18) was reported [59]. We expected a similar relationship with respect to the stationary phase surface coverage with octadecyl chains. Actually, for a given solute i chromatographed at a fixed mobile phase composition X, linearity has been found between $\log k'_{ij,x}$ and the C-18 surface, C_j, for the first three of the four phases studied (Fig. 9.15):

$$\log k'_{ij,x} = AC_j + B \tag{9.87}$$

where A and B are regression coefficients. In such a situation, it seemed probable that the coefficients a_{ij} and b_{ij} of Eq. (9.81) depend not only on the solute structure described by individual E_{Ti} and Δ_i parameters but also linearly (over a limited range) on the stationary phase C-18 coverage, C_j. Thus, the general equation describing capacity factors in terms of solute structure, mobile phase composition, and stationary phase surface properties is

$$\log k'_{ij} = (\alpha E_{Ti} + \beta \Delta_i + \gamma C_j + \delta)X + (\alpha' E_{Ti} + \beta' \Delta_i + \gamma' C_j + \delta') \tag{9.88}$$

To derive the regression coefficients α, β, γ, δ, α', β', γ', and δ', the variable matrices of $(i \times j) \times 4$ dimensions are considered:

$$\begin{bmatrix} a_{11} & E_{T1} & \Delta_1 & C_1 \\ a_{21} & E_{T2} & \Delta_2 & C_1 \\ a_{31} & E_{T3} & \Delta_3 & C_1 \\ \vdots & \vdots & \vdots & \vdots \\ a_{i1} & E_{Ti} & \Delta_i & C_1 \\ a_{12} & E_{T1} & \Delta_1 & C_2 \\ a_{22} & E_{T2} & \Delta_2 & C_2 \\ a_{32} & E_{T3} & \Delta_3 & C_2 \\ \vdots & \vdots & \vdots & \vdots \\ a_{i2} & E_{Ti} & \Delta_i & C_2 \\ \vdots & \vdots & \vdots & \vdots \\ a_{1j} & E_{T1} & \Delta_1 & C_j \\ \vdots & \vdots & \vdots & \vdots \\ a_{ij} & E_{Ti} & \Delta_i & C_j \end{bmatrix} \tag{9.89}$$

and analogously for b_{ij} coefficients.

The final equation obtained is

$$\log k'_{ij} = a_{ij}X + b_{ij} = [0.0454\,(\pm 0.0071)E_{Ti}$$
$$+\,2.6493\,(\pm 0.9187)\Delta_i - 0.1053\,(\pm 0.0672)C_j$$
$$-\,0.4946\,(\pm 0.5828)]X + [-0.0381\,(\pm 0.0039)E_{Ti}$$
$$+\,2.1659\,(\pm 0.4919)\Delta_i + 0.1696\,(\pm 0.0359)C_j$$
$$+\,1.2963\,(\pm 0.3120)]$$

The statistics are $n = 36$, $R = 0.9251$, and $s = 0.2756$ for a_{ij} and $n = 36$, $R = 0.9715$, and $s = 0.1476$ for b_{ij}.

For the sake of comparison, a similar regression analysis was done after replacing the polarity parameter Δ_i with the CNDO/2-calculated dipole moment μ_i. The statistics obtained were significantly lower. For a_{ij} calculated in terms of E_{Ti}, μ_i, and C_j, the correlation coefficient was $R = 0.8710$; for b_{ij}, the

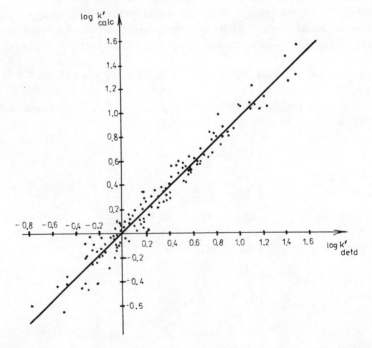

Figure 9.16. Correlation of predicted (k'_{calc}) and observed (k'_{obs}) capacity factors for 12 test solutes over all mobile and stationary phases examined. (After R. Kaliszan, K. Ośmialowski, S. A. Tomellini, S. -H. Hsu, S. D. Fazio, and R. A. Hartwick, *J. Chromatogr.*, **352**, 141, 1986. With permission.)

Table 9.11. Values of Logarithms of Capacity Factors, log k', by Coverage and Mole Fraction of Methanol MeOH[a]

Compound Number	Coverage, 1.54×10^{-4} mol/g MeOH				Coverage, 3.18×10^{-4} mol/g MeOH				Coverage, 4.96×10^{-4} mol/g MeOH				
	0.1934	0.2670	0.3524	0.4526	0.1934	0.2670	0.3524	0.4526	0.1934	0.2670	0.3524	0.4004	0.4526
1	-0.0746	-0.2672	-0.4771	-0.7782	0.1845	-0.0280	-0.2609	-0.4771	0.4141	0.1883	-0.0544	-0.1891	-0.3274
	0.0443	-0.1387	-0.3510	-0.6002	0.2890	0.0933	-0.1338	-0.4002	0.5546	0.3451	0.1021	-0.0346	-0.1832
2	0.3696	0.0946	-0.1413	-0.4424	0.6177	0.3399	0.0649	-0.1761	0.8648	0.5621	0.2679	0.1120	-0.0401
	0.3766	0.1562	-0.0995	-0.3995	0.6213	0.3883	0.1178	-0.1995	0.8870	0.6401	0.3566	0.1926	0.0175
3	0.4025	0.1799	-0.0512	-0.3259	0.6388	0.4087	0.1521	-0.0729	0.8927	0.6435	0.3716	0.2244	-0.0813
	0.3332	0.1196	-0.1284	-0.4193	0.5780	0.3516	0.0889	-0.2193	0.8436	0.6034	0.3247	0.1681	-0.0022
4	0.6532	0.3509	0.0669	-0.2553	0.9456	0.6220	0.3010	0.0142	1.2200	0.8841	0.5477	0.3716	0.2089
	0.6283	0.3804	0.0927	-0.2448	0.8730	0.6124	0.3100	-0.0448	1.1387	0.8643	0.5458	0.3669	0.1722
5	0.2527	0.0116	-0.2341	-0.5565	0.5292	0.2730	0.0000	-0.2218	—	0.5090	0.2609	0.0483	-0.1001
	0.3093	0.0960	-0.1514	-0.4417	0.5540	0.3281	0.0659	-0.2417	—	0.5799	0.3017	0.1454	-0.0247
6	0.5569	0.2706	-0.0122	-0.3259	0.8647	0.5362	0.2490	-0.0621	—	0.8233	0.4771	0.2946	0.1313
	0.5547	0.3135	0.0336	-0.2948	0.7994	0.5455	0.2509	-0.0948	—	0.7973	0.4867	0.3121	0.1223
7	0.9088	0.5712	0.2290	-0.1413	1.2467	0.8641	0.5172	0.1663	—	1.1794	0.7739	0.5619	0.3770
	0.8191	0.5398	0.2261	-0.1419	1.0548	0.7718	0.4434	0.0581	—	1.0237	0.6793	0.4857	0.2752
8	1.1461	0.7683	0.3882	0.0000	1.5091	1.0892	0.6962	0.3082	—	1.4267	0.9709	0.7450	0.5292
	1.0729	0.7726	0.4242	0.0154	1.3176	1.0047	0.6415	0.2154	—	1.2565	0.8773	0.6642	0.4324
9	—	-0.0631	-0.2341	-0.5149	0.1845	0.0134	-0.1903	-0.3632	—	0.1963	-0.0263	-0.1513	-0.2762
	—	-0.2086	-0.4110	-0.6485	0.2106	0.0234	-0.1937	-0.4485	—	0.2753	0.0421	-0.0889	-0.2315
10	—	0.2242	0.0000	-0.3010	0.5858	0.3522	0.1214	-0.1154	—	0.5798	0.3259	0.1761	0.0483
	—	0.0442	-0.1978	-0.4794	0.4938	0.2742	0.0195	-0.2794	—	0.5261	0.2553	0.1032	-0.0623
11	—	0.4652	0.1680	-0.1761	1.0855	0.7709	0.4117	0.1027	—	1.0012	0.6475	0.4597	0.2946
	—	0.5663	0.2493	-0.1225	1.0842	0.7983	0.4666	0.0774	—	1.0502	0.7025	0.5070	0.2045
12	—	0.8120	0.4649	0.0969	1.5075	1.1076	0.7339	0.3862	—	1.3979	0.9949	0.7845	0.5956
	—	0.9894	0.0692	0.1631	1.5618	1.2214	0.8265	0.3631	—	1.4732	1.0623	0.8313	0.5801

[a] For each solute in Table 9.9 the upper row gives the observed value and the lower row gives the values calculated by Eq. (9.90) (according to Kaliszan et al. [47]). With permission.

Table 9.12. Multiparameter QSSR

Number	Class of Solutes	Chromatographic System[a]	Molecular Descriptors	References
1	p-Benzoquinones and p-hydroquinones	RP HPLC	Various connectivity indices	10
2	Homologous series of alkyl esters	GC	Hammett and Taft substituent constants	18
3	Substituted benzenes	RP HPLC	Molecular connectivity, van der Waals volume, molecular area, substituent electronic and hydrophobic constants	19
4	Alkylbenzenes, substituted benzenes, phenol	RP HPLC	Substituent hydrophobic and electronic constants, partition coefficient	20
5	Alkylbenzenes, substituted benzenes, phenols, polycyclic hydrocarbons	RP HPLC	Partition coefficient, substituent hydrophobic and electronic constants, van der Waals volume, molecular area, molecular connectivity, length-to-breadth ratio	21
6	Phenols	GC	Molar refractivity, Hammett constant	24
7	Aliphatic esters	GC	Molar refractivity, refraction coefficient, dielectric constant, van der Waals volume, dipole moment	26
8	Chloro derivatives of dimethylbenzene	GC	Boiling point, van der Waals volume	27
9	Chloro derivatives of dimethylbenzene	GC	Boiling point, van der Waals volume	28
10	Sterol acetates	GC	Effective atomic charges, steric substituent constants	35
11	Cyclohexane derivatives	GC	Effective atomic charges, steric substituent constants	36
12	Olefinic and paraffinic hydrocarbons, aliphatic alcohols, methyl esters of fatty acids, aldehydes, ketones	GC	Molecular connectivities, dipole moments	37
13	Phenols	GC	Molecular refractivities, dipole moments	38

212

No.	Compounds	Method	Descriptors	Ref.
14	Phenols	GC	Molecular refractivities, Kováts indices	39
15	Aliphatic ethers, esters, alcohols, ketones, aldehydes	GC	Molecular connectivities, electronic and lipophilic substituent constants, quantum chemical indices, Wiener index, Hosoya index	43
16	Primary, secondary and heterocyclic amines	GC	Maximum excess charges, total energy, energy of highest occupied orbital	44
17	Primary, secondary, and heterocyclic amines	GC	Total energy, topological electronic index	45
18	Substituted benzene derivatives	RP HPLC	Total energy, maximum excess charge differences	46
19	Substituted benzene derivatives	RP HPLC	Total energy, maximum excess charge differences	47
20	Aliphatic nitriles, alcohols, ketones, halogenoalkanes, and nitroalkanes	GC	Molecular refractivities, dipole moments	51
21	Aromatic acids	RP HPLC	Energy effects, pK_a, van der Waals volumes	60
22	Alkylbenzenes	GC	Molecular refractivity, refraction coefficient, van der Waals volume	61
23	Alkanes, methyl esters of fatty acids, alcohols, ethers, aldehydes, ketones	GC	Molecular connectivity, hydrophobic constant, Hammett constant, Wiener number, Hosoya index	62
24	Hydrocarbons, heterocyclic compounds	GC	Taft steric constant, π-electron density, molecular weight	63
25	Alkanes, alkenes, alcohols, aldehydes, ketones, esters	GC	Molecular connectivities, lipophilic and electronic substituent constants, Wiener number, Hosoya index	64
26	cis and trans alkenes	GC	Quantum chemical indices	65
27	PAH	GC	molecular connectivity, shape parameter	66
28	PAH	RP HPLC	Molecular connectivity, length-to-breadth ratio	67

Table 9.12. (Continued)

Number	Class of Solutes	Chromatographic System[a]	Molecular Descriptors	References
29	PAH	RP HPLC	Length-to-breadth ratio, bond and group parameter F	68
30	PAH	RP HPLC	Length-to-breadth ratio, bond and group parameter F	70
31	PAH	RP and NP HPLC	Length-to-breadth ratio, bond and group parameter F	70
32	Aliphatic alcohols, ketones, ethers, and esters	GC	Molecular connectivities	71
33	Chlorinated benzenes	GC	Various molecular connectivities	72
34	Ketones	GC	Various molecular connectivities	73
35	Aliphatic hydrocarbons	GC	Various molecular connectivities	74
36	Hydrocarbons, ethers, aldehydes, ketones, esters, alcohols	GC	Various molecular connectivities	75
37	Barbiturates	HPLC	Various molecular connectivities	76
38	Arylaliphatic acids	GC	Molecular refractivities of aromatic part of molecules and of aliphatic chain	77
39	PAH	GC	Molecular connectivity, shape parameter	78
40	Pyrimidine and quinazoline derivatives	TLC, GC, HPLC	Various molecular connectivities	79
41	Polycyclic, heterocyclic, and partially saturated aromatic compounds	GC	Molecular connectivity, number of nitrogens, number of basis rings	80
42	PAH	GC, HPLC	Wiener number and its square	81
43	Aliphatic esters	GC	Molecular weight, boiling point, van der Waals volume	82
44	PAH	RP HPLC	Van der Waals volume, fragmental connectivities, moments of inertia, number of aromatic rings	

[a] Abbreviations: RP, reversed phase, NP, normal phase.

214

corresponding value was $R = 0.9040$. The correlation between dipole moment μ_i and the polarity parameter Δ_i for 12 solutes is $r = 0.7689$; this is what precludes their being used together in one regression equation. For the same reason, energies of highest occupied and lowest unoccupied molecular orbitals were not included in multiparameter regression equations.

In deriving Eq. (9.90), the data determined on the phase of coverage 6.6×10^{-4} mol/g were not included since for that phase a marked deviation from Eq. (9.87) was observed. For all the remaining data, a good correlation was found between the retention data observed experimentally and those calculated by Eq. (9.90) (Table 9.11 and Fig. 9.16):

$$\log k'_{ij}(\text{obs}) = 0.9524 \log k'_{ij}(\text{calc}) + 0.0142$$

$$n = 144; \quad r = 0.9862 \tag{9.91}$$

The benzene derivatives analyzed in our work cannot be considered congeneric from the point of view of their polarity. This becomes evident if one compares their dipole moments. The nonempirical polarity parameter Δ_i comprises that diverse set of solutes in one relationship and has convincing advantages in that respect over the dipole moment.

Fully recognizing all the approximations and limitations, one nevertheless will find the derived relationship describing reversed-phase HPLC retention in terms of nonempirical structural descriptors and properties of both stationary and mobile phases promising. The QSRRs obtained with the nonempirical structural parameters are of comparable or of better quality than the majority of multiparameter relationships reported, involving empirical data for even more closely related solutes. These multiparameter QSRRs found in the available literature are listed in Table 9.12.

References

1. A. J. P. Martin, Some theoretical aspects of partition chromatography, *Biochem. Soc. Symp.*, **3**, 4, 1950.

2. J. Green, S. Marcinkiewicz, and D. McHale, Paper chromatography and chemical structure. III. The correlation of complex and simple molecules. The calculation of R_M values for tocopherols, vitamins K, ubiquinones and ubichromenols from R_M (phenol). Effects of unsaturation and chain branching, *J. Chromatogr.*, **10**, 158, 1963.

3. E. Tomlinson, H. Poppe, and J. C. Kraak, Thermodynamics of functional groups in reversed-phase high-performance liquid–solid chromatography, *Int. J. Pharm.*, **7**, 225, 1981.

4. Cs. Horváth, W. Melander, and I. Molnar, Solvophobic interactions in liquid chromatography with nonpolar stationary phases, *J. Chromatogr.*, **125**, 129, 1976.

5. C. Hansch and A. Leo, *Substituent Constants for Correlation Analysis in Chemistry and Biology*, Wiley, New York, 1979, p. 18.

6. T. L. Hafkenscheid and E. Tomlinson, Isocratic chromatographic retention data for estimating aqueous solubilities of acidic, basic and neutral drugs, *Int. J. Pharm.*, **17**, 1, 1983.

7. S. H. Yalkowsky and S. C. Valvani, Solubilities and partitioning. II. Relationships between aqueous solubilities, partition coefficients, and molecular surface area of rigid aromatic hydrocarbons, *J. Chem. Eng. Data*, **24**, 127, 1979.

8. B.-K. Chen, and Cs. Horváth, Evaluation of substituent contributions to chromatographic retention: Quantitative structure–retention relationships, *J. Chromatogr.*, **171**, 15, 1979.

9. M. Kuchař, V. Rejholec, M. Jelinková, and O. Němeček, Parametrization of lipophilic properties of some aromatic–aliphatic acids in paper chromatography, *J. Chromatogr.*, **150**, 419, 1978.

10. J.-X. Huang, E. S. P. Bouvier, J. D. Stuart, W. R. Melander, and Cs. Horváth, High-performance liquid chromatography of substituted *p*-benzoquinones and *p*-hydroquinones. II. Retention behavior, quantitative structure–retention relationships and octanol–water partition coefficients, *J. Chromatogr.*, **330**, 181, 1985.

11. J. M. DiBussolo and W. R. Nes, Structural elucidation of sterols by reversed-phase liquid chromatography: I. Assignment of retention coefficients to various groups, *J. Chromatogr. Sci.*, **20**, 193, 1982.

12. S. Mihara and N. Enomoto, Calculation of retention indices of pyrazines on the basis of molecular structure, *J. Chromatogr.*, **324**, 428, 1985.

13. J. Osiek, J. Oszczapowicz, and W. Drzewiński, Prediction of retention indices of N^1-N^1-dialkylformamidines on a non-polar column, *J. Chromatogr.*, **351**, 177, 1986.

14. D. Sissons and D. Welti, Structural identification of polychlorinated biphenyls in commercial mixtures by gas–liquid chromatography, nuclear magnetic resonance and mass spectrometry, *J. Chromatogr.*, **60**, 15, 1971.

15. J. N. Tejedor, Prediction of retention indices of aromatic hydrocarbons, *J. Chromatogr.*, **177**, 279, 1979.

16. M. V. Budahegyi, E. R. Lombosi, T. S. Lombosi, S. Y. Mészaros, Sz. Nyiredy, G. Tarján, I. Timár, and J. M. Takács, Twenty-fifth anniversary of the retention index system in gas–liquid chromatography, *J. Chromatogr.*, **271**, 213, 1983.

17. K. Jinno, Retention prediction of phenylthiohydantoin amino acid derivatives in reversed-phase liquid chromatography, *Chromatographia*, **20**, 743, 1985.

18. J. K. Haken, A. Nguyen, and M. S. Wainwright, Gas chromatography of esters. XII. Linear extrathermodynamic relationships, *J. Chromatogr.*, **178**, 471, 1979.

19. K. Jinno and K. Kawasaki, Retention prediction of substituted benzenes in reversed-phase HPLC, *Chromatographia*, **18**, 90, 1984.

20. K. Jinno and K. Kawasaki, Automated optimization of reversed-phase liquid chromatographic separations using a computer-assisted retention prediction, *J. Chromatogr.*, **298**, 326, 1984.

21. K. Jinno and K. Kawasaki, Computer-assisted retention prediction system for

reversed-phase micro high-performance liquid chromatography, *J. Chromatogr.*, **316**, 1, 1984.

22. N. Dimov and D. Papazova, Calculation of the retention indices of C_5–C_9 cycloalkanes on squalane, *J. Chromatogr.*, **148**, 11, 1978.

23. N. Dimov and D. Papazova, Correlation equations for prediction of gas chromatographic separation of hydrocarbons on Squalane, *Chromatographia*, **12**, 720, 1979.

24. A. Radecki, J. Grzybowski, H. Lamparczyk, and A. Nasal, Relationship between retention indices and substituent constants of phenols on polar stationary phases, *J. High Resolut. Chromatogr. Chromatogr. Commun.*, **2**, 581, 1979.

25. G. Castello and G. D'Amato, Gas chromatographic separation and identification of linear and branched-chain alkyl bromides, *J. Chromatogr.*, **324**, 363, 1985.

26. J. Bermejo and M. D. Guillen, A study of Kováts retention indices of aliphatic saturated esters and their relation to the polarity of the stationary phase, *J. Chromatogr.*, **318**, 187, 1985.

27. J. Bermejo, C. G. Blanco, and M. D. Guillén, Capillary gas chromatography of chloro derivatives of 1,4-dimethylbenzene. Separation, identification and prediction of boiling points, *J. Chromatogr.*, **331**, 237, 1985.

28. J. Bermejo, C. G. Blanco, and M. D. Guillén, Gas chromatography of deuterated and protiated chloro derivatives of 1,4-dimethylbenzene, *J. Chromatogr.*, **351**, 425, 1986.

29. M. Roth and J. Novak, Utilization of the solution-of-groups concept in gas–liquid chromatography, *J. Chromatogr.*, **258**, 23, 1983.

30. A. Fredenslund, R. L. Jones, and J. M. Prausnitz, Group contribution estimation of activity coefficients in nonideal liquid mixtures, *AIChE J.*, **21**, 1086, 1975.

31. S. M. Petrović, S. Lomić, and I. Sefer, Utilization of the functional group contribution concept in liquid chromatography on chemically bonded reversed-phase, *J. Chromatogr.*, **348**, 49, 1985.

32. L. Rohrschneider, Eine Methode zur Charakterisierung von gaschromatographischen Trennflüssigkeiten, *J. Chromatogr.*, **22**, 6, 1966.

33. W. O. McReynolds, Characterization of some liquid phases, *J. Chromatogr. Sci.*, **8**, 685, 1970.

34. A. Ladon and S. Sandler, Gas-chromatographic retention and molecular structure. An extension of the James–Martin rule, *Anal. Chem.*, **45**, 921, 1973.

35. M. Gassiot, E. Fernandez, G. Firpo, R. Carbó, and M. Martin, Empirical quantum chemical approach to structure–gas chromatographic retention index relationships. I. Sterol acetates, *J. Chromatogr.*, **108**, 337, 1975.

36. G. Firpo, M. Gassiot, M. Martin, R. Carbó, X. Guardino, and J. Albaigés, Empirical quantum chemical approach to structure–gas chromatographic retention index relationships. II. Cyclohexane derivatives, *J. Chromatogr.*, **117**, 105, 1976.

37. M. Gassiot-Matas and G. Firpo-Pamies, Relationships between gas chromatographic retention index and molecular structure, *J. Chromatogr.*, **187**, 1, 1980.

38. R. Kaliszan and H.-D. Höltje, Gas chromatographic determination of molecular polarity and quantum chemical calculation of dipole moments in a group of substituted phenols, *J. Chromatogr.*, **234**, 303, 1982.

39. J. Grzybowski, H. Lamparczyk, A. Nasal, and A. Radecki, Relationship between the retention indices of phenols on polar and non-polar stationary phases, *J. Chromatogr.*, **196**, 217, 1980.

40. C. Hansch, A. Leo, S. H. Unger, K.-H. Kim, D. Nikaitani, and E. J. Lien, "Aromatic" substituent constants for structure–activity correlations, *J. Med. Chem.*, **16**, 1207, 1973.

41. R. Kaliszan, Chromatography in studies of quantitative structure–activity relationships, *J. Chromatogr.*, **220**, 71, 1981.

42. J. Halkiewicz, H. Lamparczyk, and A. Radecki, Behaviour of copper(II) and nickel(II) dialkyldithiocarbamates on polar and non-polar stationary phases, *Fresenius Z. Anal. Chem.*, **320**, 577, 1985.

43. L. Buydens, D. L. Massart, and P. Geerlings, Prediction of gas chromatographic retention indexes with topological, physicochemical, and quantum chemical parameters, *Anal. Chem.*, **55**, 738, 1983.

44. K. Ośmiałowski, J. Halkiewicz, A. Radecki, and R. Kaliszan, Quantum chemical parameters in correlation analysis of gas–liquid chromatographic retention indices of amines, *J. Chromatogr.*, **346**, 53, 1985.

45. K. Ośmiałowski, J. Halkiewicz, and R. Kaliszan, Quantum chemical parameters in correlation anlaysis of gas–liquid chromatographic retention indices of amines. II. Topological electronic index, *J. Chromatogr.*, **361**, 63, 1986.

46. R. Kaliszan, K. Ośmiałowski, S. A. Tomellini, S.-H. Hsu, S. D. Fazio, and R. A. Hartwick, Non-empirical descriptors of sub-molecular polarity and dispersive interactions in reversed-phase HPLC, *Chromatographia*, **20**, 705, 1985.

47. R. Kaliszan, K. Ośmiałowski, S. A. Tomellini, S.-H. Hsu, S. D. Fazio, and R. A. Hartwick, Quantitative retention relationships as a function of mobile and C_{18} stationary phase composition for non-cogeneric solutes, *J. Chromatogr.*, **352**, 141, 1986.

48. E. Soczewiński and C. A. Wachtmeister, The relation between the composition of certain ternary two-phase solvent systems and R_M values, *J. Chromatogr.*, **7**, 311, 1962.

49. M. Kuchař, H. Tomková, V. Rejholec, and O. Skalická, Relationships between gas–liquid chromatographic behaviour and structure of arylaliphatic acids, *J. Chromatogr.*, **333**, 21, 1985.

50. Th. Kleinert and W. Ecknig, Prediction of retention data by using parameters of intermolecular interaction. I. A model for calculation of non-polar and polar interaction parameters, *J. Chromatogr.*, **315**, 75, 1984.

51. Th. Kleinert, W. Ecknig, and J. Novák, Prediction of retention data by using parameters of intermolecular interaction. II. Application of the model to gas–liquid chromatographic systems, *J. Chromatogr.*, **315**, 85, 1984.

52. P. J. Schoenmakers, H. A. H. Billiet, and L. de Galan, Influence of organic modifiers

on the retention behaviour in reversed-phase liquid chromatography and its consequences for gradient elution, *J. Chromatogr.*, **185**, 179, 1979.

53. P. Jandera, Reversed-phase liquid chromatography of homologous series. A general method for prediction retention, *J. Chromatogr.*, **314**, 13, 1984.

54. P. Jandera, Method for characterization of selectivity in reversed-phase liquid chromatography. I. Derivation of the method and verification of the assumptions, *J. Chromatogr.*, **352**, 91, 1986.

55. P. Jandera, Method for characterization of selectivity in reversed-phase liquid chromatography. II. Possibilities for the prediction of retention data, *J. Chromatogr.*, **352**, 111, 1986.

56. V. D. Shatz, O. V. Sakhartova, V. A. Belikov, L. A. Brivkalne, and V. D. Grigor'eva, Vybor uslovii elyuirovaniya v obrashchenno-phazovoi khromatographii. Zavisimost koephphitsientov emkosti ot kontsentratsii organicheskogo komponenta podvizhnoi phazy, *Zh. Anal. Khim.*, **39**, 331, 1984.

57. L. R. Snyder, W. J. Dolan, and J. R. Gant, Gradient elution in high-performance liquid chromatography. I. Theoretical basis for reversed-phase systems, *J. Chromatogr.*, **165**, 3, 1979.

58. L. R. Snyder and J. J. Kirkland, *Introduction to Modern Liquid Chromatography*, Wiley, New York, 1974, p. 453.

59. K. Jinno, Effect of the alkyl chain length of the bonded stationary phase on solute retention in reversed-phase high-performance liquid chromatography, *Chromatographia*, **15**, 667, 1982.

60. T. Hanai, Energy effects in the retention of aromatic acids in liquid chromatography, *J. Chromatogr.*, **332**, 189, 1985.

61. J. Bermejo, J. S. Canga, O. M. Gayol, and M. D. Guillén, Utilization of physicochemical properties and structural parameters for calculating retention indexes of alkylbenzenes, *J. Chromatogr. Sci.*, **22**, 252, 1984.

62. L. Buydens and D. L. Massart, Prediction of gas chromatographic retention indices from linear free energy and topological parameters, *Anal. Chem.*, **53**, 1990, 1982.

63. C. A. Streuli and M. Orloff, Computer applications in gas–liquid chromatography. II. Prediction of retention volume for hydrocarbons and heteroatomic organic molecules using molecular parameters and multiple regression analysis, *J. Chromatogr.*, **101**, 23, 1974.

64. L. Buydens, D. Coomans, M. Vanbelle, D. L. Massart, and R. Vanden Driessche, Comparative study of topological and linear free energy-related parameters for the prediction of GC retention indices, *J. Pharm. Sci.*, **72**, 1327, 1983.

65. A. García-Raso, F. Saura-Calixto, and M. A. Raso, Study of gas chromatographic behaviour of alkanes based on molecular orbital calculations, *J. Chromatogr.*, **302**, 107, 1984.

66. A. Radecki, H. Lamparczyk, and R. Kaliszan, A Relationship between the retention indices on nematic and isotropic phases and the shape of polycyclic aromatic hydrocarbons, *Chromatographia*, **12**, 595, 1979.

67. S. A. Wise, W. J. Bonnett, F. R. Guenther, and W. E. May, A relationship between reversed-phase C_{18} liquid chromatographic retention and the shape of polycyclic aromatic hydrocarbons, *J. Chromatogr. Sci.*, **19**, 457, 1981.

68. K. Jinno and K. Kawasaki, Correlation between the retention data of polycyclic aromatic hydrocarbons and several descriptors in reversed-phase HPLC, *Chromatographia*, **17**, 445, 1983.

69. K. Jinno and M. Okamoto, Molecular-shape recognition of polycyclic aromatic hydrocarbons in reversed phase liquid chromatography, *Chromatographia*, **18**, 495, 1984.

70. K. Jinno and K. Kawasaki, Correlation of the retention data of polyaromatic hydrocarbons obtained on various stationary phases used in normal- and reversed-phase liquid chromatography, *Chromatographia*, **18**, 44, 1984.

71. L. B. Kier and L. H. Hall, Molecular connectivity analyses of structure influencing chromatographic retention indexes, *J. Pharm. Sci.*, **68**, 120, 1979.

72. A. Sabljić, Calculation of retention indices by molecular topology. Chlorinated benzenes, *J. Chromatogr.*, **319**, 1, 1985.

73. J. Raymer, D. Wiesler, and M. Novotny, Structure–retention studies of model ketones by capillary gas chromatography, *J. Chromatogr.*, **325**, 13, 1985.

74. J. S. Millership and A. D. Woolfson, The relation between molecular connectivity and gas chromatographic retention data, *J. Pharm. Pharmacol.*, **30**, 483, 1978.

75. J. S. Millership and A. D. Woolfson, Molecular connectivity and gas chromatographic retention parameters, *J. Pharm. Pharmacol.*, **32**, 610, 1980.

76. M. J. M. Wells, C. R. Clark, and R. M. Patterson, Correlation of reversed-phase capacity factors for barbiturates with biological activities, partition coefficients, and molecular connectivity indices, *J. Chromatogr. Sci.*, **19**, 573, 1981.

77. M. Kuchař, H. Tomková, V. Rejholec, and O. Skalická, Relationships between gas–liquid chromatographic behaviour and structure of arylaliphatic acids, *J. Chromatogr.*, **333**, 21, 1985.

78. H. Lamparczyk, R. Kaliszan, and A. Radecki, Correlation between retention on liquid crystalline phases and chemical structure, *J. Chromatogr.*, **361**, 442, 1986.

79. Gy. Szász, O. Papp, J. Vámas, K. Hankó-Novák, and L. B. Kier, Relationships between molecular connectivity indices, partition coefficients and chromatographic parameters, *J. Chromatogr.*, **269**, 91, 1983.

80. E. K. Whalen-Pederson and P. C. Jurs, Calculation of linear temperature programmed capillary gas chromatographic retention indices of polycyclic aromatic compounds, *Anal. Chem.*, **53**, 2184, 1981.

81. N. Adler, D. Babić, and N. Trinajstić, On the calculation of the HPLC parameters for polycyclic aromatic hydrocarbons, *Fresenius Z. Anal. Chem.*, **322**, 426, 1985.

82. G. Dahlmann, H. J. K. Köser, and H. H. Oelert, Multiple Korrelation von Retentionsindizes, *Chromatographia*, **12**, 665, 1979.

83. M. N. Hasan and P. C. Jurs, Computer-assisted prediction of liquid chromatographic retention indexes of polycyclic aromatic hydrocarbons, *Anal. Chem.*, **55**, 263, 1983.

CHAPTER

10

APPLICATION OF FACTOR ANALYSIS IN QSRR STUDIES

As pointed out by Weiner and co-workers [1] in one of the first publications concerning chromatographic applications of factor analysis, there are three main goals in the application of factor analysis in QSRR studies. One is to ascertain how many factors are necessary to account for the variance of the data. This information can be obtained without having to identify the individual factors.

Second, it is possible to attempt to identify the abstract factors with physically meaningful parameters, like molecular weight or number of carbon atoms in the solute molecule. In this manner, it is possible to gain insight into the fundamental factors that affect the measured variable of the system. The individual parameters can be tested separately for possible identification with the abstract factors of the space.

The third goal of factor analysis is the prediction of new data. One accomplishes this by finding a set of parameters (preferably physically significant) that span the space. Therefore, if one finds a set of physically significant factors that can be evaluated for a new case, it is then possible to predict the data desired without having to measure them for the new system.

The first publications reporting the application of factor analysis as a data-predicting tool appeared about 1972 [2–6]. In gas chromatography, an example in which the factors in the retention index of alcohols were identified with physically significant parameters and predictions were made for new alcohols can be found in the literature [2].

In liquid chromatography Huber et al. [5, 6] first applied factor analysis to the study of the partition of steroids and pesticides in a ternary system. These authors found that three factors were sufficient to span the space, but they did not attempt to rotate the abstract vectors into physically significant parameters. Nonetheless, Huber et al. [5, 6] proved the value of factor analysis as a good retention prediction tool.

In studies on correlation of HPLC retention volumes of substituted carboranes with molecular properties using factor analysis, Weiner and co-workers [1] considered the following relationship:

$$F_v(i, \alpha) = \sum_j U(i, j) \ V(j, \alpha) \tag{10.1}$$

221

where $F_v(i, \alpha)$ represents the function of the chromatographic retention of solute i in solvent system α; $U(i,j)$ represents the jth solute parameter for the ith solute; $V(j, \alpha)$ corresponds to the jth solvent or stationary phase factor for the αth system, and j is a dummy variable equal to the number of important factors in the space. Weiner et al. [1] measured the retention volumes of 15 solutes in 8 reversed-phase HPLC systems. Applying the factor analysis scheme to both the logarithms of the retention volume set and its transpose, it was found that three factors were sufficient to reproduce the data within experimental error.

Once the number of factors needed to span the space is determined, the next step is to try to identify these abstract factors with physically significant parameters. The advantage that factor analysis has over regression analysis is that individual factors can be tested for possible identification with the abstract factors without simultaneously identifying all the other important factors of the space. Of the several parameters associated with the solutes tested by Weiner and co-workers [1], the effective carbon number, molecular weight, and gas chromatographic retention time on a polar phase were apparently correlated with individual factors.

GC retention data have been the subject of factor analyses performed by Laffort and co-workers [7, 8]. These authors determined solubility factors for an impressive number of 240 solutes and 207 stationary phases. In their approach, the solute factors, represented by Greek letters, and the stationary phase factors, represented by Latin letters, are involved in a linear equation of predictability of retention index I as follows:

$$I = \alpha A + \omega O + \varepsilon E + \pi P + \beta B + 100 \qquad (10.2)$$

The constant 100 in Eq. (10.2) is used because all solute factors are referred to methane (values are equal to zero for this substance). The solute factors are linear combinations of retention indices measured on five selected stationary phases of different polarities. The factors of the stationary phases are calculated from retention indices of five reference solutes of different classes.

Physicochemical meanings are given to the solute factors. Thus, α is proportional to molecular volume; ω is proportional to the square of the dipole moment for simple molecules; ε is related to the ratio between molecular refraction and molecular volume; π is a proton donor factor; and β is a proton acceptor factor. Among the stationary phase factors, only E is identified. It is supposed to reflect the "compactability" of the solvent (i.e., the relative absence of "holes").

Laffort and co-workers [7, 8] related their solute factors to the Karger et al. [9] solubility parameters, partition coefficients, and biological activity.

Massart and co-workers [10–12] applied factor analysis in their studies on the correlation of GC retention indices with topological and quantum chemical parameters. For diverse solutes chromatographed on phases of different polarities, two or three important factors explained about 85–90% of the variance in the retention indices. Consistency is reported of the conclusions drawn from regression analysis and those based on factor analysis.

An interesting application of factor analysis for QSRR studies is reported by Hasan and Jurs [13]. These authors analyzed a representative set of 97 PAH, including methyl-substituted isomers. The retention behavior of the solutes in reversed-phase HPLC was quantitatively characterized by means of the logarithms of retention indices similar to Kováts retention indices in GC, except that the internal standards used were PAH instead of n-alkanes:

$$\log I_i = \log I_n + \frac{\log R_i - \log R_n}{\log R_{n+1} - \log R_n} \qquad (10.3)$$

where i represents the compound of interest, n and $n+1$ represent the bracketing lower and upper PAH standards, and the R values are the corresponding corrected retention volumes. The standards were assigned the following log I values: benzene 1; naphthalene 2; phenanthrene 3; benz[a]anthracene 4; and benzo[b]chrysene 5.

During the course of the study [13], a total of 36 molecular structure descriptors were generated for each compound in the data set. The descriptors generated can be classified into three classes: topological descriptors, geometric descriptors, and calculated physical property descriptors. Among the structural parameters considered are fragment descriptors (e.g., number of bonds), various molecular connectivity indices, substructure descriptors providing information on the presence or absence of specific substructures in the molecule, geometric descriptors (e.g., principal axes of rotation, principal moments of inertia), hydrophobicity parameters.

Since a large number of descriptors were generated, some were likely to be correlated among themselves. Simple correlations between two descriptors can be easily detected from the correlation matrix. However, correlation that involves more than two descriptors (multicollinearity) cannot be readily seen. To detect the presence of multicollinearity, Hasan and Jurs [13] applied factor analysis. It is known in mathematical statistics [14] that if any individual eigenvalues of the covariance matrix are small or the sum of their reciprocals is greater than 5 times the number of descriptors, it can be said that the variables are intercorrelated.

Of the 36 molecular structure descriptors generated by Hasan and Jurs [13], 19 appeared superfluous. Among the 17 descriptors retained for QSRRs, 5 or 4 (depending on the column applied) sufficed to derive the "best" regression equations describing log I for 97 PAH. For the descriptors used, the size of the eigenvalues of the correlation matrix indicates that no eigenvalue is small enough for the presence of unacceptably high multicollinearity among the descriptors.

Cserháti and Bordas [15] subjected a chromatographic data matrix to an analysis by computer-assisted multivariate techniques. The matrix was built of reversed-phase TLC R_f values determined for 17 substituted triazine derivatives with 27 binary solvents. The authors concluded, based on two-dimensional nonlinear mapping of the solvents, that in reversed-phase TLC solvent selectivity is determined mainly by the presence and number of free hydroxyl groups and by the presence and position of nitrogen and oxygen atoms in the solvent molecule.

Later Cserháti et al. [16] evaluated the reversed-phase TLC retention data generated for seven 5,5-dialkylsubstituted barbiturates in 21 mobile–stationary phase systems. As the result of the principal-component analysis, it was found that more than 95% of the total variance could be explained on the basis of a hidden variable. As one could expect, the determining factor in the grouping of compounds according to their chromatographic behavior is the number of carbons in the alkyl chains, which may be related directly to solute hydrophobicity.

The best orientation regarding the possibilities of obtaining valuable QSRRs by means of factor analysis may be gained from the recent publications by Chretien and co-workers [17, 18].

At first the authors [17] analyzed reversed-phase HPLC capacity factors k' of 53 chalcones (giving two series of the E-s-cis and Z-s-cis isomers) of the general formula given in Fig. 10.1 The capacity factors k' were determined in four reversed-phase chromatographic systems utilizing methanol–water 70:30 as the mobile phase and the following chemically bonded stationary phases: C-18 (Zorbax ODS), C-8 (Zorbax C_8), C-6 (Spherisorb C_6), and phenyl (μBondapak-phenyl). Numerical data are collected in Table 10.1.

Comparison of separation mechanisms in a number of chromatographic systems for the selected set of solutes was carried out in terms of the separation factor $\alpha_{a/b}$, defined as the ratio of the capacity factors k' for two separated solutes a and b. The separation factor was determined with respect to the unsubstituted chalcone as well as between the corresponding E and Z isomers.

The correspondence factor analysis, applied by Chretien and co-workers in gas chromatography [19, 20], was used again to extract information from the set of experimental reversed-phase HPLC data. By the method of factor analysis, Chrétien and co-workers [17] selected two main axes having contributions of 83.6 and 15.6% to the total inertia of the cluster. These axes defined the plane in Fig. 10.2 in which projection of the 53 compounds and four columns offers an accurate representation of their behavior in the initial cluster.

The C-18 column (51.0%) and the phenyl column (32.1%), as well as

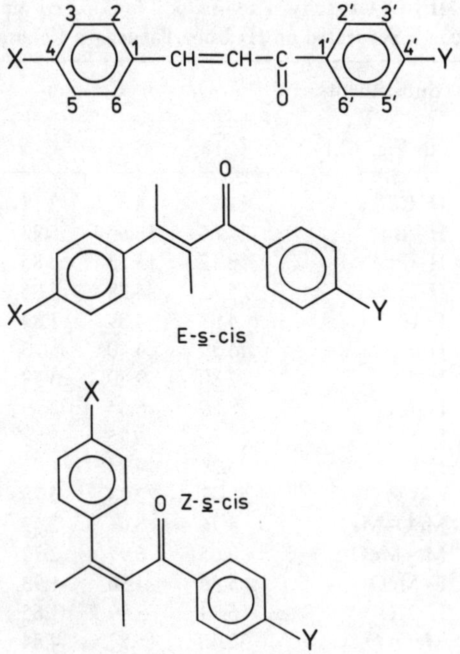

Figure 10.1. General formula of chalcones and structures of *E-s-cis* and *Z-s-cis* isomers considered by Chrétien and co-workers [17, 18].

compounds 11 (Me-Ø; 34.9%) and 16 (MeO-Ø; 11.4%) have the greatest contribution in isolating the first axis, while the phenyl column (48.4%) and the C-8 column (37.0%) as well as solutes 2* (H-*t*Bu*; 12.4%), 2 (H-*t*Bu; 12.2%), 16 (MeO-Ø; 7.9%), and 11 (Me-Ø; 6.5%) predominate in isolating the second axis.

Considering separately the arrangement of the *E-s-cis* chalcones in Fig. 10.2, one can note that the phenyl-substituted chalcones are most distinctly separated from the remaining compounds. For these chalcones Walczak et al. [17] obtained high values of the separation factor with respect to the unsubstituted derivative $\alpha_{X-Y/H-H}$ on the C-18 column. This is probably the result of high hydrophobicity and distinct donor–acceptor properties of the phenyl-substituted chalcones. The other cluster separated in Fig. 10.2 by Chrétien and co-workers [17] is the cluster a, in which the compounds with hydrophobic substituents are grouped. For the solutes in cluster a the following sequence of separation factor $\alpha_{X-Y/H-H}$ on the individual phases is characteristic: C-18 > C-8 > C-6 > phenyl. The small cluster b is formed by chalcones incorporating fluorine in their structure. Cluster c includes the

Table 10.1. HPLC Capacity Factors, k', for E-s-cis and Z-s-cis[a] Chalcones Separated on Hydrocarbonaceous Columns[b]

Number	Substituents X–Y in Fig. 10.1	Column C-18	C-8	C-6	Phenyl
1	H–CF$_3$	11.23	8.52	3.38	3.64
2	H–tBu	24.66	18.66	6.23	6.51
3	H–iPr	18.72	13.53	4.85	5.41
4	H–H	5.83	4.28	1.85	2.48
5	F–H	4.92	4.89	1.84	2.40
6	H–F	5.39	4.40	1.93	2.50
7	H–Et	12.80	9.50	3.59	4.30
8	H–Me	8.50	6.34	2.59	3.12
9	F–Me	8.83	6.85	2.53	3.14
10	F–F	5.29	4.24	1.91	2.45
11	Me–Ø	53.42	22.38	6.78	9.90
12	MeO–Me	8.76	6.64	2.59	3.53
13	Me–MeO	9.65	6.97	2.73	3.59
14	F–MeO	5.36	4.36	1.93	2.67
15	H–NO$_2$	5.89	3.89	1.65	2.85
16	MeO–Ø	32.46	15.87	4.84	8.28
17	F–NO$_2$	4.85	3.77	1.67	2.76
18	NO$_2$–Me	6.49	4.76	2.85	3.31
19	NO$_2$–H	4.02	3.25	1.48	2.62
20	MeO–MeO	5.83	4.69	1.94	3.89
21	MeO–NO$_2$	5.91	4.16	1.73	3.22
22	NO$_2$–F	4.32	3.43	1.54	2.60
23	N(Me)$_2$–NO$_2$	8.51	6.44	2.45	4.77
24	NO$_2$–MeO	4.56	3.53	1.57	2.96
25	NO$_2$–NO$_2$	4.82	3.00	1.55	2.91
26	NH$_2$–H	1.36	1.59	0.74	1.31
27	H–OH	1.88	2.00	0.96	1.46
1*	H–CF$_3$	7.19	6.53	2.81	2.68
2*	H–tBu	16.63	14.19	5.15	5.25
3*	H–iPr*	11.68	18.16	3.98	4.36
4*	H–H*	3.22	3.17	1.49	2.00
5*	F–H*	3.68	3.46	1.62	2.10
6*	H–F*	3.47	3.36	1.59	2.85
7*	H–Et*	7.85	7.87	2.90	3.46
8*	H–Me*	5.89	1.62	2.85	2.52
9*	F–Me*	5.70	5.89	2.22	2.74
10*	F–F*	3.93	3.64	1.7	2.14
11*	Me–Ø	38.93	17.42	6.78	8.84
12*	MeO–Me*	5.84	5.16	2.15	3.83

226

Table 10.1. (Continued)

Number	Substituents X–Y in Fig. 10.1	Column			
		C-18	C-8	C-6	Phenyl
13*	Me–MeO*	5.45	5.87	2.17	2.96
14*	F–MeO*	3.65	3.6	1.68	2.35
15*	H–NO$_2^*$	3.30	3.82	1.37	2.31
16*	MeO–Ø*	21.56	12.31	4.19	7.09
17*	F–NO$_2^*$	3.68	3.26	1.48	2.34
18*	NO$_2$–Me*	4.27	4.83	1.80	2.88
19*	NO$_2$–H*	2.77	2.76	1.31	2.30
20*	MeO–MeO*	3.74	3.59	1.60	2.67
21*	MeO–NO$_2^*$	4.36	3.47	1.50	2.77
22*	NO$_2$–F*	2.93	2.87	1.34	2.20
24*	NO$_2$–MeO*	2.89	2.96	1.36	2.56
25*	NO$_2$–NO$_2^*$	2.72	2.47	1.32	2.37
26*	NH$_2$–H*	1.36	1.59	0.74	1.31
27*	H–NO$_2^*$	1.15	1.46	0.75	1.15

[a] *Z-s-cis* compounds denoted by asterisks.

[b] Mobile phase, methanol–water 70:30. Numerical data from Walczak et al. [17]. With permission.

polar-substituted derivatives for which the separation factor $\alpha_{X-Y/H-H}$ is greater for the phenyl-bonded phase than for the alkyl-bonded stationary phases.

Walczak et al. [17] attempted to elucidate the division of the investigated compounds into the clusters and the respective differences in the specific selectivities by applying physicochemical interpretation to both of the discussed factors (i.e., axes 1 and 2). Accordingly, the first factor is to reflect the changes in hydrophobicity of compounds and columns, while the second describes the changes in retention caused by the different abilities of chalcones to interact specifically. The suggestions concerning the first factor are confirmed by the regression equations relating the chalcone coordinates on the first axis and the hydrophobic substituent constant, considered separately for the X and Y positions of substitution:

$$\text{Axis } 1(E\text{-}s\text{-}cis) = 0.101 - 0.193\ \pi(X) - 0.142\ \pi(Y)$$

$$n = 27; \quad R = 0.981 \tag{10.4}$$

The very interesting feature of Eq. (10.4) is that the contributions of a given substituent to the first factor axis 1 depends on the substitution site. Walczak and co-workers [17] explain this phenomenon by the different influence of the

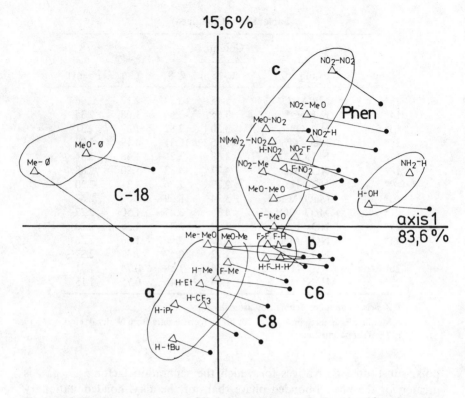

Figure 10.2. Simultaneous projection of 53 chalcones (Δ, *E-s-cis*; $\bullet$, *Z-s-cis*) and four chromatographic systems (C-18, C-8, C-6, phenyl) onto plane determined by two main inertia axes extracted from correspondence factor analysis of data given in Table 10.1. (After B. Walczak, M. Dreux, J. R. Chrétien, K. Szymoniak, M. Lafosse, L. Morin-Allory, and J. P. Doucet, *J. Chromatogr.*, **353**, 109, 1986. With permission.)

substituents in position 4 and 4′ on the electron density in the chalcone core; the stronger effect is exerted by the substituent attached to the acetophenone part of the chalcone.

Attempts to correlate chalcone coordinates on axis 2 in Fig. 10.2 with parameters such as Hammett electronic substituent constants, dipole moments, polarizabilities, electron densities on the individual chalcone atoms, and spectral data appeared unsuccessful [17]. It is suggested, instead, that some consistency of the arrangement of the investigated chromatographic systems along axis 2 is observed with their arrangement according to the number of residual silanol groups on the surface of the stationary phase used.

Correlation of the capacity factors for the *E-s-cis* and *Z-s-cis* series is high ($r = 0.989$ or more for the individual stationary phases studied). In such a situation observations and conclusions drawn for Z-s-cis chalcones are analogous to those discussed above for the E-s-cis species. The main difference concerns the dependence of the chalcone coordinates on axis 1 and the hydrophobic substituent constant:

$$\text{Axis } 1(Z\text{-}s\text{-}cis) = 0.200 - 0.133 \, \pi(X) - 0.134 \, \pi(Y)$$

$$n = 26; \quad R = 0.964 \tag{10.5}$$

Walczak et al. [17] conclude that the contributions of the 4 and 4′ substituents to axis 1 are less differentiated [actually, they are not differentiated in Eq. (10-5)] than with the E-s-cis isomers. This is due to the decrease in electron coupling in the Z-s-cis molecules, resulting from the lack of coplanarity with the constituent phenyl rings.

Continuing their studies on the application of factor analysis in QSRRs, Chrétien and co-workers [18] turned their attention to the normal-phase HPLC data obtained for chalcones on five columns of different polarities. The mobile phase applied was heptane–tetrahydrofuran 97:3. The columns used were characterized by the following functionalities: NH_2 (Zorbax NH_2), diol (LiChrospher 100 DIOL), CN (MicroPak CN), alkyl C-18 (Zorbax ODS), and alkyl C-8 (Zorbax C_8). The capacity factors k' for 47 E-s-cis and Z-s-cis chalcones separated in five chromatographic systems were investigated by the method of factor analysis.

Three main axes were extracted from factor analysis. Axes 1, 2, and 3 make contributions of 58.0, 31.2, and 8.9%, respectively, to the total cluster interia. Chrétien et al. [18] gave projections of chalcones and chromatographic systems onto the planes—the first plane is defined by axes 1 and 2 and the second plane is determined by axes 2 and 3. Next, it was attempted to detect the regularities in the behavior of the chalcones studied in different chromatographic systems based on the relative arrangement and distribution of the solutes and their subclasses on both projections. Based on the correspondence factor analysis projection, one can observe which subclasses of compounds behave specifically in defined chromatographic systems [18]. A prediction of the manner in which the relative retention of compounds, grouped in common subclasses on factor analysis projection, will change if chromatographic conditions are changed is also possible. Thus, one can choose the best chromatographic conditions for the optimization of the separation of positional or configurational isomers.

The information gained by Chrétien and co-workers [18] from the factor analysis of normal-phase HPLC data is generally of the qualitative type. Physical identification and quantitation of the factors governing retention of polar solutes on polar stationary phases in a straight-phase system have not succeeded. As expected, it appeared that there were several important factors affecting retention in normal-phase chromatography on polar phases. In contradistinction to reversed-phase partitioning liquid chromatography, where rationalization of the QSRRs observed is usually feasible, to quantify properties of solutes and systems governing normal-phase specific retention is still difficult. Certainly, the factor analysis of large sets of data that may be intercorrelated facilitates the extraction and separation of the factors determining the chromatographic phenomena. Efforts can be directed then to find physical interpretation of the individual factors in terms of the molecular structure of the species involved in the chromatographic separation process.

References

1. J. H. Kindsvater, P. H. Weiner, and T. J. Klingen, Correlation of retention volumes of substituted carboranes with molecular properties in high pressure liquid chromatography using factor analysis, *Anal. Chem.*, **46**, 982, 1974.

2. P. H. Weiner, C. J. Dack, and D. G. Howery, Retention index–structure relationships for alcohols using factor analysis, *J. Chromatogr.*, **69**, 249, 1972.

3. P. H. Weiner and D. G. Howery, Application of factor analysis to gas–liquid chromatography, *Can. J. Chem.*, **50**, 448, 1972.

4. P. H. Weiner and J. F. Parcher, A method for selecting preferred liquid phases using the technique of factor analysis, *J. Chromatogr. Sci.*, **10**, 612, 1972.

5. J. F. K. Huber, C. A. M. Meijers, and J. A. R. J. Hulsman, New method for prediction of partition coefficients in liquid–liquid systems and its experimental verification for steroids by static and chromatographic measurements, *Anal. Chem.*, **44**, 111, 1982.

6. J. F. K. Huber, E. T. Alderlieste, H. Harren, and H. Poppe, Static and chromatographic measurement and correlation of liquid–liquid partition coefficients, *Anal. Chem.*, **45**, 1337, 1973.

7. P. Laffort and F. Patte, Solubility factors in gas–liquid chromatography. Comparison between two approaches and application to some biological studies, *J. Chromatogr.*, **126**, 625, 1976.

8. F. Patte, M. Etcheto and P. Laffort, Solubility factors for 240 solutes and 207 stationary phases in gas–liquid chromatography, *Anal. Chem.*, **54**, 2239, 1982.

9. B. L. Karger, L. R. Snyder, and C. Eon, An expanded solubility parameter treatment for classification and use of chromatographic solvents and adsorbents. Parameters for dispersive, dipole and hydrogen bonding interactions, *J. Chromatogr.*, **125**, 71, 1976.

10. L. Buydens and D. L. Massart, Prediction of gas chromatographic retention indexes from linear free energy and topological parameters, *Anal. Chem.*, **53**, 1990, 1981.

11. L. Buydens, D. L. Massart, and P. Geerlings, Prediction of gas chromatographic

retention indexes with topological, physicochemical, and quantum chemical parameters, *Anal. Chem.*, **55**, 738, 1983.

12. L. Buydens, D. Coomans, M. Vanbelle, D. L. Massart, and R. Vanden Driessche, Comparative study of topological and linear free energy-related parameters for the prediction of GC retention indices, *J. Pharm. Sci.*, **72**, 1327, 1983.

13. M. N. Hasan and P. C. Jurs, Computer-assisted prediction of liquid chromatographic retention indexes of polycyclic aromatic hydrocarbons, *Anal. Chem.*, **55**, 263, 1983.

14. S. Chatterjee and B. Price, *Regression Analysis by Example*, Wiley-Interscience, New York, 1977.

15. T. Cserháti and B. Bordás, Classifying eluents in reversed-phase thin-layer chromatography by spectral mapping, *J. Chromatogr.*, **286**, 131, 1984.

16. T. Cserháti, B. Bordás, L. Ekiert, and J. Bojarski, Effect of layer and eluent characteristics on the reversed-phase thin-layer chromatographic behaviour of some barbituric acid derivatives, *J. Chromatogr.*, **287**, 385, 1984.

17. B. Walczak, M. Dreux, J. R. Chrétien, K. Szymoniak, M. Lafosse, L. Morin-Allory, and J. P. Doucet, Factor analysis and experiment design in high-performance liquid chromatography. I. Trends in selectivity of 53 chalcones in reversed-phase high-performance liquid chromatography on alkyl- or phenyl-bonded stationary phases, *J. Chromatogr.*, **353**, 109, 1986.

18. B. Walczak, J. R. Chrétien, M. Dreux, L. Morin-Allory, M. Lafosse, K. Szymoniak, and F., Membrey, Factor analysis and experiment design in high-performance liquid chromatography on columns of different polarities, *J. Chromatogr.*, **353**, 123, 1986.

19. R. F. Hirsch, R. J. Gaydosh, and J. R. Chrétien, Factor analysis of trends in selectivity in gas–solid chromatography on cation-exchange resins, *Anal. Chem.*, **52**, 723, 1980.

20. J. R. Chrétien, K. Szymoniak, C. Lion, and J. K. Haken, Gas chromatography of homologous esters. XXX. Unexpected predominancy of polar effects over steric effects in crowded aliphatic esters: A gas–liquid chromatographic appraisal of specific interactions, *J. Chromatogr.*, **324**, 355, 1985.

RELATIONSHIP BETWEEN LIQUID CHROMATOGRAPHIC RETENTION DATA AND PARTITION COEFFICIENTS

Since the turn of the century it has been recognized, due to the works of Overton [1], Meyer [2], and Baum [3], that the lipophilic properties of drugs are of importance in their pharmacological activity. The lipophilic (or hydrophobic) nature of a drug, and any other chemical substance as well, may be represented by the logarithm of the partition coefficient obtained from studies of the distribution of the substance between an immiscible polar and nonpolar solvent pair. In the early 1950s Collander [4] introduced the equation, called the general Collander equation, relating partition coefficients determined in the individual organic–water solvent systems:

$$\log P_a = m \log P_b + n \tag{11.1}$$

where P_a and P_b are partition coefficients determined in aqueous solvent systems a and b, respectively and m and n are constants that are characteristic of the two solvent systems employed.

It was already observed by Collander that the fit implied in Eq. (11.1) became poorer as the polarity differences between the organic solvents in the two aqueous partition systems became larger. Thus, a good fit can be expected for the n-octanol–water versus the pentanol–water system, while the combination of cyclohexane–water versus n-octanol–water is typical of a poorly fitting Eq. (11.1).

The n-octanol–water partitioning system proposed by Hansch and co-workers [5, 6] is the common reference system for determination of the logarithm of the partition coefficient, $\log P$, widely used in medicinal chemistry. The n-octanol–water partitioning system provides a single, continuous scale for measurement of the hydrophobicity of all drugs.

Although the choice of n-octanol as a compound reflecting the properties of the lipid components of the cell membrane has occasionally been questioned, the large amount of n-octanol–water partition data collected by the Hansch group [7] has made n-octanol a common reference standard.

Measurement of $\log P$ by the conventional "shake-flask" method is tedious and time consuming. It is difficult to determine $\log P$ for compounds that are poorly soluble in water or that cannot be detected by conventional methods. The equilibration method is not applicable to surface-active and organo-metallic substances. It is difficult to determine reliable $\log P$ data for ionic substances, volatile compounds, and solutes for which association and dissociation processes are observed. Additional problems are caused by impurities, phase separation, and formation of emulsions.

Instead of measuring $\log P$ values by equilibration methods, partition chromatographic data can be used. It has long been known that the partition chromatographic R_M value and the partition coefficient $\log P_s$ determined in a system identical with the chromatographic system are mutually related by the Martin equation [8]:

$$R_M = \log P_s + \log V_s/V_M \qquad (11.2)$$

where V_s and V_M are the volumes of the stationary and mobile phases, respectively. In practice, it is difficult to obtain a slope of Eq. (11.2) that is equal exactly to unity. The deviation may be explained according to Bird and Marshall [9] as resulting from difficulties in adjusting ionic strengths in shake-flask and chromatographic processes to precisely the same value. Generally, a slope of exactly unity could be expected only if the two chromatographic phases were identical with the two phases in the classical shake-flask experiment [10].

Assuming the extrathermodynamic linear free-energy relationship, one may expect the classical $\log P$ data to be linearly related to the partition chromatographic parameters (i.e., R_M in TLC and PC and $\log k'$ in HPLC):

$$\log P = a'R_M + b' \qquad (11.3)$$

or

$$\log P = a \log k' + b \qquad (11.4)$$

where a, b, a', and b' are constants.

Keeping the above in mind, Iwasa et al. [11] suggested the usefulness of chromatographic data for characterization of the partitioning of a substance in the n-octanol–water system. These authors obtained good correlation ($r = 0.970$) between the substituent measures of hydrophobicity determined by the classical shake-flask method (π constants) and the reversed-phase chromatographic method (ΔR_M constants). At the same time the chromato-graphic measure of hydrophobicity, R_M from reversed-phase TLC, was applied by Boyce and Milborrow [12] in quantitative structure–biological activity studies.

Reversed-phase TLC as the method of assessment of the partitioning properties of drug candidates and chemicals are popularized mainly due to the extensive studies by Biagi and co-workers initiated in 1969 [13]. Since the early 1970s reversed-phase HPLC becomes increasingly more popular as the method of hydrophobicity determination. The first papers in which the advantages of HPLC as the method of characterization of partition coefficients were pointed out were published by Huber et al. [14], Seiber [15], Haggerty and Murrill [16], and Menheere et al. [17].

To date, a great many papers have been published dealing with the chromatographic characterization of hydrophobicity and the application of chromatographically determined partitioning data in QSAR studies. The respective literature has been reviewed rather systematically since 1975 by Tomlinson [18], by this author [19–21], and by Harnisch et al. [22]. Short reviews were also published [23, 24].

Despite the large material collected, the discussion still persists concerning the reliability of chromatographic measures of hydrophobicity as well as their compatibility with the n-octanol–water log P. The discussion is somewhat academic, and sometimes individual authors advocate their views paying no attention to the results reported by others. In such a situation it seems worthwhile to bring together the more important recent findings and settlements in the field.

Correlations of chromatographic data will be discussed separately with the experimentally determined log P values and the partitioning data calculated from hydrophobic substituent and fragmental constants.

11.1. CORRELATIONS BETWEEN CHROMATOGRAPHIC PARAMETERS AND EXPERIMENTAL n-OCTANOL–WATER PARTITION COEFFICIENTS

If a precise relationship is determined between chromatographic retention data and the logarithm of n-octanol–water partition coefficients, the hydrophobicity of a given compound can be chromatographically quantified. In practice, HPLC and TLC are employed for hydrophobicity determinations. Occasionally, PC is also used. Partitioning chromatography has usually been realized by the reversed-phase mode, but there are several examples of employing of the normal-phase systems.

The advantages of the chromatographic method of lipophilicity evaluation over the classical shake-flask method are numerous. These advantages have been summarized by Kaliszan [20] and Harnisch et al. [22]. In the case of reversed-phase HPLC these authors observed that the method is fast. It is suitable for substances containing impurities and for substance mixtures. The method requires no quantitative determination and is applicable to volatile substances. The method is highly reproducible, and its measuring range is 1–8 log P units. It should be noted here that Tomlinson et al. [25] suggest that in systems where organic modifiers may be added to a mobile phase, a distribution coefficient of 10^{-12}–10^{12} is possible.

The advantages of the reversed-phase TLC method are similar to those of reversed-phase HPLC, but the former requires considerably less laboratory equipment, and many solutes can be simultaneously analyzed. However, the accuracy in TLC demands $0.2 < R_f < 0.8$ [26, 27]. This gives a range of R_M of less than one and a half decades. In such a situation, it is often necessary to change the composition of the mobile phase to obtain reliable R_M values when one deals with a series of compounds of various polarity. Obviously, TLC is less reproducible and less sensitive than HPLC. As there may be a selective adsorption of mobile phase components in TLC, a quasi-gradient elution may be observed for less retained solutes [20]. The same is true for buffer components. In contrast, HPLC provides precise control of pH and ionic strength during the separation process. Temperature control is also conveniently attained.

The main disadvantages of chromatographic methods of hydrophobicity determination are that the methods are not applicable to organometallic substances and are applicable only to a limited extent to surface-active and ionic substances. Additionally, TLC is not applicable for volatile substances. Contrary to the direct method of partition coefficient determination, chromatographic methods require a reference system.

It is important to make chromatographic determinations in systems where partition is either the sole process taking place or predominates over other methods. In partition chromatography by the reversed-phase technique, the stationary phase is hydrophobic. To get such phases, silicone oil, *n*-octanol, oleyl alcohol, liquid paraffin, and hydrocarbons chemically bonded to a support material are usually used. To check whether the R_f of compounds in reversed-phase TLC are determined exclusively by partitioning between the stationary phase and the mobile phase, Hulshoff and Perrin [28] employed the relationship

$$1/R_f = 1 + {}_sP^x k C_{ol} \tag{11.5}$$

where ${}_sP^x$ is the partition coefficient in the system oleyl alcohol/methanol–water, which is analogous to a chromatographic system; k is a constant; and C_{ol} is the oleyl alcohol concentration in the mixture used for impregnation of the plates. When no adsorption of the compounds (benzodiazepines) onto the support phase during the migration took place, straight lines were obtained when $1/R_f$ was plotted against C_{ol}, with intercepts close to unity and slopes of ${}_sP^x k$.

Special care must be taken if one deals with ionizable compounds. The reference partition coefficient P refers to the ratio of neutral, unionized compound in each phase. What one actually observes are the distribution coefficients. The distribution coefficient is defined [29] as the ratio of the concentration of a compound in the organic phase to the concentration of all species (neutral and ionized) in the aqueous phase at a given pH. The organic phase is assumed to contain only un-ionized species, although there may be some solution of an ionized compound in the hydrophobic phase as an ion pair with the buffer ions. If one knows the effects of ionization, one can simply use the distribution coefficient D in place of the partition coefficient P, but one must correct for the relative differences in hydrogen-bonding effects when comparing different experimental procedures [30]. The reference n-octanol–water partition coefficient P of neutral species is

$$P = D/(1-\alpha) \qquad (11.6)$$

where α is the degree of ionization. Thus, for acids,

$$P = D\left(\frac{K_a}{[H]} + 1\right) \qquad (11.7)$$

where K_a is the dissociation constant and $[H]$ is the hydrogen ion concentration. In the case of organic–water mobile phases a problem may arise in the determination of $[H]$, as the apparent pH may not define this quantity unequivocally.

Equation (11.7) may be written in the following form for acids:

$$\log D = \log P + \log[1/(1 + 10^{pH - pK_a})] = \log P + C_D \qquad (11.8)$$

For bases the corresponding relationship is

$$\log D = \log P + \log[1/(1 + 10^{pK_a - pH})] = \log P + C_D \qquad (11.9)$$

The second term in Eqs. (11.8) and (11.9), C_D, is a correction for dissociation. Values of C_D for various pK_a and pH differences are tabulated in Table 11.1 after Scherrer and Howard [29].

Unger et al. [31] rearranged Eq. (11.7) into the form

$$D = P + K_a(-D/[H]) \qquad (11.10)$$

Correlating the distribution coefficient D or the capacity factor k' from HPLC, with $-D/[H]$ for acids or $-D[H]$ for bases, they determined the

Table 11.1. Correction for Dissociation Factor C_D for Use with Eq. (11.8) and (11.9)^a

pK_a–pH for Acids, pH–pK_a for Bases	C_D	pK_a–pH for Acids, pH–pK_a for Bases	C_D
>2	0	−0.1	−0.35
2.0	0	−0.2	−0.41
1.0	−0.04	−0.3	−0.48
0.9	−0.05	−0.4	−0.55
0.8	−0.06	−0.5	−0.62
0.7	−0.08	−0.6	−0.70
0.6	−0.10	−0.7	−0.78
0.5	−0.12	−0.8	−0.86
0.4	−0.15	−0.9	−0.95
0.3	−0.18	−1.0	−1.04
0.2	−0.21	−2.0	−2.00
0.1	−0.25	−3.0	−3.00
0.0	−0.30		

^{a} Reprinted with permission from R. A. Scherrer and S. M. Howard, *J. Med. Chem.*, **20**, p. 53. Copyright 1977 American Chemical Society.

intercept P and the slope K_a (or $1/K_a^b$ for bases). To determine it most accurately, measurements should be made near pH 2 and approximately the pK_a for acids or near pH 8 and the pK_a^b for bases. The method is capable of simultaneously determining partition, distribution, and ionization coefficients (P, D, and K_a, respectively), all of which are interesting from the point of view of QSRRs. The method has further been developed by Unger and Feuerman [32] based on the work by Horváth et al. [33] to model bulk phase partitioning for lipophilic acids, including ion pair partitioning. The equation used has the form

$$\log D = \log(P + P'K_a/[\mathrm{H}]) - \log(1 + K_a/[\mathrm{H}]) \qquad (11.11)$$

where P' is the partition coefficient for the anion. The constants P, P', and K_a were derived from the nonlinear least-squares fit of Eq. (11.11). Unger and Feuerman [32] supported Wahlund and Beijersten [34] in concluding that anions of lipophilic acids can be considerably more lipophilic than is commonly assumed and their partitioning should be taken into consideration for QSRR purposes.

11.1.1. Chromatographic Systems with Octanol like Properties

Much effort has been directed to getting chromatographic systems (mainly the HPLC systems) to mimic the conventional n-octanol–water partition system. To obtain such an HPLC system, Mirrlees et al. [35] coated a silanized Kieselguhr support with water-saturated n-octanol and used n-octanol-saturated buffer as an eluent. For a set of seven solutes an excellent correlation ($r = 0.999$) has been reported between the logarithm of the corrected elution time and log P from the literature. The slope of this relationship is indistinguishable from unity, as it should be in the case of two identical partitioning systems. By varying the column length and flow rate, Mirrlees et al. [35] attained the range in log P of -0.3 to $+3.7$. They have also claimed that at a flow rate of 2 mL/min, recoating of the column is not normally required in less than 50 h of actual use. More recently these authors [36] reported some modifications of their procedure aimed at determinations of heterocyclic partition coefficients.

Concurrent with Mirrlees et al. [35] Henry et al. [37] prepared columns using the pellicular silica (Corasil II) material with 1% loadings of n-octanol. The eluents were acetate buffers of pH 4.0 and 5.0 and a phosphate buffer of pH 6.5 presaturated with the stationary phase. Analogous experiments were done with the same pellicular silica material but coated with 1% squalene and using the chemically bonded octadecylsilica column (Bondapak C-18). Two groups of solutes were chromatographed (i.e., sulfonamides and barbiturates) for which the log P data were known from literature. Surprisingly enough, the best have been the correlations between log P and the logarithm of the retention volume on the ordinary C-18 column. Correlation of the retention data derived from the supposedly octanol-like systems with log P have been significantly worse.

Miyake and Terada [38] examined the in situ n-octanol-coating method described by Mirrlees et al. [35]. They did not obtain satisfactory results because the n-octanol dislodged and the retention time decreased gradually during the chromatography. In such a situation Miyake and Terada [38] proposed a method for preparing an n-octanol-coated column that had high stability. To prepare the column, a silica material (Corasil I) was heated at 110°C overnight and then, while still hot, was mixed with n-octanol. The slurry was packed in a PTFE tube, and excess n-octanol was removed by elution with the mobile phase until a stable baseline was obtained on the recorder. With the mobile phase HCl–KCl buffer of pH 2.0 saturated with n-octanol, the logarithm of the capacity factor, log k', for a set of aromatic acids and phenol derivatives was linearly related (slope 0.965; $r = 0.996$) to the literature log P.

In a series of papers, Unger and co-workers [31, 32, 39] described the optimization of HPLC conditions for determination of n-octanol–aqueous partition. Their method involves the use of the persililated octadecylsilica phase with a n-octanol-saturated buffer of adjusted ionic strength. These authors are interested in obtaining chromatographic data corresponding to the distribution of a solute between n-octanol and water (log D) at physiological pH and ionic strength. Therefore, they have carried out their chromatographic experiments using phosphate buffer saline of ionic strength $\mu = 0.15$ and pH 7.4 saturated with n-octanol. To determine the lipophilicity of neutral and acidic compounds, Unger and co-workers [31, 32] used phosphate buffer 0.15 μ PO_4. However, these conditions were unsatisfactory in the case of lipophilic amines. To improve the correlation between log k' and log D, a phosphate buffer with added NaCl (0.01 μ PO_4 + 0.14 μ NaCl) was applied to which N,N-dimethyloctylamine was added at a concentration of 4 mmol/L. The lipophilic N,N-dimethyloctylamine is added to the eluent to swamp out the binding of amines chromatographed to residual silanol sites on the stationary phase material. If N,N-dimethylamine is not added, compounds such as phenothiazines and tricyclic antidepressants are retained too long (appear too lipophilic) and give unsymmetrical peak shapes. With the procedure proposed, Unger and Chiang [39] obtained a convincing correlation between log k' and log D data for a series of neurotropic drugs (Fig. 11.1).

In the case of TLC, an octanol like system was reported by Bird and Marshall [9], who impregnated microcrystalline cellulose layers with

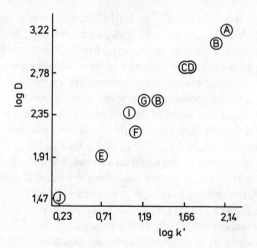

Figure 11.1. Relationship between shake-flask distribution coefficient, log D, and capacity factors, log k', determined on octadecylsilica columns with phosphate buffer containing NaCl and N,N-dimethyloctylamine as eluent. Solutes: (A) chloropromazine; (B) promazine; (C) promethazine; (D) amitriptyline; (E) mepacrine; (F) doxepin; (G) imipramine; (H) cyproheptadine; (I) indoramine; (J) tilorone. (Reprinted with permission from S. H. Unger and G. H. Chiang, *J. Med. Chem.*, **24**, 262. Copyright 1981 American Chemical Society.)

n-octanol and used an n-octanol-saturated buffer as the solvent. For a group of penicillins, plots of R_M versus $\log P$ were straight lines with a slope close to unity when the pH was below the pK_a of the solutes. However, the TLC system applied did not give meaningful results for the very hydrophilic penicillins α-aminobenzylpenicillin (ampicillin) and α-carboxybenzylpenicillin (carbenicillin). These compounds ran very close to the solvent front at the pH studied. This is a realistic result because the presence of an ionized group in the side chain makes it very unlikely that any significant amount of these penicillins would partition into the octanol phase. This also illustrates one of the limitations of the octanol like systems, especially evident in the case of TLC.

There are other disadvantages of n-octanol like chromatographic systems. It is time consuming to prepare such systems. There is some doubt as to whether the column characteristics do not change during use. Thus, reproducibility of the results may not be sufficiently high. As practically no organic modifier can be added to the aqueous mobile phase (because of column stability), the chromatography of more hydrophobic solutes is impossible. Actually, the distribution coefficient range determined in octanol like chromatographic systems may not be wider than 10^{-2}–10^6 (10^4), as in shake-flask methods.

11.1.2. Stable Liquid Chromatographic Partition Systems

Retention parameters obtained from HPLC measurements ($\log k'$) as well as those obtained from TLC or PC measurements (R_M) can be taken as hydrophobic indices in three different ways. First, the retention data obtained with the hydrophobic stationary phase and pure water as a mobile phase may be used directly as experimentally determined. However, in practice, too long retention times usually preclude the employment of this simplest procedure. Moreover, the retention mechanism on the now commonly used highly hydrophobic chemically bonded stationary phases changes at high water content in the mobile phase ($>90\%$), as compared to the practically applied region of 30–90% of water in the aqueous solvent. A second approach is to use isocratic data measured at a certain eluent composition X ($\log k'_x$, R_M^x) and to calculate n-octanol–water partition coefficients ($\log P$) by means of the Collander-type Eq. (11.3) or (11.4). A third possibility is to employ chromatographic data extrapolated to 0% of the organic modifier in mobile phase.

In practice, one must decide whether to use isocratic data or do a series of determinations of several different isocratic data, required for extrapolation of chromatographic parameters to the hypothetical value corresponding to 100% water in the mobile phase. Intuitively, the extrapolated data appear to be more reliable, and at present these are preferred by a majority of the workers in the field of QSARs. However, the problem usually arises what range of composition of binary aqueous–organic solvent should serve for extrapolation of retention data to pure water. As pointed out in Chapter 4, the dependence of retention parameters on mobile phase composition is complex [40, 41]. This complexity is illustrated in Fig. 4.8 [42] concerning reversed-phase TLC data.

Usually, there is only a limited range of concentration X of an organic component in the aqueous mobile phase for which a linear relationship is observed between $\log k'$ or R_M values of a given solute and the organic solvent concentration X. To identify unequivocally such a region of linearity of plots given in Fig. 4.8 seems to be a rather difficult task. Yet, Pietrogrande et al. [42] advocate the method of linear extrapolation of retention data experimentally derived at rather narrow solvent concentration ranges to pure water. For example, in the case of benzodiazepines chromatographed in a reversed-phase HPLC system employing a C-18 stationary phase and methanol-buffer (pH 7.4) mixtures, the logarithm of the capacity factor extrapolated to 100% water, $\log k'_w$, was derived from isocratic capacity factors determined at methanol concentrations ranging from 50 to 60% and from 45 to 55% for the majority of solutes studied. The same authors [42] also analyzed the capacity factors corresponding to the 100% water mobile phase but calculated from the Schoenmakers et al. [49] equation

$$\ln k' = \ln k'_w + AX^2 + BX + C\sqrt{X} \tag{11.12}$$

where A, B, and C are regression coefficients; X is the volume fraction of the organic modifier in the binary mobile phase with water; and k'_w is the solute capacity factor in pure water. The $\log k'_w$ data calculated by means of Eq. (11.12) appeared to be poorly correlated to $\log P$, contrary to the $\log k'_w$ values extrapolated from the linear part of the $\log k'$ versus X curve.

A limited region of linearity of the plots of $\log k'$ (or R_M) versus X observed for a majority of compounds is the reason for discarding the $\log k'_w$ (R_M^w) by several authors in favor of isocratic $\log k'_x$ (R_M^x) values. Hafkenscheid and Tomlinson [43] have proposed to use isocratic $\log k'_x$ since extrapolated values to 100% water composition might contain errors due to regression. Based on arguments that at high concentrations of organic modifier (a) large changes in eluent pH may occur and (b) the order of retention may be reversed due to specific effects of the modifier on retention, Pietrogrande et al. [43] suggested that with octadecylsilane as the stationary phase, an eluent of methanol–water 50:50 can be used to obtain correlation of chromatographic parameters with n-octanol–water distribution coefficients.

For pure water as the eluent the extrapolated $\log k'_w$ (R_M^w) of an individual solute should be constant for a given stationary phase, independent of the organic solvent used in mixtures with water for particular isocratic data required for extrapolation. However, the reported retention parameters k'_w (R^w_M), obtained by extrapolation of Eq. (11.12) or its simple linear transformation, to pure water depend on the nature of the organic modifier present in the aqueous mobile phase. For a given set of solutes, $\log k'_w$ (R^w_M) values derived from different solvent systems are more or less intercorrelated. For example, for a group of phenylurea herbicides, the relationship between the extrapolated capacity factors from the methanol–water system, $\log k'_w$ (CH_3OH), and those obtained in the acetonitrile–water system, $\log k'_w$ (CH_3CN), is [44]

$$\log k'_w\,(CH_3OH) = 1.435 \log k'_w\,(CH_3CN) + 0.279 \qquad (11.13)$$

with correlation coefficient $r = 0.939$. Braumann et al. [44] concluded that in the chromatographic system studied, the methanol–water mobile phase produced retention data k'_w that were strongly related to $\log P$, whereas acetonitrile–water and tetrahydrofuran–water yielded k'_w data poorly correlated with $\log P$. However, even the data derived in methanol–water systems, $\log k'_w$ (CH_3OH), gave separate correlation plots for the three families of solutes studied (Fig. 11.2). Assuming the reliability of the $\log P$ data used for the correlation, shifts of individual $\log P$ versus $\log k'_w$ (CH_3OH) plots in Fig. 11.2 suggest differences in specific interactions of the particular classes of solutes with the stationary phase (no attempt to deactivate free silanol sites was undertaken).

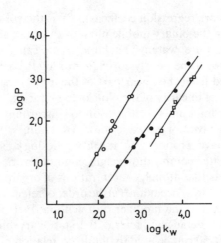

Figure 11.2. Relationship between $\log P$ and $\log k'_w$ (CH_3OH) for phenylureas (●), phenoxycarbonic acids (○), and phenoxycarbonic acid methyl esters (□). Octadecylsilica column used without any special treatment. Mobile phase consisted of different volume fractions of methanol in water (0.5 M acetate buffer pH 2.9 was used in the case of phenoxycarbonic acids). (After T. Braumann, G. Weber, and L. H. Grimme, *J. Chromatogr.*, **261**, 329, 1983. With permission.)

In TLC, the extrapolated R_M^w data are used for QSRR studies involving log P since the first publication by Boyce and Milborrow [12]. Also in TLC, the extrapolated R_M^w have initially been reported to depend on the mobile phase used. Riley et al. [45] reported that R_M^w values of acetophenone determined on the paraffin oil-impregnated plates with acetone–water, methanol–water, and dimethylformamide–water solvents differed significantly. However, similar values of methylene group contribution were found for a series of n-alkylphenyl ketones. In 1984 Biagi and co-workers [46] reported R_M^w values of a series of demorphin-related oligopeptides determined in two reversed-phase TLC systems. In one system the mobile phase formed an aqueous buffer mixed with various amounts of methanol. In the other system the organic modifier used was acetone. A nonpolar stationary phase was obtained by impregnating the silica gel layer with silicone oil DC 200 (Applied Sciences Labs.). The following relationship between the R_M values extrapolated to pure water, R_M^w, for the methanol–water and acetone–water systems was reported:

$$R_M^w \ (CH_3OH) = 1.76 - 0.482 \ R_M^w \ (CH_3COCH_3) \tag{11.14}$$

For the 23 solutes considered, the correlation coefficient was $r = 0.987$. Biagi et al. [46] suggest, however, that the R_M^w values extrapolated from the R_M data determined at low concentrations of the organic modifier in the mobile phase were very similar in both systems and close to the experimental R_M^w available for five compounds of the series. This situation well illustrates the ambiguities in deciding which part of the R_M versus X curve is linear. The Italian group preferred R_M^w from methanol–water determined at low concentrations of methanol in the solvent, and these data were used in subsequent publications concerning the measurement of lipophilicities of the demorphin-related oligopeptides mentioned above [47, 48]. The impressive experience of the Italian group allowed them more recently to extrapolate R_M data for a series of prostaglandins determined for the methanol–water and acetone–water systems in such a manner that the resulting R_M^w values derived from the two systems did not differ significantly [49]. Such a result would theoretically be expected if R_M^w is to be a measure of the partitioning between water and the silicone oil impregnating the silica gel layer.

The unambiguities encountered at the determinations of the extrapolated to 100% water retention data are certainly a reasonable explanation of the opposing opinions concerning the validity of the extrapolated data. The results reported by Kuchař et al. [50], for example, cast doubt on the reliability of extrapolated retention parameters. These authors determined R_M and log k' values in different reversed-phase TLC and HPLC systems for alkoxy and phenylalkoxy arylacetic acids. It was found that when an aqueous mobile phase containing an organic solvent (50% acetone or 60% methanol) was used, the changes in hydrophobicity corresponded to the changes in $\log P$ measured in the reference system n-octanol–water. Extrapolation of retention indices to pure water was not advantageous and negatively influenced the calculation of hydrophobic substituent parameters from the corresponding retention indices. Similar conclusions were drawn by Jinno [51, 52] based on QSRR studies of $\log k'$ values determined by the micro-HPLC technique for alkylbenzenes. The data analyzed by Jinno were determined on C-2, C-8, and C-18 hydrocarbonaceous columns with acetonitrile–water mixtures as mobile phases.

On the other hand, Hammers et al. [53] and El Tayar et al. [54] strongly advocate the extrapolation to pure water procedures, claiming better correlations between $\log P$ and $\log k'_w$ from reversed-phase HPLC than analogous correlations involving isocratic $\log k'_x$ values measured at a given composition X of methanol in the mobile phase. In the most recent publications by Kraak et al. [55] and de Biasi et al. [56], the isocratic $\log k'_x$ parameters are correlated with $\log P$, however.

An interesting approach to the problem of selection of the mobile phase composition that provides the best correlation of the chromatographic and the shake-flask partition data has been proposed by Valkó [57, 58]. Assuming the Collander-type relationship between n-octanol–water partition coefficient, $\log P$, and capacity factor, $\log k'$, measured in a chromatographic partition system, one can write, for the ith solute,

$$\log P_i = a \log k'_i + b \qquad (11.15)$$

For individual un-ionized solutes, i, the linear relationship between $\log k'_i$ and the percentage X of the organic modifier in the aqueous mobile phase usually holds true. Then,

$$\log k'_i = A_i X + B_i \qquad (11.16)$$

where the coefficients A_i and B_i may be determined for the ith solute by regression analysis of its capacity factors observed at several compositions X of the eluent used for chromatography. Combining Eqs. (11.15) and (11.16) yields

$$\log P_i = a A_i X + a B_i + b \qquad (11.17)$$

where the coefficient B_i corresponds to the capacity factor of the ith solute extrapolated to 100% water in the mobile phase (i.e., $\log k'_{w,\,i}$. Having determined individual A_i and B_i parameters for a set of solutes, one can correlate them with the corresponding $\log P_i$ values by means of a two-parameter regression equation:

$$\log P_i = a'A_i + aB_i + b \tag{11.18}$$

Thus, the regression coefficients a', a, and b can be calculated using the least-squares method. Comparing Eqs. (11.18) and (11.17),

$$a' = aX \quad \text{or} \quad X = a'/a \tag{11.19}$$

Knowing a' and a from Eq. (11.18), one can calculate optimum eluent composition a'/a by which the best correlation can be found between $\log P$ and reversed-phase HPLC retention data. For a set of neutral (at chromatographic conditions) solutes chromatographed on a C-18 phase with acetonitrile–KH_2PO_4 buffer (0.05 M; pH 4.6), the slopes and intercepts of Eq. (11.16) are given in Table 11.2 [57]. The regression equation calculated with these data and the literature $\log P$ values is of the form

$$\log P_i = 90.23A_i + 1.854B_i + 1.911$$
$$n = 26; \quad R = 0.949; \quad s = 0.413; \quad F = 104.7 \tag{11.20}$$

Correlation between experimental $\log P$ values and those calculated by Eq. (11.20) is illustrated in Fig. 11.3.

The quotient of the regression coefficients at the A_i and B_i variables in Eq. (11.20) gives 48.6. It means that a relationship like Eq. (11.20) should be obtained between $\log P_i$ and $\log k'_{X,\,i}$ determined at 48.6% acetonitrile in the mobile phase. Valko [57] concluded that the composition of eluent and the C-18 column used showed the greatest similarity to the n-octanol–water partitioning system.

Comparing the calculated and the known $\log P$ values of each compound (Fig. 11.3), one agrees that none can be considered a significant outliner. Although the residual error is 10 times as high as the usual error of $\log P$ determination by the shake-flask method, it is not too high in view of the deviation of the $\log P$ values obtained from different laboratories.

The chromatographic techniques employed for hydrophobicity characterization deserve some characterization here. In TLC, according to the method employed since 1969 [13, 49], glass plates (20 × 20 cm) are coated with silica gel G. A nonpolar stationary phase is obtained by impregnating the silica gel G layer with Silicone DC 200 oil. The impregnation is carried out by developing the plates in a 5% silicone solution in diethyl ether. The plates are left in the developing chamber for several hours after the silicone solution has reached the top of the plates. Next, the plates are withdrawn and left to evaporate the diethyl ether. A migration of 10 cm is obtained on all plates by

Table 11.2. Slopes A_i and Intercepts B_i of Linear Relationship
$\log k_i' = f(X)^a$

Solute	A_i	B_i	r
Resorcin	−0.0150	0.259	0.990
Sulfadimidine	−0.0280	0.854	0.997
Sulfamethoxypyridazine	−0.0285	0.892	0.990
Barbital	−0.0402	1.063	0.981
Phenobarbital	−0.0319	1.341	0.999
Chloramphenicol	−0.0414	1.625	0.997
Salicylamide	−0.0255	0.871	0.984
Phenacetin	−0.0226	1.002	0.981
Vanillin	−0.0244	0.866	0.999
Benzaldehyde	−0.0303	1.575	0.999
Acetanilide	−0.0270	1.021	0.991
Nicotinamide	−0.0382	0.251	0.941
Benzoic acid	−0.0284	1.252	0.987
Salicylic	−0.0301	1.425	0.988
Acetylsalicylic acid	−0.0272	1.077	0.974
Coffein	−0.0299	0.552	0.979
Hydrochlorothiazide	−0.0456	0.887	0.912
Cortexolone	−0.0138	0.757	0.993
Dexamethasone	−0.0139	0.568	0.997
Desoxycortone	−0.0147	1.120	0.990
Sulfaguanidine	−0.0272	0.011	0.932
Isoniazide	−0.0382	0.060	0.947
Methylsalicylate	−0.0244	1.727	0.997
Hydrocortisone	−0.0129	0.436	0.991
Progesterone	−0.0192	1.831	0.995
Testosterone	−0.0143	1.085	0.998

^{a} X is percentage of acetonitrile in aqueous eluent; r is correlation coefficient of the relationship for individual solutes. Reprinted from Valko [57], p. 1419, by courtesy of Marcel Dekker, Inc.

cutting the layer at 12 cm and spotting the compounds analyzed on a line 2 cm from the lower edge of the plate. The mobile phase saturated with silicone is an aqueous buffer (sodium acetate–Veronal buffer, 1/7 mol/L at pH 7.0), alone or mixed with various amounts of acetone or methanol. The procedure is similar to that introduced by Boyce and Milborrow [12], who impregnated the plates with 5% paraffine oil solution in n-hexane.

It should be noted here that in the reversed-phase TLC system employed by the Italian group, specific adsorption interactions were reported with

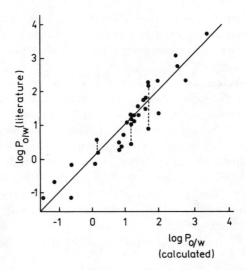

Figure 11.3. Plot of *n*-octanol–water partition coefficients calculated by Eq. (11-20) versus that in literature for solutes listed in Table 11.2. Broken lines join different literature log *P* data for same solutes. (Reprinted from K. Valkó *J. Liq. Chromatogr.*, 7, 1405, 1984. Courtesy of Marcel Dekker, Inc.)

arylaliphatic acids by Kuchař et al. [59]. These interactions explain the lack of correlation of R_M^w data derived in the reversed-phase TLC system described above and log *P* for a set of 5-nitroimidazoles, reported by Guerra et al. [60]. For the same solutes an apparent correlation was observed in the same work [60] between log *P* and log k_w' determined on a C-18 column.

Hulfshoff and Perrin [28] obtained the TLC stationary nonpolar phase covering the plates with a Kieselguhr G layer, which was directly impregnated with oleyl alcohol. These authors coated the glass plates with a slurry of Kieselguhr G in a mixture of oleyl alcohol, acetone, and dioxane. More recently Grünbauer et al. [61] applied the now classical method of impregnation of silica plates with liquid paraffin, introduced by Boyce and Milborrow [12]. Of the three organic co-solvents studied (i.e., acetone, methanol, and dimethylformanide), acetone appeared to be the least suitable for preparation of aqueous mobile phases. The solutes studied by these authors were *n*-alkyl phenyl ketones.

Since thin-layer plates precoated with nonpolar chemically bonded phases [62] became commercially available, they have often been employed for hydrophobicity determinations. Certainly, all comments concerning the limitations and shortcomings of reversed-phase HPLC with chemically bonded phases are also valid when the plates are precoated with octadecylsilica TLC. Eight types of such plates were compared [63]. As reported, the advantage of the RP-18-coated plates (Merck) of unlimited use without added NaCl is partly offset by the shorter migration time observed with KC_{18} end capped with C_2 plates (Whatman) in the often used 30–70% methanol in water concentration range.

Certainly, there are some advantages of TLC on chemically bonded phases (see p. 235). Unfortunately, however, as opposed to the HPLC stationary phases, the chromatographic plates cannot be subjected to the special treatments aimed at the improvement of retention data as the hydrophobicity descriptors. That improvement has supposedly been realized through suppressing the activity of the residual, free, unreacted silanol sites on the surface of the modern chemically bonded liquid chromatographic phases.

In preparing a reversed-phase material, the surface coverage by hydrophobic groups is usually maximized to yield the maximum hydrophobic character. Even with maximized surface coverage and 'capping' techniques, approximately 50% of the surface hydroxyls (or 3.5–5 $\mu mol/m^2$) remain unreacted [64, 65].

It is commonly assumed (see Chapter 4) that the free silanol groups on the surface of chemically bonded reversed-phase material are responsible for specific interactions between solutes and a stationary phase during the chromatographic separation process. Thus, it has been argued that due to these interactions, the correlations between $\log k'$ and $\log P$ hold only as long as the solutes considered belong to the same congeneric class. For different congeneric groups, the plots of $\log P$ versus $\log k'$ form separate parallel lines, similar to those illustrated in Fig. 11.2.

The term *congeneric* is used rather intuitively. Homologues of a given series are undoubtedly congeneric. As noted by Koopmans and Rekker [66], in many instances noncongenericity may be hidden in a series of compounds consisting of outwardly congeneric-looking structures. On the other hand, with regard to their behavior in the system considered, congenerics may be a class of apparently diverse solutes.

Interactions with the silanol sites of reversed-phase stationary phases are especially pronounced in the case of highly polar ionizable solutes. An illustration of the role of free silanol groups on the retention of aromatic acids on the ordinary C-18 phase (ODS-Hypersil) at varying compositions of the ethanol–water mobile phase, is given in Fig. 11.3 [67]. As evident from Fig. 11.4, the anionic solutes are excluded (k' is negative) from a reversed-phase packing material. Maximal exclusion occurs at a composition of about 50% water in the mobile phase. With more than 80% water benzoic acid and salicylic acid become retained, but sulfanilic acid is still excluded.

The U-shaped plots of $\log k'$ versus methanol content in the aqueous mobile phase are attributed to the strong polar interaction between the solutes and the residual silica hydroxyls [68, 69]. The increase in retention as the

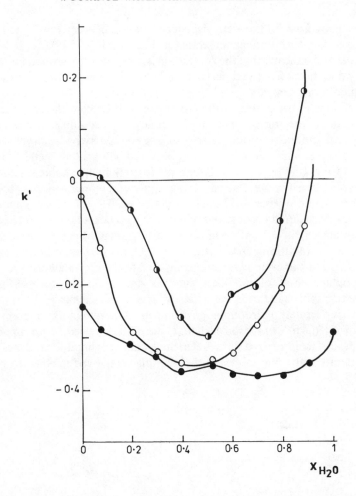

Figure 11.4. Dependence composition of capacity factors k' for acids eluted from ODS-Hypersil by water–ethanol mixtures (eluents contained 10^{-3} M $NaNO_3$): ($\bigcirc$) benzoic acid; ($\bigcirc$) salicylic acid; ($\bullet$) sulfanilic acid. (After J. H. Knox, R. Kaliszan, and G. J. Kennedy, *Faraday Symp. Chem. Soc.*, **15**, 113, 1980. With permission.)

concentration of the organic modifier increases beyond the value corresponding to the minimum is thought to be due to the column's taking on some normal-phase character.

That part of the parabola describing $\log k'$ versus the volume fraction of the organic modifier in the eluent, which corresponds to water-rich eluents, can be regarded as linear and permit the linear extrapolation of capacity factors to $\log k'_w$. Such extrapolation in the case of protonated basic compounds

(pK_a ranging from 7.48 to 9.50) yielded correlation between $\log P$ and $\log k'_w$ (corrected for ionization) as illustrated in Fig. 11.5 [70, 71].

Figure 11.5 indicates that the correlation is good despite the uncertainties in $\log P$ values taken from various sources, but deviations are large for the less lipophilic compounds.

When amines are added to the mobile phase it is often the case that the minimum on the $\log k'$ versus the organic modifier concentration plots disappears, and the retention appears to be governed by hydrophobic effects [68, 69, 72]. Similar effects have been observed when alkali metal salts are added to the mobile phase, although moderately hydrophobic quaternary amines are much more efficient at an equivalent concentration [65, 73]. Tramposch and Weber [65] suggest that the effect of added amines in reversed-phase liquid chromatography is not to block all chemical effects of residual silica hydroxyls. Although quaternary amines mask the interaction of the support with cationic species, the total dipolarity of the stationary phase is not reduced (it can even increase as compared to the protonated silica form). The primary effect of modifiers such as amines in the reversed-phase chromatographic systems is their effectiveness as ion exchange site counter-ions in a moderately hydrophobic environment. According to Tramposch and Weber [65], the pK_a of the ion exchange sites of the reversed-phase support is not typical of underivitized silica, but a small number of highly acidic sites exist that interact with ionic solutes at the pH range employed in reversed-phase HPLC.

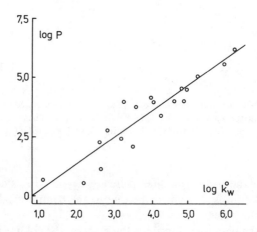

Figure 11.5. Relationship between published n-octanol–water partition coefficients, $\log P$, and reversed-phase HPLC capacity factors, $\log k'_w$, for a series of protonated basic compounds (neuroleptics drugs). extrapolated to pure water. (After N. El Tayer, H. van de Waterbeemd, and B. Testa, *J. Chromatogr.*, **320**, 305, 1985. With permission.)

The above-discussed complexity of the retention processes in the reversed-phase systems employing chemically bonded stationary phases makes interpretation of experimental chromatographic data difficult. Certainly, the convenience of hydrophobicity determinations by means of reversed-phase HPLC with chemically bonded stationary phases is unquestionable. The determined chromatographic data are generally well intercorrelated with log *P*—the less diverse the set of solutes analyzed, the better the intercorrelation. In practice, series of related compounds are studied in order to optimalize some of their properties (e.g., bioactivity). In such instances log *k'* well characterizes the relative lipophilicity of an individual compound of the series. Thus, there is not much need from a practical point of view to strive for obtaining a chromatographic system that provides a continuous hydrophobicity scale analogous to that offered by *n*-octanol–water partitioning systems. It is important from the practical point of view, however, that some diversity is allowed. The log *k'* versus log *P* correlations, whose validity is limited to homologous series, would be of little value. Now, the approaches will be briefly presented, aiming at the broadening of the range of structures for which reliable parameterization of partitioning properties may be gained from an individual reversed-phase HPLC system, employing chemically bonded stationary phases.

The first attempts to improve correlations between log *P* and the reversed-phase HPLC capacity factors, log *k'*, determined on chemically bonded alkylsilica columns included the reduction of free silanol sites in the column by additional silylation [74]. More recently, an exhaustive silylation of the C-18 columns has been recommended by Brent et al. [75]. These authors, however, have not attempted to duplicate log *P* by chromatographic data but rather have developed a new method of reliable lipophilicity measurement for a broad spectrum of compounds.

In spite of the special silylation procedures applied, unreacted residual silanol groups exist on the surface of reversed-phase matrices, which causes extra retention and asymmetric peak shapes, especially in the case of ionizable solutes. To diminish the interaction with acids, mobile phases of low pH are used, providing the presence of an acid in neutral form. The use of mobile phases is restricted to pH values below ~8 for the alkyl-bonded silicas commonly used in reversed-phase liquid chromatography, however. Thus, it is often difficult to suppress completely the ionization of basic solutes. In such a situation the extra interaction with stationary phases can be reduced by blocking the silanol sites with an amine added to the mobile phase [32, 72, 76, 77]. Whereas Yamana et al. [77] added ammonium salt to block the active silanol sites, at present, most authors use the hydrophobic alkylamines (e.g., *n*-decylamine or *N,N*-dimethyloctylamine).

Another method of reduction of the effect of free silanol groups on chromatographic measurements of basic solute partitioning properties consists in masking the basic site of the solute molecules by ion pair formation [78–81]. This may be realized by adding an anionic surfactant to the mobile phase. Riley et al. [81] proposed a system using sodium dodecylsulfate as the pairing ion and methanol as the organic modifier for the determination of indices of hydrophobicity in a group of 1,3-5-s-triazines chromatographed on an octadecylsilica column. Kraak et al. [55] also used sodium dodecylsulfate as the pairing agent but added 0.1 M sodium perchlorate to the methanol–phosphate buffer (pH 4.0; 0.005 M) 55:45 (v/v) used as the mobile phase.

The composition of a buffer and/or the type of substance used for adjustment of ionic strength may also affect retention of ionizable solutes. According to Wang and Lien [82], for acidic and neutral solutes, phosphate buffer appears to give partition coefficients closer to the values obtained from the n-octanol–water system than acetate and bicarbonate buffers. Adding a cation to the mobile phase decreases the exclusion of acids [67] and decreases the retention of basic solutes [83]. This effect also depends on the nature of the cation applied.

The variety of factors affecting retention of different classes of solutes is the cause of the lack of a universal liquid chromatographic method for determination of the n-octanol–water partition coefficients that would be suitable for every compound. The recommended methods comprise neutral solutes or individual classes of polar compounds but are uncertain when dealing with highly diverse structures. It should be noted here that the n-octanol–water partition system is not an ideal one. Chromatographically derived measures of hydrophobicity can be even more reliable than the classical log P. Some more interesting recent propositions regarding chromatographic methods of estimation of log P values will now be briefly discussed.

In 1984 Garst [84] recommended an "accurate, wide-range, automated" HPLC method of hydrophobicity measurements. The method suggested is based on the linear relationship between log P and log k'. The procedure applied by Garst [84] utilizes a temperature- and column-length-corrected retention volume. The method does not apply for zwitterionic compounds and is not accurate for compounds having multiple substituents or high log P values [85]. The acids and bases are treated separately according to the method, and the authors [84, 85] employ 0.004 M trifluoroacetic acid for acidic solutes and 0.035 M triethylamine for bases.

Haky and Young [86] used unmodified commercial octadecylsilica column and methanol–0.05 M ammonium phosphate buffer 55:45 to determine log k' values for a set of 68 compounds of varying functionality and structure type. Correlation between log P and log k' has been characterized by the coefficient $r = 0.966$. For the determination of the partition coefficients of unknowns,

Haky and Young [86] used one of two sets of standards to calibrate the system (i.e., phenolic calibration standards or nonphenolic calibration standards). The choice of calibration standard depends on the hydrogen-bonding character of the compounds being evaluated.

An up-to-date approach to lipophilicity characterization by means of reversed-phase HPLC has been reported by El Tayar et al. [54]. These authors determined capacity factors for a set of mono- and disubstituted benzenes chromatographed on an octadecylsilica column with the methanol buffer pH 7.4 eluent at different fractions of methanol (10–70% v/v).

El Tayar et al. [54] used *n*-decylamine (0.2% v/v) to eliminate interactions with free silanol groups. The pH of the mobile phase was adjusted by addition of HCl to the solution of *n*-decylamine in pure distilled water. 3-Morpholino-propane sulfonic acid, a zwitterionic buffer not forming ion pairs, was used throughout. In correlation analysis with literature log P data, the capacity factors, log k'_w, obtained by extrapolation to 100% water eluent were used along with the log k'_{50} values, corresponding to 50% concentration of water in the mobile phase. These data are collected in Table 11.3. Both type of capacity factors strongly correlated to log P values. Linear equation log $k'_{50} = f(\log P)$ for 49 solutes has been characterized by correlation coefficient $r = 0.967$, whereas the respective number for the relationship log $k'_w = f(\log P)$ is higher, namely $r = 0.982$. These differences are graphically illustrated in Fig. 11.6. El Tayar et al. [54] concluded from their results that water eluent capacity factors extrapolated to 100% are the accurate estimates of lipophilicity.

Table 11.3. Reversed-Phase HPLC Capacity Factors k' and Octanol–Water Partition Coefficients P of Mono- and Disubstituted Benzenes

Compound	log k'^b_{50}	log k'^c_w	log P
C_6H_6	0.810	1.987	2.13
$C_6H_5CH_3$	1.129	2.603	2.69
C_6H_5Cl	1.151	2.702	2.84
$C_6H_5NO_2$	0.659	1.814	1.86
$C_6H_5NH_2$	0.175	0.977	0.90
C_6H_5OH	0.243	1.274	1.46
$CH_3C_6H_42CH_3$	1.394	3.088	3.12
$CH_3C_6H_43CH_3$	1.446	3.186	3.20
$CH_3C_6H_44CH_3$	1.457	3.197	3.15
$CH_3C_6H_42Cl$	1.501	3.363	3.42
$CH_3C_6H_43Cl$	1.495	3.359	3.28

Table 11.3. (Continued)

Compound	$\log k_{50}^b$	$\log k_w^{\prime c}$	$\log P$
$CH_3C_6H_44Cl$	1.480	3.336	3.33
ClC_6H_42Cl	1.459	3.318	3.38
ClC_6H_43Cl	1.551	3.410	3.38
ClC_6H_44Cl	1.462	3.315	3.39
$NH_2C_6H_42Cl$	0.598	1.713	1.90
$NH_2C_6H_43Cl$	0.600	1.782	1.88
$NH_2C_6H_44Cl$	0.595	1.753	1.83
$CH_3C_6H_42NO_2$	0.894	2.319	2.30
$CH_3C_6H_43NO_2$	0.962	2.401	2.45
$CH_3C_6H_44NO_2$	0.930	2.345	2.42
$ClC_6H_42NO_2$	0.893	2.372	2.24
$ClC_6H_43NO_2$	1.035	2.466	2.46
$ClC_6H_44NO_2$	0.920	2.307	2.39
OHC_6H_42Cl	0.566	2.191	2.19
OHC_6H_43Cl	0.729	2.401	2.48
OHC_6H_44Cl	0.700	2.329	2.40
$OHC_6H_42CH_3$	0.583	1.748	1.96
$OHC_6H_43CH_3$	0.491	1.798	1.99
$OHC_6H_44CH_3$	0.495	1.801	1.94
$NO_2C_6H_42NO_2$	0.602	1.961	1.58
$NO_2C_6H_43NO_2$	0.613	1.663	1.49
$NO_2C_6H_44NO_2$	0.489	1.481	1.46
$OHC_6H_42NO_2$	0.717	2.372	1.79
$OHC_6H_43NO_2$	0.499	1.906	2.00
$OHC_6H_44NO_2$	0.529	1.996	1.91
$NH_2C_6H_42NO_2$	0.518	1.753	1.83
$NH_2C_6H_43NO_2$	0.348	1.423	1.37
$NH_2C_6H_44NO_2$	0.266	1.391	1.39
$NH_2C_6H_42NH_2$	-0.037	0.632	0.15
$NH_2C_6H_43NH_2$	-0.310	0.273	-0.36
$OHC_6H_42NH_2$	0.235	0.916	0.52
$OHC_6H_43NH_2$	-0.204	0.485	0.16
$OHC_6H_44NH_2$	-0.308	0.123	0.04
$NH_2C_6H_42CH_3$	0.402	1.397	1.32
$NH_2C_6H_43CH_3$	0.423	1.477	1.41
$NH_2C_6H_44CH_3$	0.441	1.506	1.39
OHC_6H_43OH	-0.128	0.868	0.77
OHC_6H_44OH	-0.247	0.395	0.59

[a] From Tayar et al. [54]. With permission.
[b] At eluent composition methanol–water 50:50.
[c] Extrapolated to pure water.

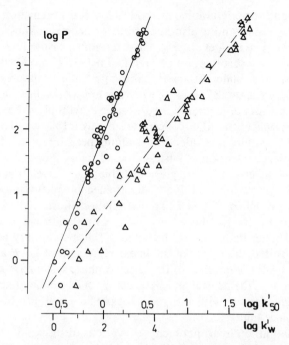

Figure 11.6. Relationships between *n*- octanol–water partition coefficients, log *P*, and reversed-phase HPLC capacity factors measured for mobile phase containing 50% methanol (Δ) and capacity factors extrapolated to 100% water mobile phase (○). (After N. El Tayar, H. van de Waterbeemd, and B. Testa, *Quant. Struct. Act. Relat.*, **4**, 69, 1985. With permission.)

The correlations between log *P* and reversed-phase liquid chromatographic data are often found for neutral solutes. There are still difficulties, however, involved in the evaluation of retention data of organic bases with pK_a in water greater than about 7. This is mainly so because the use of mobile phases is restricted to pH values below ~ 8 for the alkyl-bonded silicas commonly used in reversed-phase liquid chromatography. Having that in mind, de Biasi et al. [56] applied the styrene-divinylbenzene copolymeric stationary phase PRP-1 for the direct measurement of retention data of un-ionized organic bases. The PRP-1 phase was chosen because it is stable over a wide range of mobile phase pH, and it contains no silanol groups. De Biasi et al. [56] analyzed 12 drugs of basic character using methanol–0.5 *M* aqueous sodium hydroxide 90:10 as the eluent. The method is feasible, but the resulting correlation log *P* versus log *k'* is not impressive with correlation coefficient $r = 0.906$. In the same paper the authors reported the lack of correlation ($r = 0.216$) between log *P* and capacity factors determined on porous graphitic carbon, a new stationary phase material recently introduced by Knox and co-workers [87].

The ion pair chromatographic method that permits prediction of the log P of basic compounds (and is also applicable for neutral solutes) has recently been described by Kraak et al. [55]. These authors measured the retention of neutral and basic test solutes on phenyl- and octadecyl-modified silica using methanol–aqueous buffer (0.005 M phosphate buffer) mixtures containing sodium dodecylsulfate (SDS) as the pairing ion as the mobile phase. The volume percentages of methanol used were 55% with phenylsilica and 65% with octadecylsilica; the SDS concentration was 0.1% w/v. To the mobile phase, 0.1 M sodium was added as sodium perchlorate to diminish the unfavorably large retention of basic solutes. The log P data and the solutes studied by Kraak et al. [55] are given in Table 11.4. Correlation between the literature log P data and the log k' determined on phenylsilica stationary phase is illustrated in Fig. 11.7. The correlation illustrated in Fig. 11.7 is characterized by the coefficient $r = 0.987$. Correlation on phenylsilica is significantly better than that on the octadecylsilica stationary phase. Moreover, it seems that the slopes of the linear regression of log P and log k' of neutral and basic compounds in the selected phase system are about equal. Thus, the log P of the neutral and basic compounds can be predicted from a common regression line with an error in the prediction of ± 0.10 log units at a 95% confidence limit.

Determination of hydrophobicity of various drugs and chemicals by chromatographic methods has been the subject of numerous publications. A number of publications concern environmental pollutants. With that research area has been connected one of the first works dealing with reversed-phase HPLC as a method of evaluation of log P published by Carlson et al. [88] in 1975. Other log k', or R_M, versus log P relationships for various series of solutes of interest for environmental sciences, the interested reader is referred to the literature [89–98].

The problem of reversed-phase HPLC determination of the lipophilicity of individual classes of solutes as expressed by their log P values is discussed by Jinno and co-workers [51, 52, 99] and Hanai and co-workers [99–102]. Other more recent publications dealing with the relationships between the chromatographic and the shake-flask lipophilicity data are available [103–114].

Nearly all workers in the field of partition chromatography share the conviction that the Collander relationship comprises both the shake-flask and liquid chromatographic systems. In such a situation these workers seek high log P versus log k' (R_M) correlation by all possible means. It seems worthwhile then to turn the attention of the uncritical enthusiasts to the publication by Huang et al. [113]. These authors report the lack of correlation between log k' data determined for a set of 12 compounds on three alkylsilica stationary phases using three mobile phases and the log P data determined for these solutes.

Table 11.4. Solutes Used in Correlation Studies of Kraak et al[a] and Literature Values of
n-Octanol–Water Partition Coefficients, log P

	Neutral Solutes			Basic Solutes	
Number	Compound	log P	Number	Compound	log P
1	Linurol	3.00	21	Clonidine	1.49
2	Acetophenone	1.63	22	4-Chloroaniline	1.64
3	Benzene	2.02	23	Baygon	1.52
4	Toluene	2.65	24	Atropine	1.80
5	Diuran	2.85	25	Droperidol	3.50
6	Dichlorobenzil	2.65	26	Diphenylamine	3.37
7	Ethylbenzene	3.13	27	Diphenhydramine	3.30
8	p-Xylene	3.18	28	Propranolol	3.56
9	Naphthalene	3.33	29	Halopemide	3.92
10	1,3-Dichlorobenzene	3.52	30	Haloperidol	4.31
11	Chloroxuron	4.00	31	Bromoperidol	4.72
12	Biphenyl	3.91	32	Pimozide	6.23
13	Diphenyl ether	4.20	33	Chlopimozide	7.05
14	Butylbenzene	4.28	34	Chlorpromizine	5.30
15	Phenanthrene	4.52	35	Indole	2.14
16	4,4'-Dichlorobiphenyl	5.28	36	Pyridinebenzimidazolon	0.85
17	Theobromine	−0.80	37	Piperidinebenzimidazolon	1.14
18	Aniline	0.90	38	Chloropiperidinebenzimidazolon	1.59
19	Morphine	0.76	39	Phenylhydroxypiperidine	1.98

[a] Reference 55. With permission.

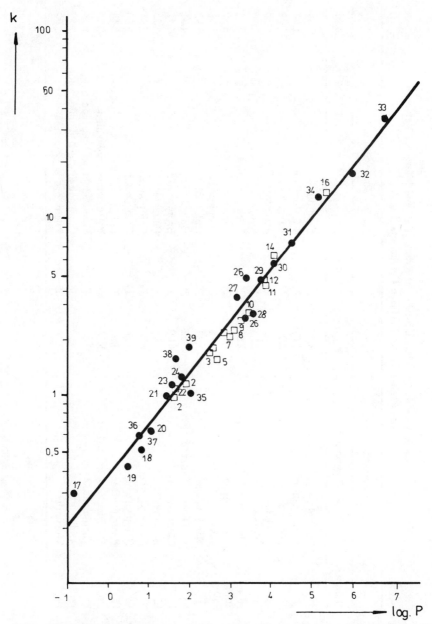

Figure 11.7. Relationship between n-octanol–water partition coefficients of neutral (□) and basic (●) test solutes and their capacity ratios measured in ion pair system with phenylsilica as stationary phase; mobile phase of pH 4.0 and of the composition as described in text. (After J. C. Kraak, H. H. van Rooij, and J. L. G. Thus, *J. Chromatogr.*, **352**, 455, 1986. With permission.)

Occasionally the normal-phase partition chromatography is applied for lipophilicity evaluation. There are reports published [115, 116] on paper chromatography or TLC on cellulose layers, where the support has been impregnated with a polar solvent. Such an impregnating solvent may be formamide in ethanol, whereas cyclohexane is the mobile phase. In that context, it is interesting that Cserháti [117] reported that the lipophilic character of 3,5-dinitrobenzoic acid esters measured on unimpregnated cellulose layers correlated well with similar values determined on silica and alumina impregnated with paraffin oil.

In reversed-phase HPLC Baker et al. [118] report the correlation between n-octanol–water partition coefficients and the Baker liquid chromatographic retention index based on the relative retention of a homologous series of 2-ketoalkanes (see Chapter 4 for details). The relationship holds true for large series of barbiturates, propranolol, and anthranilic acid analogs, narcotic analgesics, and glucuronide metabolites. Based on that relationship, one can predict the liquid chromatographic retention index of a given substance from its log P value.

The findings by El Tayar et al. [119] concerning isotopic effects on the lipophilicity measured by reversed-phase HPLC method are interesting. For a series of deuterated and protiated aromatic isotopomers (benzene, toluene, p-xylene, and aniline) the capacity factors determined indicate that deuterated compounds are less lipophilic than their protium isomers by about -0.006 per deuterium atom on the log P scale. According to El Tayar et al. [119], the classical shake-flask method also reveals isotopic effects. But the HPLC technique confirmed these observations with a greater precision. These results confirm the earlier observations by Tanaka and Thornton [25] that deuterated compounds have 0.2–0.5% smaller capacity factors (log k') than their protium isomers.

Although the majority of workers look for a chromatographic system yielding retention data precisely linearly related to the standard n-octanol–water partition coefficients, the other approaches to the chromatographic evaluation of lipophilicity should also be noted. For example, Clark and co-workers [120, 121] suggest the use of the quadratic and cubic relationships between log k' and log P for predictive purposes. Interestingly, Rowland and co-workers [122, 123] proposed a reversed-phase HPLC-based index of hydrophobicity, R_Q, defined as

$$R_Q = \log(1 - R_0/R_T) \tag{11.21}$$

where R_0 and R_T are retention times of unretained solute and the compound studied, respectively. Contrary to log k', R_Q gives a linear relation when plotted against acetonitrile concentration in the aqueous mobile phase for homologous barbiturates. By definition, R_Q cannot exceed zero [123]. Yet, several of the R_Q^w values extrapolated to pure water are numerically greater than zero. These values form excellent correlations with the logarithm of the n-octanol–water partition coefficients taken from the literature for the series of barbiturates studied.

11.1.3. Applications of Gas–Liquid Chromatography for Determination of Partitioning Characteristics

At present, when the more convenient HPLC and TLC techniques are available, there are not many convincing reasons to use gas chromatography directly for the evaluation of the lipophilicity of solutes. Certainly, GC can be employed for determination of the concentration of a volatile solute in one of the phases of the n-octanol–water partition system [124]. However, its use for immediate lipophilicity determinations may only be justified in the case of agents highly sensitive to hydrolysis in aqueous TLC or HPLC mobile phases or when detection problems arise in TLC or HPLC.

In 1979 Boček [125] developed a GC method of hydrophobicity assessment using water and oleyl alcohol as the liquid stationary phases. The method is based on the assumption that the partition coefficient of a compound in the oleyl–water system is equal to the ratio of partition coefficients in the oleyl alcohol (saturated with water)-nitrogen and water-nitrogen which can be determined by GC. Unfortunately, the method is complicated as care must be taken to eliminate the influence of adsorption in order to determine true partition coefficients.

Rehn et al. [126] reported correlation between log P and the logarithm of the experimental GC retention times, measured on a 3% OV-17 phase, for a set of O-acylated derivatives of p-hydroxybenzoic acid.

A systematic approach to the problem of application of GC data for evaluation of solute lipophilicity has been developed by Valkó and co-workers [127–129]. The approach is based on the hypothesis that the ratio of two gas–liquid partition coefficients is equal to the liquid–liquid partition coefficient. As the retention index I of a compound measured on a given stationary phase changes with the logarithm of its gas–liquid partition coefficient, it follows that the weighted difference between the retention indices of a compound measured on two stationary phases $(aI_1 - bI_2)$ is directly related to the log P values of that compound

$$\log P = aI_1 - bI_2 + c \qquad (11.22)$$

where a, b, and c are regression coefficients. In order to make Eq. (11.22) suitable for the practical determination of log P, one should use the gas chromatographic stationary phase pair that models best the n-octanol–water partition system. Valkó et al. [129] analyzed retention indices of seven model compounds obtained on 61 stationary phases. These indices and the corresponding log P values were introduced into Eq. (11.22). The correlation coefficient of that equation was regarded as a measure of the suitability of a given stationary phase pair for modeling the n-octanol–water system. On the basis of the total 1830 equations covering all combinations of the 61 phases, the best stationary phase pair chosen was 15% OV-1 (Merck, Darmstadt) and 15% EPON 1001 (Supelco) loaded on Gas-Chrom Q support. However, equally good as EPON 1001 for replacing the water phase in the n-octanol–water system were diethylene glycol succinate (DEGS) and neopentyl glycol succinate (NPGS). Instead of OV-1, several less polar paraffin-type or methylsilicon polymer stationary phases can be used. If the group of test solutes was enlarged to 15 compounds given in Table 11.5, the following regression equation was derived:

$$\log P = 0.0111 I_1 - 0.0065 I_2 + 0.764$$

$$n = 15; \quad R = 0.868; \quad s = 0.4666; \quad F = 18.4 \qquad (11.23)$$

Table 11.5. n-Octanol–Water Distribution Coefficients P^a

		log P	
Number	Compound	Observed	Calculated
1	Toluene	2.69	2.35
2	n-Pentanol	1.48	1.26
3	Cyclohexane[b]	0.81	1.97
4	Isobutyl methyl ketone[b]	1.38	1.81
5	4-Methylpyridine	1.22	0.75
6	2-Ethylpyridine	1.69	1.59
7	Isopentanol	1.29	1.23
8	3-Methyl-1-pentin-3-ol	1.36	1.04
9	Benzene	2.15	2.08
10	n-Butanol	0.88	0.89
11	2-Pentanone[b]	0.91	1.48
12	Nitropropane	0.87	0.96
13	Pyridine	0.65	0.52
14	2-Methyl-2-pentanol	1.91	1.81
15	2-Octyne	3.97	3.51

[a] Measured and calculated from Eq. (11.23) for test solutes analyzed by Valkó et al. [129]. With permission.
[b] Compounds not used to derive Eq. (11.23).

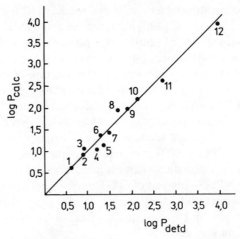

Figure 11.8. Plot of measured values of log P versus the same values calculated from Eq. (11.23). Compounds: (1) pyridine; (2)n-butanol; (3) nitropropane; (4) 4-methylpyridine; (5) 3-methyl-pent-1-in-3-ol; (6) isopentanol; (7) n-pentanol; (8) 2-ethylpyridine; (9) 2-methyl-2-pentanol; (10) benzene; (11) toluene; (12) 2-octyne. (After K. Valkó, O. Papp, and F. Darvas, *J. Chromatogr.*, **301**, 355, 1984. With permission.)

where I_1 and I_2 are retention indices on the EPON 1001 and OV-1 phases, respectively. The correlation in Eq. (11.23) is not high. Valkó et al. [129] excluded 3 compounds of the set of 15 from the correlation and again derived a regression equation that they recommend for calculation of log P values. The correlation of that new regression equation for the reduced set of solutes is

$$\log P = 0.0127 I_1 - 0.0068 I_2 + 0.058$$
$$n = 12; \quad R = 0.988; \quad s = 0.159; \quad F = 181.0 \qquad (11.23)$$

Equation (11.23) was next used to calculate log P for several diverse solutes. Correlation of the measured and calculated log P for these solutes is illustrated in Fig. 11.8.

The results obtained by Valkó et al. [129] give evidence of the applicability of GC for lipophilicity calculation. The method has no evident advantage over liquid chromatographic methods used for that purpose, however.

11.2. SUBSTITUENT AND FRAGMENTAL HYDROPHOBIC CONSTANTS AS RETENTION DESCRIPTORS

The substituent hydrophobic parameter most often used in correlation studies in pharmacochemistry and in chromatography is the π lipophilic constant introduced by Hansch and co-workers in early 1960 [130]. The π value for

a given substitutent is defined as the logarithm of the ratio of the
n-octanol–water partition coefficients of a substituted to an unsubstituted
derivative,

$$\pi_X = \log \frac{P_{R-X}}{P_{R-H}} \tag{11.24}$$

where X denotes the substituent; R–H is the unsubstituted and R–X is the
substituted compound, for which the n-octanol–water partition coefficients P
is experimentally determined. The standard π values are derived from benzene
compounds. The important feature of π constants is their additivity.

Already in 1965 Hansch and co-workers [11] reported significant correlation between π and the term ΔR_M from TLC. The paramater ΔR_M accounts
for the effect of a substituent upon the R_M value of a parent, unsubstituted
compound.

It should be noted here that in 1967 correlation was reported between the
Zahradnik β constants and R_M values [131]. These β constants are regarded as
analogous to the π parameters.

Chromatographic retention data are related to π values of individual
substituents derived from the n-octanol–water partition data of a model
aromatic system. The quality of such relationships depends not only on the
chromatographic data but also on the reliability of an individual universal π
constant as the measure of the contribution, which a given substituent affords
to log P. Thus, in addition to all the limitations discussed in the preceding
section, one must take into consideration the limitations resulting from the
nature of the π constant.

As has been the case for liquid chromatographic retention versus log P
relationships, equally good correlations are reported between reversed-phase
HPLC and TLC retention data and π values for separate classes of related
solutes. In TLC such correlations are observed and reported long since, and
the representative early works may be those by Biagi et al. [132–134] and by
others [135, 136]. In HPLC similar results were reported since the mid-1970s
[88, 137]. Carlson et al. [88] got good correlations between the substituent
contributions to the logarithm of the HPLC capacity factor [i.e., log k'
(substituted) minus log k' (unsubstituted) and π values for the respective
substituents]. The correlations were good only if the two classes of solutes
studied (i.e., phenols and anilines) were analyzed separately.

In a study of the effect of ionization on the chromatographic behavior of
some β-aryl-n-butyric acids, Kuchař et al. [138] obtained equations relating
R_M values to π values with R_M data derived from a chromatographic system (A)
where the acids would be ionized and another system (B) where they are not. In
the case of the system A, the introduction of the Hammett electronic constant

σ improves the correlation over the straight π versus R_M, whereas that term is not needed for system B. The continuing their studies on the parameterization of the lipophilicity of arylaliphatic acids, Kuchař and co-workers [59] found significant departures from linearity in the relationship between π and R_M when the solutes studied represented the broader region of lipophilicity. To account for the nonlinearity observed, these authors [59] replaced the linear relationship $\pi = f(R_M)$ by a quadratic dependence $\pi = f(R_M, R_M^2)$.

In another paper of the series Kuchař et al. [139] observed that lipophilicities of benzyloxyarylaliphatic acids were better described by the tabulated parameters π than by the experimental values obtained from reversed-phase TLC on silica gel impregnated with silicone oil with 50% v/v acetone–water as the mobile phase. Quite recently Kuchař and co-workers [50] analyzed reversed-phase TLC and HPLC data of their group of arylaliphatic acids. Ready-made TLC plates were used, either silanized or impregnated with silicone oil; buffer–acetone mixtures were the mobile phases. In reversed-phase HPLC, an octadecylsilica column was used with methanol–buffer eluent. Kuchař et al. [50] found that extrapolation of retention parameters to pure water was not advantageous and negatively influenced the calculation of π parameters for the dialkoxy and phenylalkoxy derivatives from the corresponding retention indices.

In studies of structure–retention relationships in a series of O-alkyl-O-aryl-phenylphosphanothioates Steurbaut et al. [140] analyzed TLC R_M values determined in both a reversed-phase system and on polyamide layers as well as GC relative retention times as compared with the unsubstituted compound. Some correlations with π constants were found only for the reversed-phase TLC system.

Bachrata et al. [141] reported correlation between π values and the substituent contributions, ΔR_M, to the TLC retention measured in a normal-phase partition system. Chromatography was carried out on the ready-made cellulose or silica layers impregnated with formamide; n-heptane was the eluent. The correlations reported hold true among the individual homologous series of solutes studied. However, the ortho-, meta-, and para-substituted homologues give different relationships of the form $\pi = f(\Delta R_M)$. Such results are due to the assumption of the same π values for given substituent independent of the position of substitution. In fact, the π values assumed by Bachrata et al. [141] reflect nothing more but the number of carbon atoms in individual homologues.

Rittich et al. [142] found some relationship between $\log k'$ obtained in reversed-phase HPLC on octadecylsilica stationary phase with dioxan–water eluent and the π constant for a group of variously substituted phenols. However, the relationship reported holds true only when no more than 18 compounds of a total of 26 studied are arbitrarily selected. On the other hand, in the case of closely congeneric solutes, like alkylbenzenes, the expected strong correlation between $\log k'$ values and the π constants of the alkyl substituents has been confirmed [143, 144], with capacity factors determined either on C-8 or C-18 alkylsilica columns at different concentrations of acetonitrile in the aqueous mobile phase.

When the solutes chromatographed differ structurally, the correlations between chromatographic data and π usually deteriorate. For example, the relationships between reversed-phase HPLC capacity factor and π for different phenyl-substituted phenylcarbamoylmethyliminodiacetic acids divide the solutes into three groups whose membership is determined by the degree of ortho substitution [145]. Similarly, separate correlation equations had to be derived, describing the relationship between the π parameter and the contribution of a given substituent to the reversed-phase HPLC capacity factor, for C^2-, C^6-, and C^8-substituted cyclic nucleotides (analogs of adenosine 3',5'-monophosphate) [146]. The lack of linearity of the plot of $\log k'$ versus π has been reported for α-melanotropin analogs [147]. The compounds were chromatographed on octadecylsilica with four mobile phases. However, the series of structures analyzed is too small (three substituents) for general conclusions.

Side-chain hydrophobic constants, π_R, determined for 20 amino acids were related to TLC and HPLC retention data [148, 149]. Correlations were good when the side-chain contributions to the chromatographically determined hydrophobicity were considered. The correlations of overall amino acid hydrophobicity with the retention coefficients were much lower but still significant.

Cserháti [117] found linear relationship between R_M of 3,5-dinitrobenzoic acid esters chromatographed on unimpregnated TLC layers and concentration X of methanol in aqueous eluent. Both the slope and, intercept of the linear relationship obtained by Cserháti [117] correlated well with the respective π values of the solutes studied. One would expect intercorrelation between the slope and intercept of the $R_M = f(X)$ relationship in reversed-phase liquid chromatography. In such a situation both coefficients could be related to π. The results obtained by Cserháti suggest that unimpregnated cellulose may be used for lipophilicity characterization of the group of 3,5-dinitrobenzoic acid esters.

By summation of π values of the substituents present in a given multiple-substituted parent structure, the lipophilicities of a series of derivatives may be compared. Guerra et al. [47] obtained moderate correlation of the reversed-phase HPLC capacity factors and $\Sigma\pi$ for a series of demorphin-related oligopeptides. On the other hand, an excellent correlation of the $\log k'$ versus $\Sigma\pi$ relationship was reported by Macek et al. [150] for a series of phenylaliphatic acids. These authors, however, corrected their reversed-phase

HPLC capacity factors for dissociation and calculated log k' values for neutral molecules.

Both from the chromatographic and pharmacological point of view, it is highly desirable to be able to calculate partition coefficients a priori, directly from structural formula of a given compound.

A number of shortcomings in the π substituent constants system prompted Rekker [151] and later Hansch and Leo [130] to develop fragmental methods in which the hydrophobic fragmental constant of hydroden, f_H, is different from zero. Whereas the lipophilic substituent constant π is based on differentiation in the solute group, the Rekker f constant was designed to reflect differentiation in the solute structure. Arguing that adding a fragmental constant of hydrogen, f_H, to the π constant for any substituent gives the fragment constant, f_X, for that structure, Hansch and Leo [130] proposed their own fragment method of hydrophobicity calculation. The general equation to calculate log P in the two fragmental systems is

$$\log P = \sum (\text{fragmental values}) + \text{correction terms} \qquad (11.25)$$

Calculation of log P by fragmental methods often requires a proper choice among the individual fragmental values and correction terms. In fact, the correctness of calculations should be tested experimentally for representatives of the series of compounds studied. The lack of correlations between reversed-phase liquid chromatographic retention data and log P calculated by fragmental methods may result from errors in calculations of the latter data. In any case, Hammers et al. [53] interpret the lack of correlation of log k' and log P (calculated) as resulting from the lack of reliability of the Rekker f method [151] for prediction of log P values of highly substituted aromatic compounds. Not impressive have also been the correlation of TLC R_M values and log P values calculated by the Hansch and Leo f method [130] for a series of crotonolactones [116]. Dadáková et al. [116] carried out the chromatographic determinations in a straight-phase system consisting of cellulose sheets impregnated with formamide; cyclohexane was the mobile phase. Correlations between relative R_M values and the calculated hydrophobicity were $r = 0.877$ and $r = 0.919$ at low and high coverage of the stationary phase with formamide, respectively. Of roughly similar value has been the relationship between hydrophobicity calculated by the f method of Hansch and Leo [130] and R_M^0 values determined for a series of pyrazine CH and NH acids [152]. In that case TLC was carried out on ready-made silanized silica plates with ethanol–buffer pH 7.4 mixtures. The R_M values were extrapolated to 100% buffer eluent yielding the R_M^0 parameter.

According to El Tayar et al. [54], a correction term must be added to the relationship between the observed partition data and the calculated ones from fragmental or substituent contributions, which accounts for substituent interactions.

Yamagami et al. [153] used the Hammett sigma of the substituent for corrections of the relationship they observed between reversed-phase HPLC capacity factors and the partition parameter calculated from π values of meta- and para-substituted phenylacetanilides.

In recent years Rekker has actively propagated the application in chromatography of his fragmental method of lipophilicity calculation. At first, Koopmans and Rekker [66] reported correlation of $\log k'$ determined on an octadecylsilica HPLC column with methanol–water 70:30 as the eluent, with the calculated n-octanol–water partition coefficient Σf for a series of alkylbenzenes. The correlation was better than that obtained with the experimental $\log P$ data. However, as the compounds studied were closely congeneric, equally good correlations was found [66] between $\log k'$ and molecular connectivity indices. Based on their own and literature TLC data for other sets of congeners, Rekker and co-workers [154–156] proposed R_M fragmentation similar to that applied for the partition coefficient. These authors concluded that into the equation $\Sigma f = f(R_M)$ a correction factor must be introduced to improve the relationship to a statistical level comparable to that attained for alkylbenzenes. The nature of that correction term is not clear, but it is assumed to reflect resonance, steric, and systemic peculiarities of the structure under consideration.

References

1. E. Overton, Über die osmotischen Eigenschaften der Zelle in ihrer Bedeutung für die Toxikologie und Pharmakologie, Z. Physikal. Chem., **22**, 189, 1897.

2. H. Meyer, Zur Theorie der Alkoholnarkose. Erste Mitteilung. Welche Eigenschaft der Anästhetica bedingt ihre narkotische Wirkung? Arch. Exp. Pathol. Pharmakol., **42**, 109, 1899.

3. F. Baum, Zur Theorie der Alkoholnarkose. Zweite Mitteilung. Ein physikalisch-chemischer Beitrag Zur Theorie der Narkotica, Arch. Exp. Pathol. Pharmakol., **42**, 119, 1899.

4. R. Collander, The partition of organic compounds between higher alcohols and water, Acta Chem. Scand., **5**, 774, 1951.

5. C. Hansch and T. Fujita, ρ-σ-π analysis. A method for the correlation of biological activity and chemical structure, J. Am. Chem. Soc., **86**, 1616, 1964.

6. A. J. Leo, C. Hansch, and D. Elkins, Partition coefficients and their uses, Chem. Rev., **7**, 525, 1971.

7. C. Hansch and A. Leo, Log P and parameter database: A tool for the quantitative prediction of bioactivity, Pomona College Medicinal Chemisty Project, 1983, issues 22 and 23.

8. A. J. P. Martin, Some theoretical aspects of partition chromatography, *Biochem. Soc. Symp.*, **3**, 4, 1950.

9. A. E. Bird, and A. C. Marshall, Reversed-phase thin-layer chromatography and partition coefficients of penicillins, *J. Chromatogr.*, **63**, 313, 1971.

10. J. K. Baker, Estimation of high pressure liquid chromatographic retention indices, *Anal. Chem.*, **51**, 1693, 1979.

11. J. Iwasa, T. Fujita, and C. Hansch, Substituent constants for aliphatic functions obtained from partition coefficients, *J. Med. Chem.*, **8**, 150, 1965.

12. C. B. C. Boyce and B. V. Milborrow, A simple assessment of partition data for correlating structure and biological activity using thin-layer chromatography, *Nature*, **208**, 537, 1965.

13. G. L. Biagi, A. M. Barbaro, M. F. Gamba, and M. C. Guerra, Partition data of penicillins determined by means of reversed-phase thin-layer chromatography, *J. Chromatogr.*, **41**, 371, 1969.

14. J. F. K. Huber, C. A. M. Meijers, and J. A. R. J. Hulsman, New method for prediction of partition coefficients in liquid–liquid systems and its experimental verification for steroids by static and chromatographic measurements, *Anal. Chem.*, **44**, 111, 1972.

15. J. N. Seiber, Reversed-phase liquid chromatography of some pesticides and related compounds. Solubility–retention relationships, *J. Chromatogr*, **94**, 151, 1974.

16. W. J. Haggerty, Jr, and E. A. Murrill, Applications of HPLC in contract research, *Res. Dev.*, **25**, 30, 1974.

17. P. Menheere, C. Devillez, C. Eon, and G. Guiochon, Method for prediction of partition coefficients in liquid–liquid chromatography, *Anal. Chem.*, **46**, 1375, 1974.

18. E. Tomlinson, Chromatographic hydrophobic parameters in correlation analysis of structure–activity relationships, *J. Chromatogr.*, **113**, 1, 1975.

19. R. Kaliszan, Chromatography in studies of quantitative structure–activity relationships, *J. Chromatogr.*, **220**, 71, 1981.

20. R. Kaliszan, High performance liquid chromatography as a source of structural information for medicinal chemistry, *J. Chromatogr. Sci.*, **22**, 362, 1984.

21. R. Kaliszan, Quantitative relationships between molecular structure and chromatographic retention. Implications in physical, analytical, and medicinal chemistry, *Crit. Rev. Anal. Chem.*, **16**, 323, 1986.

22. M. Harnisch, H. J. Möckel, and G. Schulze, Relationship between log P_{ow} shake-flask values and capacity factors derived from reversed-phase high-performance liquid chromatography for *n*-alkylbenzenes and some OECD reference substances, *J. Chromatogr.*, **282**, 315, 1983.

23. K. K. Butani, Chromatography in structure–activity correlationship, *Pharmacos*, **20**, 12, 1975.

24. B. Rittich and H. Dubský, Využiti vysoce účinne kapalinové chromatografie ke stanoveni hydrofobnich parametrú, *Cesk. Farm.*, **33**, 125, 1984

25. N. Tanaka and E. R. Thornton, Isotopic effect in hydrophobic binding measured by HPLC, *J. Am. Chem. Soc.*, **98**, 1617, 1976.

26. E. Tomlinson, H. Poppe, and J. C. Kraak, Thermodynamics of functional groups in reversed-phase high-performance liquid–solid chromatography, *Int. J. Pharm.*, **7**, 225, 1981.

27. I. E. Bush, *The Chromatography of Steroids*, Pergamon, London, 1961.

28. A. Hulshoff and J. Perrin, A comparison of the determination of partition coefficients of 1,4-benzodiazepines by high-performance liquid chromatography and thin-layer chromatography, *J. Chromatogr.*, **129**, 263, 1976.

29. R. A. Scherrer and S. M. Howard, Use of distribution coefficients in quantitative structure–activity relationships, *J. Med. Chem.*, **20**, 53, 1977.

30. T. Fujita, T. Nishioka, and M. Nakajima, Hydrogen-bonding parameter and its significance in quantitative structure–activity studies, *J. Med. Chem.*, **20**, 1071, 1977.

31. S. H. Unger, J. R. Cook, and J. S. Hollenberg, Simple procedure for determining octanol–aqueous partition, distribution, and ionization coefficients by reversed-phase high-pressure liquid chromatography, *J. Pharm. Sci.*, **67**, 1364, 1978.

32. S. H. Unger and T. F. Feuerman, Octanol–aqueous partition, distribution and ionization coefficients of lipophilic acids and their anions by reversed-phase high-performance liquid chromatography, *J. Chromatogr.*, **176**, 426, 1979.

33. Cs. Horváth, W. Melander, and I. Molnár, Liquid chromatography of ionogenic substances with nonpolar stationary phases, *Anal. Chem.*, **49**, 142, 1977.

34. K.-G. Wahlund and I. Beijersten, Stationary phase effects in reversed-phase liquid chromatography of acids and ion pairs, *J. Chromatogr.*, **149**, 313, 1978.

35. M. S. Mirrlees, S. J. Moulton, C. T. Murphy, and P. J. Taylor, Direct measurement of octanol–water partition coefficients by high-pressure liquid chromatography, *J. Med. Chem.*, **19**, 615, 1976.

36. S. J. Lewis, M. S. Mirrlees, and P. J. Taylor, Rationalisations among heterocyclic partition coefficients. Part 1: The π-value of phenyl, *Quant. Struct.–Act. Relat.*, **2**, 1, 1983.

37. D. Henry, J. H. Block, J. L. Anederson, and G. R. Carlson, Use of high-pressure liquid chromatography for quantitative structure–activity relationship studies of sulfonamides and barbiturates, *J. Med. Chem.*, **19**, 619, 1976.

38. K. Miyake and H. Terada, Preparation of a column with octanol-like properties for high-performance liquid chromatography. Direct measurements of partition coefficients in an octanol–water system, *J. Chromatogr.*, **157**, 386, 1978.

39. S. H. Unger and G. H. Chiang, Octanol-physiological buffer distribution coefficients of lipophilic amines by reversed–phase high-performance liquid chromatography and their correlation with biological activity, *J. Med. Chem.*, **24**, 262, 1981.

40. P. J. Schoenmakers, H. A. H. Billiet, and L. de Galan, Description of solute retention over the full range of mobile phase compositions in reversed-phase liquid chromatography, *J. Chromatogr.*, **282**, 107, 1983.

41. W. R. Melander, B.-K. Chen, and Cs. Horváth, Mobile phase effects in reversed-phase chromatography. VII. Dependence of retention on mobile phase composition and column temperature, *J. Chromatogr.*, **318**, 1, 1985.

42. M. C. Pietrogrande, C. Bighi, P. A. Borea, A. M. Barbaro, M. C. Guerra, and G. L. Biagi, Relationship between log k' values of benzodiazepines and composition of the mobile phase, *J. Liq. Chromatogr.*, **8**, 1711, 1985.

43. T. L. Hafkenscheid and E. Tomlinson, Correlations between alkane/water and octanol/water distribution coefficients and isocratic reversed-phase liquid chromatographic capacity factors of acids, bases and neutrals, *Int. J. Pharm.*, **16**, 225, 1983.

44. T. Braumann, G. Weber, and L. H. Grimme, Quantitative structure–activity relationships for herbicides. Reversed-phase liquid chromatographic retention parameter, log k_w, versus liquid–liquid partition coefficient as a model of the hydrophobicity of phenylureas, s-triazines and phenoxycarbonic acid derivatives, *J. Chromatogr.*, **261**, 329, 1983.

45. C. M. Riley, E. Tomlinson, and T. M. Jefferies, Functional group behaviour in ion-pair reversed-phase high-performance liquid chromatography using surface-active pairing ions, *J. Chromatogr.*, **185**, 197, 1979.

46. A. M. Barbaro, M. C. Pietrogrande, M. C. Guerra, G. Cantelli-Forti, P. A. Borea, and G. L. Biagi, Relationship between the chromatographic behaviour of demorphin-related oligopeptides and the composition of the mobile phase in reversed-phase thin-layer chromatography: Comparison of extrapolated R_M values, *J. Chromatogr.*, **287**, 259, 1984.

47. M. C. Guerra, A. M. Barbaro, G. Cantelli Forti, M. C. Pietrogrande, P. A. Borea, and G. L. Biagi, Determination of lipophilic character of a series of demorphin-related oligopeptides by means of reversed-phase HPLC, *J. Liq. Chromatogr.*, **7**, 1495, 1984.

48. G. L. Biagi, A. M. Barbaro, M. C. Guerra, G. Cantelli Forti, P. A. Borea, M. C. Pietrogrande, S. Salvadori, and R. Tomatis, Chromatographic parameters of lipophilicity of a series of demorphin related oligopeptides, in H. Kalász and L. S. Ettre, (Eds), *Chromatography, the State of the Art*. Akadémiai Kiadó, Budapest, 1985, p. 179.

49. A. M. Barbaro, M. C. Guerra, G. L. Biagi, M. C. Pietrogrande, and P. A. Borea, Influence of the mobile phase composition on the reversed-phase thin-layer chromatographic behaviour of a series of prostaglandins. Comparison of the extrapolated R_M values, *J. Chromatogr.*, **347**, 209, 1985.

50. M. Kuchař, V. Rejholec, E. Kraus, V. Miller, and V. Rábek, Influence of intramolecular interactions on chromatographic behaviour of arylaliphatic acids. I. Comparison of reversed-phase thin-layer and high-performance liquid chromatography, *J. Chromatogr.*, **280**, 279, 1983.

51. K. Jinno, Utilization of micro-HPLC technique for estimating hydrophobic substituent parameters in QSAR works, *Anal. Lett.*, **15**, 1533, 1982.

52 K. Jinno, Utilization of micro-HPLC for the direct measurements of the partition coefficients needed in studying the quantitative structure–activity relationships, *Chromatographia*, **15**, 723, 1982.

53. W. E. Hammers, H. J. Meurs, and C. L. De Ligny, Correlations between liquid chromatographic capacity ratio data on Lichrosorb RP-18 and partition coefficients in the octanol–water system, *J. Chromatogr*, **247**, 1, 1982.

54. N. El Tayar, H. van de Waterbeemd, and B. Testa, The prediction of substituent interactions in the lipophilicity of disubstituted benzenes using RP-HPLC, *Quant. Struct.–Act. Relat.*, **4**, 69, 1985.

55. J. C. Kraak, H. H. van Rooij, and J. L. G. Thus, Reversed-phase ion-pair systems for the prediction of n-octanol–water partition coefficients of basic compounds by high-performance liquid chromatography, *J. Chromatogr.*, **352**, 455, 1986.

56. V. de Biasi, W. J. Lough, and M. B. Evans, Study of the lipophilicity of organic bases by reversed-phase liquid chromatography with alkaline eluents, *J. Chromatogr.*, **353**, 279, 1986

57. K. Valkó, General approach for the estimation of octanol–water partition coefficient by reversed-phase high-performance liquid chromatography, *J. Liq. Chromatogr.*, **7**, 1405, 1984.

58. K. Valkó, Microcomputer analysis of HPLC reversed-phase retention behaviour of several pharmaceutical agents depending on the composition of eluent, in H. Kalász and L. S. Ettre, (Eds.), *Chromatography, the State of the Art*, Akadémiai Kiadó, Budapest, 1985, p. 739.

59. M. Kuchař, V. Rejholec, M. Jelinkova, V. Rábek, and O. Nemecek, Reversed-phase thin-layer chromatography in the parametrization of lipophilicity of some series of arylaliphatic acids, *J. Chromatogr.*, **162**, 197, 1979.

60. M. C. Guerra, A. M. Barbaro, G. Cantelli-Forti, G. L. Biagi, and P. A. Borea, R_M values, retention times and octanol–water partition coefficients of a series of 5-nitroimidazoles, *J. Chromatogr.*, **259**, 329, 1983.

61. H. J. M. Grünbauer, G. J. Bijloo, and T. Bultsma, Influence of eluent composition on lipophilicity measurements using reversed-phase thin-layer chromatography, *J. Chromatogr.*, **270**, 87, 1983.

62. A. M. Siouffi, T. Wawrzynowicz, F. Bressolle, and G. Guiochon, Problems and applications of reversed-phase thin-layer chromatography, *J. Chromatogr.*, **186**, 563, 1979.

63. U. A. Th. Brinkman and G. de Vries, Thin-layer chromatography on chemically bonded phases: A comparison of precoated plates, *J. Chromatogr.*, **258**, 43, 1983.

64. P. Roumeliotis and K. K. Unger, Structure and properties of n-alkyldimethylsilyl bonded silica reversed-phase packings, *J. Chromatogr.*, **149**, 211, 1978.

65. W. G. Tramposch and S. G. Weber, Effect of some amines on the ionic, dipolar, and hydrogen bonding interactions between test solutes and support in reversed-phase liquid chromatography, *Anal. Chem.*, **56**, 2567, 1984.

66. R. E. Koopmans and R. Rekker, High-performance liquid chromatography of alkylbenzenes. Relationship with lipophilicities as determined from octanol–water partition coefficients or calculated from hydrophobic fragmental data and connectivity indices: Lipophilicity predictions for polyaromatics, *J. Chromatogr.*, **285**, 267, 1984.

67. J. H. Knox, R. Kaliszan, and G. J. Kennedy, Enthalpic exclusion chromatography, *Faraday Symp. Chem. Soc.*, **15**, 113, 1980.

68. W. R. Melander, J. Stoveken, and Cs. Horváth, Stationary phase effects in reversed-phase chromatography, I. Comparison of energetics of retention on alkyl-silica bonded phases, *J. Chromatogr.*, **199**, 35, 1980.

69. A. Nahum and Cs. Horváth, Surface silanols in silica-bonded hydrocarbonaceous stationary phases. I. Dual retention mechanism in reversed-phase chromatography, *J. Chromatogr.*, **203**, 53, 1981.

70. N. El Tayar, H. van de Waterbeemd, and B. Testa, Lipophilicity measurements of protonated basic compounds by reversed-phase high-performance liquid chromatography. I. Relationship between capacity factors and the methanol concentration in methanol–water eluents, *J. Chromatogr.*, **320**, 293, 1985.

71. N. El Tayar, H. van de Waterbeemd, and B. Testa, Lipophilicity measurements of protonated basic compounds by reversed-phase high-performance liquid chromatography. II. Procedure for the determination of a lipophilic index measured by reversed-phase high-performance liquid chromatography, *J. Chromatogr.*, **320**, 305, 1985.

72. A. Sokolowski and K.-G. Wahlund, Peak tailing and retention behaviour of tricyclic antidepressant amines and related hydrophobic ammonium compounds in reversed-phase ion-pair liquid chromatography on alkyl-bonded phases, *J. Chromatogr.*, **189**, 299, 1980.

73. O. A. G. J. Van der Houwen, R. H. A. Sorel, A. Hulshoff, J. Teeuwsen, and A. W. M. Indemans, Ion-exchange phenomena and concomitant pH shifts on the equilibration of reversed-phase packings with ion-pairing reagents, *J. Chromatogr.*, **209**, 393, 1981.

74. J. M. McCall, Liquid–liquid partition coefficients by high-pressure liquid chromatography, *J. Med. Chem.*, **18**, 549, 1975.

75. D. A. Brent, J. J. Sabatka, D. J. Minick, and D. W. Henry, A simplified high-pressure liquid chromatography method for determining lipophilicity for structure–activity relationships, *J. Med. Chem.*, **26**, 1014, 1983.

76. J. C. Kraak and P. Bijster, Determination of amitriptyline and some of its metabolites in blood by high-performance liquid chromatography, *J. Chromatogr.*, **143**, 499, 1977.

77. T. Yamana, A. Tsuji, E. Miyamoto, and O. Kubo, Novel method for determination of partition coefficients of penicillins and cephalosporins by high-pressure liquid chromatography, *J. Pharm. Sci.*, **66**, 747, 1977.

78. J. H. Knox and G. R. Laird, Soap chromatography: A new high-performance liquid chromatographic technique for separation of ionizable materials. Dyestuff intermediates, *J. Chromatogr.*, **122**, 17, 1976.

79. J. C. Kraak, K. M. Jonker, and J. F. K. Huber, Solvent-generated ion-exchange systems with anionic surfactants for rapid separations of amino acids, *J. Chromatogr.*, **142**, 671, 1977.

80. J. P. Crombeen, J. C. Kraak, and H. Poppe, Reversed-phase systems for the

analysis of catecholamines and related compounds by high-performance liquid chromatography, *J. Chromatogr.*, **167**, 219, 1978.

81. C. M. Riley, E. Tomlinson, and T. M. Jefferies, Functional group behaviour in ion-pair reversed-phase high-performance liquid chromatography using surface-active pairing ions, *J. Chromatogr.*, **185**, 197, 1979.

82. P.-H. Wang and E. J. Lien, Effects of different buffer species on partition coefficients of drugs used in quantitative structure–activity relationships, *J. Pharm. Sci.*, **69**, 662, 1980.

83. Cs. Horváth, W. Melander, I. Molnar, and P. Molnar, Enhancement of retention by ion-pair formation in liquid chromatography with nonpolar stationary phases, *Anal. Chem.*, **49**, 2295, 1977.

84. J. E. Garst, Accurate, wide-range, automated, high-performance liquid chromatographic method for the estimation of octanol/water partition coefficients I: Effect of chromatographic conditions and procedure variables on accuracy and reproducibility of the method, *J. Pharm. Sci.*, **73**, 1616, 1984.

85. J. E. Garst and W. C. Wilson, Accurate, wide-range, automated, high-performance liquid chromatographic method for the estimation of octanol/water partition coefficients II: Equilibrium in partition coefficients measurements, additivity of substituent constants, and correlation of biological data, *J. Pharm. Sci.*, **73**, 1623, 1984.

86. J. E. Haky and A. M. Young, Evaluation of a simple HPLC correlation method for the estimation of the octanol–water partition coefficients of organic compounds, *J. Liq. Chromatogr.*, **7**, 675, 1984.

87. J. H. Knox, B. Kaur, and G. R. Millward, Structure and performance of porous graphitic carbon in liquid chromatography, *J. Chromatogr.*, **352**, 3, 1986.

88. R. M. Carlson, R. E. Carlson, and H. L. Kopperman, Determination of partition coefficients by liquid chromatography, *J. Chromatogr.*, **107**, 219, 1975.

89. B. McDuffie, Estimation of octanol–water partition coefficients for organic pollutants using reversed-phase HPLC, *Chemosphere*, **10**, 73, 1981.

90. H. Ellgehausen, C. D'Hondt, and R. Fuerer, Reversed-phase chromatography as a general method for determining octanol/water partition coefficients, *Pestic. Sci.*, **12**, 219, 1981.

91. W. W. Bruggeman, J. Van der Steen, and O. Hutzinger, Reversed-phase thin-layer chromatography of polynuclear aromatic hydrocarbons and chlorinated biphenyls. Relationship with hydrophobicity as measured by aqueous solubility and octanol–water partition coefficient, *J. Chromatogr.*, **238**, 335, 1982.

92. C. V. Eadsforth and P. Moser, Assessment of reversed-phase chromatographic methods for determining partition coefficients, *Chemosphere*, **12**, 1459, 1983.

93. R. L. Swann, D. A. Laskowski, P. J. McCall, K. Vander Kuy, and H. J. Dishburger, A rapid method for the estimation of the environmental parameters octanol/water partition coefficient, soil sorption constant, water to air ratio, and water solubility, *Resid. Rev.*, **85**, 17, 1983.

94. S. P. Wasik, M. M. Miller, Y. B. Tewari, W. E. May, W. J. Sonnefeld, H. DeVoe,

and W. H. Zoller, Determination of the vapor pressure, aqueous solubility and octanol/water partition coefficient of hydrophobic substances by coupled generator column/liquid chromatographic methods, *Resid. Rev.*, **85**, 29, 1983.

95. R. A. Rapaport and S. J. Elsenreich, Chromatographic determinations of octanol–water partition coefficients (K_{ow}'s) for 58 polychlorinated biphenyl congeners, *Environ. Sci. Technol.*, **18**, 163, 1984.

96. L. P. Sarna, P. E. Hodge, and G. R. B. Webster, Octanol–water partition coefficients of chlorinated dioxins and dibenzofurans by reversed-phase HPLC using several C_{18} columns, *Chemosphere*, **13**, 975, 1984.

97. H. Govers, C. Ruepert, and H. Aiking, Quantitative structure–activity relationships for polycyclic aromatic hydrocarbons: Correlation between molecular connectivity, physico-chemical properties, bioconcentration and toxicity in Daphnia Pulex, *Chemosphere*, **13**, 227, 1984.

98. C. Ruepert, A. Grinwis, and H. Govers, Prediction of partition coefficients of unsubstituted polycyclic aromatic hydrocarbons from C_{18} chromatographic and structural properties, *Chemosphere*, **14**, 279, 1985.

99. K. Jinno and K. Kawasaki, Correlations between retention data of isomeric alkylbenzenes and physical parameters in reversed-phase micro high-performance liquid chromatography, *Chromatographia*, **17**, 337, 1983.

100. T. Hanai, K. C. Tran, and J. Hubert, Prediction of retention times for aromatic acids in liquid chromatography, *J. Chromatogr.*, **239**, 385, 1982.

101. T. Hanai and J. Hubert, Prediction of retention time of phenols in liquid chromatography, *J. High Res. Chromatogr. Chromatogr. Commun.*, **6**, 20, 1983.

102. T. Hanai, Y. Arai, M. Hirukawa, K. Noguchi, and Y. Yanagihara, Reversed-phase liquid chromatography on rigid vinyl alcohol copolymer gels, *J. Chromatogr.*, **349**, 323, 1985.

103. T. Schettler, B. Aufm'Kolk, M. J. Atkinson, H. Radeke, C. Enters, and R. D. Hesch, Analysis of immunoreactive and biologically active human parathyroid hormone-peptides by high-performance liquid-chromatography, *Acta Endocrinol. (Copenh.)*, **107**, 60, 1984.

104. R. M. Arendt and D. J. Greenblatt, Liquid chromatographic retention of β-adrenoceptor antagonists: An index of lipid solubility, *J. Pharm. Pharmacol.*, **36**, 400, 1984.

105. L. Cellai, D. Corradini, C. Corradini, M. Brufani, and E. Marchi, A chromatographic approach to the study of structure–activity relationships in 3-substituted rifamycin S derivatives, *Il Farmaco, Ed. Sc.*, **39**, 575, 1984.

106. H. Sunada, K. Furukawa, R. Ishino, A. Otsuka, M. Sugiura, and Y. Hamada, Adsorption of amino-acetamide derivatives from 1-octanol solution by silica gel, *Chem. Pharm. Bull.*, **32**, 4084, 1984.

107. B. Rittich, J. Štohandl, R. Žaludová, and H. Dubský, Relationship between retention parameters (log k') and octanol–water partition coefficients (log P) of a series of phenyl carbomates and phenylureas. Part III, *J. Chromatogr.*, **294**, 344, 1984.

108. N. Ronzani, Ch. Julien-Larose, and A. La Berre, Hydrophobicity measurement by reversed phase thin-layer chromatography of potent insecticides (carbamates), *Analusis*, **12**, 151, 1984.

109. H. Könemann, R. Zelle, F. Busser, and W. E. Hammers, Determination of log P_{oct} values of chloro-substituted benzenes, toluenes and anilines by high performance liquid chromatography on ODS-silica, *J. Chromatogr.*, **178**, 559, 1979.

110. P. H. Hinderling, O. Schmidlin, and J. K. Seydel, Quatitative relationships between structure and pharmacokinetics of beta-adrenoceptor blocking agents in man, *J. Pharmacokin. Biopharm.*, **12**, 263, 1984.

111. S. Lembo, V. Sasso, C. Silipo, and A. Vittoria, Physicochemical properties and drug design in a set of analgesics, *Il Farmaco, Ed. Sc.*, **38**, 750, 1984.

112. J. C. Caron and B. Shroot, Determination of partition coefficients of glucocortico-steroids by high-performance liquid chromatography, *J. Pharm. Sci.*, **73**, 1703, 1984.

113. J.-X. Huang, E. S. P. Bouvier, J. D. Stuart, W. R. Melander, and Cs. Horváth, High-performance liquid chromatography of substituted *p*-benzoquinones and *p*-hydroquinones. II. Retention behaviour, quantitative structure–retention relationships and octanol–water partition coefficients, *J. Chromatogr.*, **330**, 181, 1985.

114. M. Kuchař, E. Kraus, M. Jelinková, V. Rejholec, and V. Miller, Thin-layer and high-performance liquid chromatography in the evaluation of the lipophilicity of aryloxoalkanoic and arylhydroxyalkanoic acids, *J. Chromatogr.*, **347**, 335, 1985.

115. M. Bachratá, M. Blešová, A. Schultzová, L. Grolichová, Ž. Bezáková, and A. Lukáš, Chromatographic study of local anaesthetics-basic esters of substituted carbanilic acids. II. The relationships between chromatographic values, other physico-chemical parameters and activity, *J. Chromatogr.*, **171**, 29, 1979.

116. K. Dadáková, J. Hartl, and K. Waisser, Relationship between R_M values and hydrophobicity. Chromatographic behaviour of crotonolactones, *J. Chromatogr.*, **254**, 277, 1983.

117. T. Cserháti, Lipophilicity determination of some 3,5-dinitro-benzoic-acid esters on unimpregnated cellulose layer, *Chromatographia*, **18**, 18, 1984.

118. J. K. Baker, G. J. Hite, M. Reamer, and P. Salva, Effect of stereochemistry on the high-performance liquid chromatographic retention index of azabicycloalkanes, *Anal. Chem.*, **56**, 2932, 1984.

119. N. El Tayar, H. van de Waterbeemd, M. Gryllaki, B. Testa, and W. F. Trager, The lipophilicity of deuterium atoms. A comparison of shake-flask and HPLC methods, *Int. J. Pharm.*, **19**, 271, 1984.

120. M. J. M. Wells, C. R. Clark, and R. M. Patterson, Correlation of reversed-phase capacity factors for barbiturates with biological activities, partition coefficients, and molecular connectivity indices, *J. Chromatogr. Sci.*, **19**, 573, 1981.

121. M. J. M. Wells, and C. R. Clark, Study of the relationship between dynamic and static equilibrium methods for the measurement of hydrophobicity. Comparison of capacity factors and partition coefficients for some 5,5-disubstituted barbituric acids, *J. Chromatogr.*, **284**, 319, 1984.

122. S. Toon and M. Rowland, Simple method for the optimisation of mobile phase composition for HPLC analysis of multi-component mixture, *J. Chromatogr.*, **208**, 391, 1981.

123. S. Toon, J. Mayer, and M. Rowland, Liquid chromatographic determination of lipophilicity with application to a homologous series of barbiturates, *J. Pharm. Sci.*, **73**, 625, 1984.

124. T. M. Xie and D. Dyrssen, Simultaneous determination of partition coefficients and acidity constants of chlorinated phenols and guaiacols by gas chromatography, *J. Chromatogr.*, **160**, 21, 1984.

125. K. Boček, Water and oleyl alcohol as stationary liquid phases in gas–liquid chromatography. Determination of partition coefficients of volatile substances for analysis of quantitative structure–activity relationships, *J. Chromatogr.*, **162**, 209, 1979.

126. D. Rehn, W. Zerling, and U. Püst, Experimental evaluation of relative partition coefficients by gas chromatography, *Pharm. Acta Helv.*, **58**, 144, 1983.

127. O. Papp, K. Valkó, Gy. Szász, I. Hermecz, J. Vámos, K. Hankó-Novák, and Zs. Ignáth-Halász, Determination of partition coefficient of pyrido-[1,2]-pyrimidin-4-one derivatives by traditional shake, thin-layer chromatographic and gas–liquid chromatographic methods, *J. Chromatogr.*, **252**, 67, 1982.

128. K. Valkó and A. Lopata, Applicability of gas–liquid chromatography in determining liquid–liquid partition data, *J. Chromatogr.*, **252**, 77, 1982.

129. K. Valkó, O. Papp, and F. Darvas, Selection of gas chromatographic stationary phase pairs for characterization of the 1-octanol–water partition coefficient, *J. Chromatogr.*, **301**, 355, 1984.

130. C. Hansch and A. Leo, *Substituent Constants for Correlation Analysis in Chemistry and Biology*, Wiley, New York, 1979.

131. J. Kopecky and K. Boček, A correlation between constants used in structure–activity relationships, *Experientia*, **23**, 125, 1967.

132. G. L. Biagi, M. C. Guerra, and A. M. Barbaro, Relationship between lipophilic character and hemolytic activity of testosterone and testosterone esters, *J. Med. Chem.*, **13**, 944, 1970.

133. G. L. Biagi, A. M. Barbaro, M. C. Guerra, G. C. Forti, and M. E. Francasso, Relationship between π and R_m values of sulfonamides, *J. Med. Chem.*, **17**, 28, 1974.

134. G. L. Biagi, A. M. Barbaro, O. Gandolfi, M. C. Guerra, and G. Cantelli-Forti, R_m values of steroids as an expression of their lipophilic character in structure–activity studies, *J. Med. Chem.*, **18**, 873, 1975.

135. E. J. Lien, L. L. Lien, and G. L. Tong, Central nervous system activities of thiolactams and lactams. Structure–activity correlations, *J. Med. Chem.*, **14**, 846, 1971.

136. J. H. Tubby, Hydrophobic substituent constants from thin–layer chromatography on polyamide plates, *J. Pharm. Pharmacol. Suppl.*, **26**, 74P, 1974.

137. K. Callmer, L.-E. Edholm, and B. E. F. Smith, Study of retention behaviour of

alkylphenols in straight- and reversed-phase liquid chromatography and application to the analysis of complex phenolic mixtures in conjunction with gas chromatography, *J. Chromatogr.*, **136**, 45, 1977.

138. M. Kuchař, B. Brůnová, V. Rejholec, and V. Rábek, Relationship between paper chromatographic R_M values and Hansch's π parameters in dissociable compounds, *J. Chromatogr.*, **92**, 381, 1974.

139. M. Kuchař, V. Rejholec, Z. Roubal, and O. Maloušowá, The effects of lipophilicity on the inhibition of denaturation of serum albumin and on the activation of fibrynolysis observed with a series of benzyloxyarylaliphatic acids, *Coll. Czech. Chem. Commun.*, **48**, 1077, 1983.

140. W. Steurbaut, W. Dejonckheere, and R. H. Kips, Relationships between chromatographic properties, partition data and chemical structure of O-alkyl-O-arylphenylphosphonothioates, *J. Chromatogr.*, **160**, 37, 1978.

141. M. Bachratá, M. Blešová, A. Schultzová, L. Grolichová, Ž. Bezáková, and A. Lukaš, Chromatographic study of local anaesthetics- basic esters of substituted carbanilic acids, II. The relationship between chromatographic values, other physiocochemical parameters and activity, *J. Chromatogr.*, **171**, 29, 1979.

142. B. Rittich, M. Polster, and O. Králik, Reversed-phase high-performance liquid chromatography. I. Relationship between R_M values and Hansch's π parameters for a series of phenols, *J. Chromatogr.*, **197**, 43, 1980.

143. K. Jinno, Utilization of micro-HPLC technique for estimating hydrophobic substituent parameters in QSAR works, *Anal. Lett.*, **15**, 1533, 1982.

144. K. Jinno and K. Kawasaki, A retention prediction system in reversed-phase high-performance liquid chromatography based on the hydrophobic parameter for alkylbenzene derivatives, *Anal. Chim. Acta*, **152**, 25, 1983.

145. A. Nunn, Structure–distribution relationships of radiopharmaceuticals. Correlation between the reversed-phase capacity factors for Tc-99 m phenylcarbamoylmethyliminodiacetic acids and their renal elimination, *J. Chromatogr.*, **255**, 91, 1983.

146. T. Braumann and B. Jastroff, Physico-chemical characterization of cyclic nucleotides by reversed-phase high-performance liquid chromatography. II. Quantitative determination of hydrophobicity, *J. Chromatogr.*, **350**, 105, 1985.

147. W. L. Cody, B. C. Wilkes, and V. J. Hruby, Reversed-phase high-performance liquid chromatography studies of α-MSH fragments, *J. Chromatogr.*, **314**, 313, 1984.

148. J.-L. Fauchere, Partition coefficients of amino acids and hydrophobic parameters π of their side-chains as measured by thin-layer chromatography, *J. Chromatogr.*, **216**, 79, 1981.

149. J.-L. Feuchere and V. Pliška, Hydrophobic parameters π of amino-acid side chains from the partitioning of N-acetyloamino-acid amides, *Eur. J. Med. Chem.*, **18**, 369, 1983.

150. J. Macek and E. Smolková-Keulemansová, Correlation of the chromatographic retention of some phenylacetic and phenylpropionic acid derivatives with molecular structure, *J. Chromatogr.*, **333**, 309, 1985.

151. R. F. Rekker, *The Hydrophobic Fragmental Constant: Its Derivatization and Application. A Means of Characterizing Membrane Systems*, Elsevier, Amsterdam, 1977.

152. R. Kaliszan, B. Pilarski, K. Ośmiałowski, H. Strzałkowska-Grad, and E. Hać, Analgesic activity of new pyrazine CH and NH acids and their hydrophobic and electron donating properties, *Pharm. Weekbl.*, **7**, 141, 1985.

153. C. Yamagami, H. Takami, K. Yamamoto, K. Miyoshi, and N. Takao, Hydrophobic properties of anticonvulsant phenylacetanilides. Relationship between octanol–water partition coefficient and capacity factor determined by reversed-phase liquid chromatography, *Chem. Pharm. Bull.*, **32**, 4994, 1984.

154. A. Kakoulidou and R. F. Rekker, A critical appraisal of log *P* fragmental procedures and connectivity indexing for reversed-phase thin layer chromatographic and high-performance liquid chromatographic data obtained for a series of benzophenones, *J. Chromatogr.*, **295**, 341, 1984.

155. R. F. Rekker, The position of reversed-phase thin-layer chromatography amongst solvent partitioning techniques, *J. Chromatogr.*, **300**, 109, 1984.

156. G. J. Bijloo and R. F. Rekker, Influence of stationary phase modifications on lipophilicity measurements of benzophenones using reversed-phase thin-layer chromatography, *J. Chromatogr.*, **351**, 122, 1986.

CHAPTER

12

CHROMATOGRAPHIC RETENTION DATA IN STUDIES OF QUANTITATIVE STRUCTURE–BIOLOGICAL ACTIVITY RELATIONSHIPS

Drug action is the end result of several branched or consecutive reaction steps, any of which can be rate limiting. In the sequence of processes at the basis of drug action, three main phases can be distinguished, pharmaceutical, pharmacokinetic, and pharmacodynamic (Fig. 12.1) [1].

The pharmaceutical phase comprises the processes involved in the release (liberation, L) of the active agent from the dosage form, thus determining the concentration of the drug available for absorption at an absorption site. The processes in the pharmacokinetic phase can be divided into absorption into general circulation (A), distribution (D) among various body compartments, biotransformation (metabolism, M) and excretion (E). These processes determine eventually the concentration of the drug in the extracellular fluid and therewith at the sites of actions in the target tissue.

The pharmacodynamic phase of drug action comprises the molecular interactions of the drug with its site of action (pharmacological receptor), which initiates the sequence of processes leading to the biological effect observed. The term *drug* is used here to mean various biologically active agents, natural and synthetic, like real drugs, toxins, environmental pollutants, hormones, ferromones, pesticides.

Considering the complex nature of drug action, it is a difficult task to find quantitative structure–biological activity relationships (QSARs), which are the starting condition for rational drug design.

Leaving aside the pharmaceutical phase, it may be assumed that the biological effect is determined by processes of transport (usually passive diffusion) of drug to the site of action and by intermolecular interactions between drug molecules and receptors [2].

The QSAR methods commonly applied are based on the assumption that the linear free-energy relationships (LFERs) exist between the chemical and biological reactions. Hansch [3] attributed the free-energy change in a standard biological response to hydrophobic, electronic, and steric contributions. Electronic and steric properties of biologically active substances are of importance for highly specific interactions with the receptor. Hydrophobic properties of drugs are rather assumed to determine their transport to the

279

receptor, although these properties also influence the drug–receptor binding. At any rate, if compounds of a given group possess a moiety (pharmacophore) as a requisite of the stimulation of the receptor, the effect will depend on the number of molecules at the range of the receptor. Many of the reported QSARs that are successful for homologous series are in fact related to the pharmacokinetics of the drugs.

To derive the QSAR equations of a possible value for the design of an agent of the required bioactivity, one must at first be able to describe the properties of the structures considered in quantitative terms. Since the beginning of QSAR studies (the early 1960s), chromatography has been extensively exploited as a tool for quantitation of physicochemical properties of drugs. Until the early 1970s, TLC and PC were applied [4]. Later, HPLC and GC became increasingly popular [1, 5, 6]. Chromatography is now one of the most popular methods of characterizing biologically active substances. It is commonly, but often mechanically, applied in medicinal chemistry.

The earliest reported study relating the chromatographic parameter (R_M from RP TLC) to biological activity is that by Boyce and Millborrow [7]. These authors determined R_M values on a stationary phase impregnated with liquid paraffin using an acetone–water mobile phase. A parabolic relation was obtained when the molluscicidal activities of some N-n-alkyltritylamines were plotted against their R_M values.

Another quadratic relationship between the R_M values (extrapolated to pure

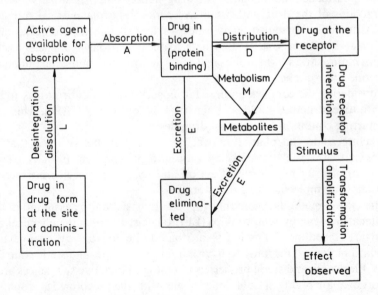

Figure 12.1. Essential processes at basis of drug action.

water) and the logarithm of the reciprocal of the minimum lethal dose in cats was found by Biagi [8] for some cardiac glycosides. In the same year, 1967, Biagi [9], in a study of the lipid solubility and human serum binding of various penicillins, showed that for some of the penicillins the correlation between the R_M from reversed-phase chromatography and human serum binding was greater than that obtained using the substituent lipophilic constant π. Soon, another quadratic relationship was reported for a group of 4-(1-cyclopentyl-n-alkyl)-2,6-dinitrophenols between their substituent ΔR_M values and their fungicidal activities by Clifford et al. [10].

According to Hansch and Fujta [11], the parabolic relationship between an index of hydrophobicity and bioactivity may be related to the probabilistic movement of a drug from an extraceludar phase to its site of action. However, for a given series of biologically active agents, the range of hydrophobicity among the members of the series need not to be wide enough to define both the ascending and the descending branch of the parabola. In the quadratic relationships relating bioactivity and the chromatographic measures of lipophilicity, the square term has not always been statistically significant because it is based on few data points only.

In studies of the influence of the hydrophobic character on the activities of a series of penicillins and cephalosporins, Biagi et al. [12] found a linear relationship between R_M and the logarithm of the reciprocal of minimum antibacterial concentration against *Escherichia coli*. The negative sign for the slope coefficient in the correlation equation derived suggests that the activity increases with a decrease in the hydrophobic character of the penicillins. If the relationship is statistically significant and the R_M data are the reliable measures of the hydrophobicity of the penicillins studied, one may assume that the hydrophobicities of the drugs under consideration are higher than optimum for the inhibition of the growth of *E. coli*. Activity of penicillins against *Staphylococcus aureus* and *Trypanosoma pallidum*, as well as the activities of cephalosporins against all the three bacterial species studied, were related to R_M values by typical quadratic functions. The same was the case with rifamycins studied separately by the same authors [13]. Another early report by Biagi et al. [14] on the influence of the lipophilic character on the bioactivity of some oligosaccharide antibiotics provided additional arguments for the effectiveness of using R_M data in QSARs.

In vitro hemolytic activity of a set of testosterone esters was described by two-parameter quadratic equations comprising R_M and $(R_M)^2$ terms [15]. The R_M values were used corresponding to 54% concentration of acetone or methanol in the mobile phase. The plates were impregnated with silicone oil. The QSAR equations obtained were of a quality comparable to that obtained with the lipophilic substituent constant π.

The effect of some testosterone esters in the capon's comb test was linearly related to the R_M values in another study by Biagi et al. [16]. A linear relationship was also found between the percentage of binding of cortico-steroids to serum albumin and R_M values [17].

Dearden and Tomlinson [18] found a linear correlation between human buccal absorption data of some acetanilide drugs and their ΔR_M substituent constants derived from R_M measurements in TLC systems using either n-octanol or liquid paraffin as the stationary phase. The n-octanol–water solvent pair acted as a better model reference system than the liquid paraffin system. Similar differences in linear correlations were found by Dearden and co-workers [19, 20] in studies on the protein binding of acetanilides to bovine serum albumin. Equally good as lipophilicity descriptors were R_M values obtained with the n-octanol–water TLC system [19] and those determined in a polyamide/acetone–water–dioxan 1:2:1 system [20].

Dearden and Tomlinson [21] correlated also the analgesic potencies of a group of p-substituted acetanilides in mice with substituent ΔR_M values. A two-parameter equation involving ΔR_M and $(\Delta R_M)^2$ terms was of a higher statistical value when chromatography was carried out on n-octanol-impregnated plates as compared to the liquid paraffin stationary phase.

Biliary excretion of penicillins in the rat was related ($r = 0.84$) to the R_M data measured in a reversed-phase silicone oil–water TLC system [22].

For N-alkyl-substituted normeperidine homologues, the R_M values obtained on cellulose TLC plates with an ethanol–water 40:60 solvent system were linearly related to butyrylcholinesterase–drug dissociation constants [23]. Linear relationships were not observed, however, with acetyl-cholinesterase.

Concentrations of a homologous series of N,N'-bis-dichloroacetyl diamines inhibiting in vitro mitochondrial electron transport were related to R_M and $(R_M)^2$ by two-parameter regression equations [24]. The R_M values on the silicone oil stationary phase were determined by extrapolating the data to 50% acetone in the mobile phase.

For a limited set of five thiolactam compounds, the lethal toxicities in mice were better correlated with R_M values than with n-octanol–water partition coefficients [25].

Plá-Delfina et al. [26–28] found that for a group of barbituric acids studied, the gastric absorption rate constants were correlated ($r \approx 0.9$) with R_M values determined in several reversed-phase TLC systems. Later, that research group found a hyperbolic relationship between the rat gut in situ absorption rate constants, k_a, and reversed-phase TLC data for 18 antibacterial sulfonamides [29]

$$k_a = [3.9396(1/R_f - 1)^{2.6364}]/[0.0719 + (1/R_f - 1)^{2.6364}] \qquad (12.1)$$

The correlation as described by Eq. (12.1) was better than that observed when $\log k_a$ was related to R_M and $(R_M)^2$.

Chandry and James [30], using literature R_M values of some nandrolone esters, have related the anabolic activities of these compounds to the chromatographic parameter R_M determined in a straight-phase system. A moderate correlation ($r = 0.84$) was obtained.

Correlations between R_M values obtained from TLC and the bioactivity of steroids [31], phenols [32], naphtols and acetophenones [33], and benzodiazepines [34] were reported by the Biagi group. With steroids, there is a parabolic relationship between the data for hemolytic activity and the membrane binding of the drugs with regard to their lipophilicity expressed by R_M values. On the other hand, introduction of the $(R_M)^2$ term does not improve the correlation between either the protein binding or the duration of action of testosterone esters and hydrophobicity. Hemolytic activity, antibacterial activity against *S. aureus*, and acute toxicity to mice for a group of phenols have been shown to be linearly dependent on R_M values [32]. The data for phenols were later combined with the corresponding data for naphtols and acetophenones [33]. In spite of the fact that the R_M values for phenols, naphtols, and acetophenones were determined at different times, the linear relationships between either acute toxicity or hemolytic activity and R_M values for a group of nearly 60 compounds were satisfactory (for five halogenated acetophenones an indicator variable was additionally used). The Biagi group preferred silicone oil as the stationary phase. They found the R_M values from the silicone system to be better correlated with the bioactivity of benzodiazepines than those obtained on *n*-octanol-impregnated TLC plates [34].

Fujii and co-workers [35, 36] applied R_M values for the description of in vivo antistaphylococcal activity in nice of ω-amino acids and their L-histidine dipeptides and carboxylic acids.

Hulshoff and Perrin [37] obtained a good correlation of the R_M values (extrapolated to 100% water in the mobile phase) with the bovine-serum-albumin-binding constants, antihemolytic activity, and the inhibition of Na^+K^+-activated adenosine triphosphatase activity for a series of phenothiazine derivatives. The biological activity data used were corrected for the state of ionization.

The R_M data from the liquid paraffin/acetone–water system gave no significant correlation with antibacterial activity against *Mycobacterium tuberculosis* in a series of isonicotinic acid hydrazide derivatives [38]. Correlation improved when pK_a data were included, thus providing a correction for ionization to the R_M data. A similar situation has been observed in the case of the in vitro tuberculostatic activity of pyrazine carbothioamide derivatives [39] and 2-cyanomethylbenzimidazole derivatives [40]. Here, too, TLC data from the liquid paraffin/water system gave satisfactory correlation with bioactivity when used together with the spectroscopic data related to polarity and thus to the ionization ability of individual derivatives.

A parabolic dependence of activity against eight bacterial species on R_M values determined on Kieselgel 60 silanisiert plates with acetone–water 60:40 as the mobile phase for a series of α, β-unsaturated γ-lactones was observed by Dal Pozzo et al. [41].

The normal-phase TLC data determined on silica gel and a cellulose support impregnated with formamide were linearly correlated with the surface local anesthetic activity of a series of 2-morpholinoethyl esters of 2-, 3-, and 4-alkoxycarbanilic acids [42]. The correlations were derived separately for the subgroups of ortho-, meta-, and para-substituted derivatives.

TLC partition data proved useful for QSAR studies of a series of rifamycins as inhibitors of viral RNA-directed DNA polymerase and mammalian α- and β-DNA polymerases [43].

Partitioning into erythrocytes of potential antimalarial sulfonamides was also related to R_M values obtained in TLC in the liquid paraffin/phosphate buffer pH 5 system [44].

Ferguson and Denny [45, 46] based on R_M values from partition chromatography, concluded that for the tumor-active but mutagenic anilino-acridines, separation of the two classes of bioactivity was possible by simple manipulation of the agent lipophilic–hydrophilic balance.

Maksay et al. [47] found a good correlation between pharmacokinetic constants characterizing oxazepam brain levels observed after the intravenous administration of prodrugs (oxazepam esters) and the chromatographic R_M values. They found that an increase in hydrophobicity of the esters decreased oxazepam brain penetration. To explain this rather unexpected finding, these authors suggested that hydrolysis precedes brain penetration and hydrophobicity might primarily influence the hydrolysis rate. The amount of tissue storage, total excertion rates, and serum binding were also correlated with hydrophobicity in the subsequent publication of Maksay et al. [48].

Areas under the effect–time curves were correlated with R_M values (extrapolated to pure water) for some new hypoglycemic sulfonamides [49].

Bieganowska [50] analyzed QSARs in a group of N-phenylamides of benzoylacetic acid using the reversed-phase TLC and HPLC retention parameters. TLC was carried out on the ready-made plates precoated with octadecyl silica. Also, HPLC was done on octadecyl silica material. In both cases the same eluent was used (i.e., aqueous solution of methanol containing acetic acid). Chromatographic retention parameters R_M and log k' were determined at specific compositions of the mobile phase and also extrapolated to pure water as the eluent. The dependent variables were bioactivity data, related to the inhibition of the prostaglandin synthetase by the compounds studied and to their binding to albumins. Both bioactivity measures considered by Bieganowska [50] have been better described by the HPLC data, as compared to the R_M values from TLC (or high-performance thin-layer chromatography, HPTLC, as insists the author). She also reports the lack of advantage in the QSARs studied of the retention parameters extrapolated to pure water.

Takeya and co-workers [51] found significant correlation between R_M values determined on the precoated silanized plates and the measure of the speed of positive inotropic effect of a group of ericaceous toxins. The correlation ($r = 0.82$) was negative (i.e., toxins with a more lipophilic nature caused faster development of the positive inotropic effect). In a subsequent publication Takeya and co-workers [52] included analogous QSARs for a group of cardenolides. For that group a positive and close correlation ($r = 0.98$) was observed between the time required for half maximum positive inotropic effect and R_M; that is, the more lipophilic the cardenolides, the more time was required to reach the full development of the positive inotropic effect. These authors [52] found the silica plates impregnated with octanol as unsuitable for lipophilicity measurements, whereas the commercially available silanized gel plates gave the lipophilicity measures comparable to those derived from reversed-phase HPLC (intercorrelation of the R_M and log k' values was $r = 0.87$). High intercorrelation between R_M and log k' was reported by Govers et al. [53] for a series of polycyclic aromatic hydrocarbons. For that closely congeneric group of solutes nearly all structural parameters are intercorrelated. In such a situation some scepticism is justified as far as the observation by the authors is concerned that R_M data provide the best prediction of bioconcentration of PAH in *Daphnia pulex*.

The lipophilicity determined by reversed-phase TLC showed a good correlation with the ability of some neuroleptics to enhance striatal dopamine release [54].

In studies of fibrynolysis and inhibition of denaturation of serum albumin by a large series of substituted benzyloxyarylaliphatic acids, Kuchař et al. [55] found that the lipophilicities of the compounds were better described by the substituent lipophilic constant π than by experimental TLC values obtained on silica gel impregnated with silicone oil.

Dadakova et al. [56] determined ΔR_M values for a series of crotonolactones in a normal-phase partitioning TLC system. A stationary phase was a formamide-impregnated cellulose, and the development was carried out by cyclohexane. The correlation of the chromatographic parameters with in vitro antituberculotic activity has been rather low ($r = 0.74$).

The R_M data determined on a cellulose support were related to antitumor activity of derivatives of ansacrine by Denny et al. [57]. A good parabolic relationship has been reported by Spengler et al. [58] between R_M values and the H_2-antihistaminic activity of a series of 5,6-substituted 4-pyrimidones

(Fig. 12.2). The R_M data used were determined on the paraffin-coated silica plates and extrapolated to pure water.

In recent years HPLC has been usually employed for chromatographic characterization of lipophilicities for the purposes of QSAR analysis instead of TLC.

Already in 1975, Carlson et al. [59] correlated toxicity against *Daphnia magna* with the reversed-phase HPLC capacity factor for a series of nine phenols. The correlation coefficient of a linear relationship ($r = 0.68$) was similar to that obtained with the π constant ($r = 0.76$).

The ability of a short series of 1,3,5-triazine herbicides to inhibit the Hill reaction has been correlated with HPLC retention times determined on octadecylsilica columns with water–methanol 95:5 as the mobile phase [60].

An extensive study on the application of HPLC data in correlation with the

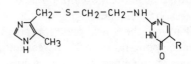

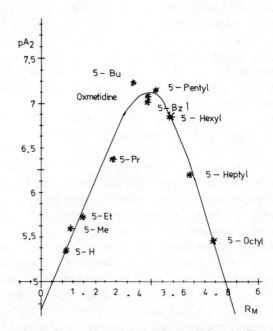

Figure 12.2. Dependence of antagonist activity (pA_2) toward histamine H_2 receptor on R_M from reversed-phase TLC for series of 4-pyrimidones. (J. -P. Spengler, K. Wegner, and W. Schunack, Structure-activity relationships in H_2-receptor antagonists containing a 4-pyrimidone moiety, *Agents Actions*, **14**, 566, 1984. Birkhäuser Verlag, AG, Basel, Switzerland.)

activity of sulfonamides against *E. coli* and inhibitory potencies of barbiturates on rat brain oxygen uptake and *Arbacia* egg cell division was described by Henry et al. [61].

These early applications of the HPLC-derived structural parameters in QSARs met only moderate success. Later in 1979, Riley et al. [62], based on extensive studies of ion pair reversed-phase HPLC, proposed a system using sodium dodecylsulfate as the pairing ion and methanol as the organic modifier for the determination of indices of hydrophobicity. They successfully applied their hydrophobicity indices to describe the antibacterial activity of a group of 1,3,5-*s*-triazines.

Fungicidal activity of mono- and disubstituted phenols was correlated with reversed-phase HPLC data [63]. The capacity factors were corrected for ionization at the environmental pH. Nonetheless, the pK_a parameter was used by the authors along with the logarithm of the capacity factor in two-parameter QSAR equations.

Rabbit hypnotic dose, rat hypnotic dose, and concentration for 50% inhibition of rat brain were related to the log k' values by means of a quadratic regression [64]. Correlations equivalent to those derived with *n*-octanol–water log P were obtained.

Unger and Chiang [65] proved the usefulness of their procedure of chromatographic determination of hydrophobicity by deriving QSAR equations relating the literature bioactivity data to log k' for a series of lipophilic amines.

Davydov et al. [66] observed a relationship between toxicity and the chromatographic behavior of 17 cardiac glycosides. Jinno [67] recalculated the data of Davydov et al. [66] using the logarithmic scale. A more convincing QSAR was obtained (Fig. 12.3).

Reversed-phase HPLC data of ligands obtained on octadecylsilica columns with a buffer–methanol mobile phase were related to renal clearance, protein binding, and in vivo distribution of substituted phenylcarbamoylmethyl-

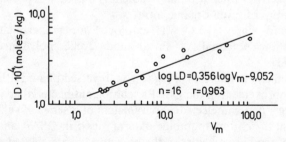

Figure 12.3. Correlation between toxicity, LD, of series of cardiac glycosides and their retention volumes, V_m. G-strophantin is not included in correlation. (After K. Jinno, *J. Chromatogr.*, **264**, 485, 1983. With permission.)

iminodiacetic acid technetium-99m complexes by Nunn and co-workers [68, 69]. Moderate correlation ($r = 0.82$) was observed when mono- and ortho-substituted derivatives were excluded.

Jinno [70] used capacity factors determined by reversed-phase micro-HPLC on octadecylsilica phase with methanol–water solvent as the molecular descriptors in QSAR studies of phenols. Several literature bioactivity data sets were related to log k' with correlation coefficients ranging from 0.90 to 0.99.

Valkó et al. [71] correlated the bioactivity data of a group of new analgesic azidomorphines with reversed-phase HPLC retention parameters. The independent variable was the so-called offset rate of inhibitory action measured on the muscle strip of the guinea pig ileum. The capacity factor determined at 40% acetonitrile in aqueous eluent, log k'_{40}, gave best linear correlation with bioactivity. The authors attempted to rationalize that observation, suggesting that at 40% acetonitrile in the eluent the biological partition system was best modeled by the chromatographic system used. To prove their hypothesis, Valkó et al. [71] derived linear relationships between log k' of individual azidomorphines and concentrations of acetonitrile in the mobile phase. Next, the slope values and the intercept values derived from those linear relationships were used jointly in a two-parameter regression equation as the independent variables describing bioactivity. Using an argument similar to that used with the log P prediction [see Eqs. (11.15)–(11.20)], Valkó et al. [71] calculated the optimum concentration of acetonitrile in the solvent from the ratio of the regression coefficients referring to the slope and to the intercept variables in the two-parameter QSAR equation. The calculated value, 38.7% of acetonitrile, is really close to the composition of eluent (40%), yielding capacity factors that best predict the bioactivity studied. Surprisingly, for the compounds analysed, there is no correlation ($r = 0.086$) between the slopes and the intercepts of the relationship log k' versus the acetonitrile concentration in the eluent.

Unusually high correlation ($r = 0.99$) between the reversed-phase HPLC data and bioactivity (the vasodilator potency) has been reported by Noack [72] for a group of seven organic nitrates.

The reversed-phase HPLC was employed for the estimation of the bioaccumulation potential of linear and cyclic polydimethylsiloxane oligomers [73].

The lipophilicity in the form of log k' from reversed-phase HPLC provided significant improvement in the QSARs relating antibacterial activity against *Mycobacterium lufu* and electronic parameters of a series of sulfones [74].

Most often, the chromatographic data are used in QSAR analysis as the descriptors of the lipophilicities of the bioactive agents, related more or less directly to the *n*-octanol–water partition system. There is no reason, however, to assume that the *n*-octanol–water system is the best modeling partition process in a living organism. On the other hand, one can imagine a specific chromatographic system that would mimic the biological transport and the receptor-binding phenomena. In such chromatographic model system, the hydrophobic, electronic, and steric interactions would certainly be effective. So far, there are no such systems described, and the majority of the authors seem to strive to prepare a purely hydrophobic system (with specific intermolecular interactions minimized) or a system precisely reflecting n-octanol–water partition processes.

For several years the reversed-phase technique has dominated both the theory and practice of liquid chromatography. Much has been achieved, but a further development of chromatography requires the widening of the set of factors affecting the separation processes. Studies of the factors determining molecular interactions during the chromatography should also provide structural information useful for QSAR studies. A few reports announcing the new approach to chromatography as the tool for generating structural data for QSAR purposes are briefly discussed below.

Baker et al. [75] claimed the superiority of their HPLC retention index over the n-octanol–water partitioning data for the quantitative description of biological activity of propranolol and barbiturate analogs. However, in the case of the antiinflammatory activity of the anthranilic acid derivatives, the same authors found a better description of pharmacological data when using the n-octanol–water partition data. The Baker retention index was successfully applied in QSAR studies of 4-hydroxyquinoline-3-carboxylic acids as inhibitors of cell respiration [76]. The Baker index proved useful also for prediction of the antihypertensive activity of quinazolinamine derivatives [77].

The log k' data determined on a highly deactivated octadecyl-bonded silica column with a methanol–buffer mobile phase were used in QSAR studies concerning several types of bioactivity data of sulfonamides and barbiturates [78]. Considering the quality of QSAR equations, the results obtained were indistinguishable from those produced by the Baker retention index.

According to Braumann et al. [79], log k' determined on octadecylsilica columns can be used to model the behavior of s-triazines in thylakoid membranes, whereas log P cannot.

Applications of gas–liquid chromatographic data in QSAR studies are unique. A gas chromatographic measure of solute polarity was applied in QSAR studies of the olfactory activity of phenols [80]. The activity, expressed by human detection thresholds, has been satisfactorily described by hydrophobicity parameters, and the introduction of no electronic data gave a statistically significant improvement. The chromatographic polarity measure proved to be useful as a correction factor for ionization to the hydrophobicity parameter. Previously, Laffort and Patte [81] reported correlation ($r = 0.79$)

between experimental olfactory thresholds and the calculated ones based on their gas chromatographic solubility factors.

Data concerning the induction of an enzyme N-demethylase by a group of polycyclic aromatic hydrocarbons have been correlated with the shape parameter calculated from GC retention data determined on nematic phases [82]. In the same work, the chromatographic shape parameter was used for quantitative description of mutagenicity of the compounds.

References

1. R. Kaliszan, High performance liquid chromatography as a source of structural information for medicinal chemistry, *J. Chromatogr. Sci.*, **22**, 362, 1984.

2. J. K. Seydel and K.-J. Schaper, Quantitative structure–pharmacokinetic relationships and drug design, *Pharmacol. Ther.*, **15**, 131, 1981.

3. C. Hansch, Quantitative structure–activity relationships in drug design, in E. J. Ariens (Ed.), *Drug Design*, Vol. 1, Academic, New York and London, 1971, p. 271.

4. E. Tomlinson, Chromatographic hydrophobic parameters in correlation analysis of structure–activity relationships, *J. Chromatogr.*, **113**, 1, 1975.

5. R. Kaliszan, Chromatography in studies of quantitative structure–activity relationships, *J. Chromatogr.*, **220**, 71, 1981.

6. R. Kaliszan, Quantitative relationships between molecular structure and chromatographic retention. Implications in physical, analytical, and medicinal chemistry, *Crit. Rev. Chem.*, **16**, 323, 1986.

7. C. B. C. Boyce and B. V. Millborrow, A simple assessment of partition data for correlating structure and biological activity using thin-layer chromatography, *Nature (London)*, **208**, 537, 1965.

8. G. L. Biagi, Correlation between silicone oil–water partition coefficients of cardiac glycosides and their biological activity, *Fitoterapia*, **38**, 110, 1967; after *Chem. Abstr.*, **69**, 92985a, 1967.

9. G. L. Biagi, Lipid solubility and serum-protein binding of various penicillins, *Antibiotica*, **5**, 198, 1967.

10. D. R. Clifford, A. C. Deacon, and M. E. Holgate, Fungicidal activity and chemical constitution. XVI. Use of partition data for an analysis of the activities of alkyldinitrophenols fungicides against apple mildew, *Ann. Appl. Biol.*, **64**, 131, 1969.

11. C. Hansch and T. Fujita, ρ-σ-π Analysis. A method for the correlation of biological activity and chemical structure, *J. Am. Chem. Soc.*, **86**, 1616, 1964.

12. G. L. Biagi, M. C. Guerra, A. M. Barbaro, and M. F. Gamba, Influence of lipophilic character on the antibacterial activity of cephalosporins and penicillins, *J. Med. Chem.*, **13**, 511, 1970.

13. G. L. Biagi, M. C. Guerra, and A. M. Barbaro, Relation between R_m values and antibiotic activity of rifamycins against *E. coli* and *S. aureus*, *Farmaco, Ed. Sc.*, **25**, 755, 1970.

14. G. L. Biagi, A. M. Barbaro, and M. C. Guerra, The influence of lipophilic characters on the biological activity of oligosaccharides antibiotics in *Eschericia coli*, *Pharmacol. Res. Commun.*, **2**, 121, 1970.

15. G. L. Biagi, M. C. Guerra, and A. M. Barbaro, Relationship between lipophilic character and hemolytic activity of testosterone and testosterone esters, *J. Med. Chem.*, **13**, 944, 1970.

16. G. L. Biagi, A. M. Barbaro, and M. C. Guerra, Relationship between R_m values and biological activity of steroid esters, *Experientia*, **27**, 918, 1971.

17. O. Gandolfi, A. M. Barbaro, and G. L. Biagi, Relationship between R_m values and protein binding of corticosteroids and androgens, *Experientia*, **29**, 689, 1973.

18. J. C. Dearden and E. Tomlinson, Buccal absorption as a parameter of analgesic activity of some *p*-substituted acetanilides, *J. Pharm. Pharmacol. Suppl.*, **23**, 73S, 1971.

19. J. C. Dearden, A. M. Patel, and J. H. Tubby, Hydrophobic substituent constants from thin-layer chromatography on polyamide plates, *J. Pharm. Pharmacol. Suppl.*, **26**, 74P, 1974.

20. E. Tomlinson and J. C. Dearden, Chromatographic substituent constants and their use in quantitative structure–activity relationships. The binding of acetanilides to serum albumin, *J. Chromatogr.*, **106**, 481, 1975.

21. J. C. Dearden and E. Tomlinson, Correlation of chromatographically obtained substituent constants and analgesic activity, *J. Pharm. Pharmacol. Suppl.*, **24**, 115P, 1972.

22. A. Ryrfeldt, Biliary excretion of penicillins in the rat, *J. Pharm. Pharmacol.*, **23**, 463, 1971.

23. S. T. Christian, C. W. Gorodetzky, and D. V. Lewis, Structure–activity relationships of normeperidine congeners on cholinesterase systems *in vitro* and analgesia *in vivo*, *Biochem. Pharmacol.*, **20**, 1167, 1971.

24. J. D. Turnbull, G. L. Biagi, A. J. Merola, and D. O. Cornwell, Structure–activity relationships for *N,N'-bis*-dichloroacetyl diamines and substituted naphthoquinones in the inhibition of mitochondrial electron transport, *Biochem. Pharmacol.*, **20**, 1383, 1971.

25. E. J. Lien, L. L. Lien, and G. L. Tong, Central nervous system activities of thiolactams and lactams. Structure–activity correlations, *J. Med. Chem.*, **14**, 846, 1971.

26. J. M. Plá-Delfina, J. Moreno, and A. del Pozo, Use of chromatographic R_m values as a possible approach to calculation of the absorption rate constants for some related drugs, *J. Pharmacokinet. Biopharm.*, **1**, 243, 1973.

27. J. Duran and J. M. Plá-Delfina, Relaciones entre cromatografia, estructura y velocidad de absorcion en los barbiturates 5-sustitudos. I. Planteamiento general: Constantes de absorción y parámetros cromatográficos, *Cienc. Ind. Farm.*, **6**, 83, 1974.

28. J. M. Plá-Delfina, J. Moreno, J. Duran, and A. del Pozo, Calculation of the gastric absorption rate constants of 5-substituted barbiturates through the R_m values of

substituent. ΔR_m constants in reversed-phase partition chromatography, *J. Pharmacokinet. Biopharm.*, **3**, 115, 1975.

29. J. M. Plá-Delfina, J. Moreno, P. Riera, V. Frías, R. Obach, and A. del Pozo, An experimental assessment of the Wagner–Sedman extraction theory for the intestinal absorption of antibacterial sulfonamides, *J. Pharmacokinet. Biopharm.*, **8**, 297, 1980.

30. M. A. Q. Chandry and K. C. James, A Hansch analysis of the anabolic activities of some nandrolone esters, *J. Med. Chem.*, **17**, 157, 1974.

31. G. L. Biagi, A. M. Barbaro, O. Gandofli, M. C. Guerra, and G. Cantelli-Forti, R_m values of steroids as an expression of their lipophilic character in structure–activity studies, *J. Med. Chem.*, **18**, 873, 1975.

32. G. L. Biagi, O. Gandolfi, M. C. Guerra, A. M. Barbaro, and G. Cantelli-Forti, R_m values of phenols. Their relationship with log P values and activity, *J. Med. Chem.*, **18**, 868, 1975.

33. G. L. Biagi, A. M. Barbaro, M. C. Guerra, G. Hakim, G. C. Solaini, and P. A. Borea, R_M values of naphtols and acetophenones in structure–activity studies, *J. Chromatogr.*, **177**, 35, 1979.

34. G. L. Biagi, A. M. Barbaro, M. C. Guerra, M. Babbini, M. Gaiardi, M. Bartoletti, and P. A. Borea, R_m values and structure–activity relationship of benzodiazepines, *J. Med. Chem.*, **23**, 193, 1980.

35. A. Fujii, J. H. Bush, K. E. Shores, R. G. Johnson, R. J. Garascia, and E. S. Cook, Probiotics: Antistaphylococcal activity of 4-aminocyclohexanecarboxylic acid, aminobenzoic acid, and their derivatives and structure–activity relationships, *J. Pharm. Sci.*, **66**, 844, 1977.

36. A. Fujii, K. E. Shores, J. H. Bush, R. J. Garascia, and E. S. Cook, Comparison of stationary phases in reversed-phase TLC for correlation between structure and biological response of probiotics, *J. Pharm. Sci.*, **67**, 713, 1978.

37. A. Hulshoff and J. H. Perrin, Quantitative correlations between albumin binding constants and chromatographic R_M values of phenothiazine derivatives, *J. Med. Chem.*, **20**, 430, 1977.

38. J. K. Seydel, K.-J. Schaper, E. Wempe, and H. P. Cordes, Mode of action and quantitative structure–activity correlations of tuberculostatic drugs of the iso-nicotinic acid hydrazide type, *J. Med. Chem.*, **19**, 483, 1976.

39. R. Kaliszan, H. Foks, and M. Janowiec, Studies on the quantitative structure–activity relationships in pyrazine carbothioamide derivatives, *Pol. J. Pharmacol. Pharm.*, **30**, 579, 1978.

40. R. Kaliszan, B. Milczarska, B. Łega, P. Szefer, and M. Janowiec, Studies on quantitative relationships between the structure and *in vitro* tuberculostatic potency of 2-cyanomethylbenzimidazole derivatives, *Pol. J. Pharmacol. Pharm.*, **30**, 585, 1978.

41. A. Dal Pozzo, A. Dansi, and M. Biassoni, α, β-Unsaturated γ-lactones corre-lations between lipophilicity and biological activity, *Arzneim. Forsch.*, **29**, 877, 1979.

42. M. Bachratá, M. Blešová, A. Schultzová, L. Grolichová, Ž. Bezáková, and

A. Lukáš, Chromatographic study of local anaesthetics-basic esters of substituted carbanilic acids. II. The relationships between chromatographic values, other physico-chemical parameters and activity, *J. Chromatogr.*, **171**, 29, 1979.

43. R. S. Wu, M. K. Wolpert-DeFilippes, and F. R. Quinn, Quantitative structure–activity correlations of rifamycins as inhibitors of viral RNA-directed DNA polymerase and mammalian α- and β-DNA polymerases, *J. Med. Chem.*, **23**, 256, 1980.

44. A. K. Saxena and J. K. Seydel, QSAR studies of potential antimalarial sulfonamide partitioning into erythrocytes, *Eur. J. Med. Chem.*, **15**, 241, 1980.

45. L. R. Ferguson and W. A. Denny, Potential antitumor agents. XXX. Mutagenic activity of some 9-anilinoacridines: Relationship between structure, mutagenic potential, and antileukemic activity, *J. Med. Chem.*, **22**, 251, 1979.

46. L. R. Ferguson and W. A. Denny, Potential antitumor agents. XXXIII. Quantitative structure–activity relationships for mutagenic activity and antitumor activity of substituted 4'-(9-acridinylamino)methanesulfonanilide derivatives, *J. Med. Chem.*, **23**, 269, 1980.

47. G. Maksay, Z. Tegyey, V. Kemény, I. Lukovits, L. Ötvös, and E. Palosi, Oxazepan esters. I. Correlation between hydrolysis rates and brain appearance of oxazepan, *J. Med. Chem.*, **22**, 1436, 1979.

48. G. Maksay, Z. Tegyey, and L. Ötvös, Oxazepam esters. II. Correlation of hydrophobicity with serum binding, brain penetration, and excretion, *J. Med. Chem.*, **22**, 1443, 1979.

49. R. Kaliszan, I. Kozakiewicz, F. Gajewski, and A. Madrala, QSAR studies in a series of new hypoglycemic sulfonamides, *Pharmazie*, **34**, 246, 1979.

50. M. L. Bieganowska, Relationships between biological activity of *N*-phenylamides of benzoylacetic acid and their capacity ratios in reversed-phase systems, *J. Liq. Chromatogr.*, **5**, 39, 1982.

51. N. Shirai, J. Sakakibara, T. Kaiya, S. Kobayashi, Y. Hotta, and K. Takeya, Quantitative structure–inotropy relationship applied to substituted grayanotoxins, *J. Med. Chem.*, **26**, 851, 1983.

52. Y. Hotta, H. Ando, N. Shirai, J. Sakakibara, and K. Takeya, Relationship between the inotropy speeds in guinea-pig myocardium and lipophilic character of cardenolides and ericaceous toxins, *Japan J. Pharmacol.*, **36**, 205, 1984.

53. H. Govers, C. Ruepert, and H. Aiking, Quantitative structure–activity relationships for polycyclic aromatic hydrocarbons: Correlation between molecular connectivity, physico-chemical properties, bioconcentration and toxicity in *Daphnia Pulex*, *Chemosphere*, **13**, 227, 1984.

54. M. W. Goosey and N. S. Doggett, Relationship between ability of some neuroleptics to enhance striatal [^{3}H]-dopamine release and their lipophilicity, *Biochem. Pharmacol.*, **32**, 2411, 1983.

55. M. Kuchař, V. Rejholec, Z. Roubal, and O. Maloušová, The effects of lipophilicity on the inhibition of denaturation of serum albumin and on the activation of fibrynolysis observed with a series of benzyloxyarylaliphatic acids, *Coll. Czech. Chem. Commun.*, **48**, 1077, 1983.

56. K. Dadákova, J. Hartl, and K. Waisser, Relationship between R_M values and hydrophobicity. Chromatographic behaviour of crotonolactones, *J. Chromatogr.*, **254**, 277, 1983.

57. W. A. Denny, G. J. Atwell, and B. C. Baguley, Potential antitumor agents. XXXVIII. 3-Substituted 5-carboxamido derivatives of ansacrine, *J. Med. Chem.*, **26**, 1619, 1983.

58. J.-P. Spengler, K. Wegner, and W. Schunack, H_2-Antihistaminics. 20. Structure–activity relationships in H_2-receptor antagonists containing a 4-pyrimidone moiety, *Agents Actions*, **14**, 566, 1984.

59. R. M. Carlson, R. E. Carlson, and H. L. Kopperman, Determination of partition coefficients by liquid chromatography, *J. Chromatogr.*, **107**, 219, 1975.

60. T. Vitali, E. Gaetani, C. F. Laureri, and C. Branca, Cromatografia liquido–liquido ad alta pressione. Correlazione fra tempi di ritenzione ed activitá biologica di derivativi 1,3,5-triazinici, *Farmaco, Ed. Sci.*, **31**, 58, 1976.

61. D. Henry, J. H. Block, J. L. Anderson, and G. R. Carlson, Use of high-pressure liquid chromatography for quantitative structure–activity relationships studies of sulfonamides and barbiturates, *J. Med. Chem.*, **19**, 619, 1976.

62. C. M. Riley, E. Tomlinson, and T. M. Jefferies, Functional group behaviour in ion-pair reversed-phase high-performance liquid chromatography using surface-active pairing ions, *J. Chromatogr.*, **185**, 197, 1979.

63. B. Rittich, M. Polster, and O. Kralik, Reversed-phase high-performance liquid chromatography. I. Relationship between R_M values and Hansch's π parameters for a series of phenols, *J. Chromatogr.*, **197**, 43, 1980.

64. M. J. M. Wells, C. R. Clark, and R. M. Patterson, Correlation of reversed-phase capacity factors for barbiturates with biological activities, partition coefficients, and molecular connectivity indices, *J. Chromatogr. Sci.*, **19**, 573, 1981.

65. S. H. Unger and G. H. Chiang, Octanol-physiological buffer distribution coefficients of lipophilic amines by reversed-phase high-performance liquid chromatography and their correlation with biological activity, *J. Med. Chem.*, **24**, 262, 1981.

66. V. Ya. Davydov, M. Elizalde Gonzales, and A. V. Kiselev, Correlation between the retention of cardiac glycosides in reversed-phase high-performance liquid chromatography with diphenylsilyl stationary phase, the structure of their molecules and their biological activity, *J. Chromatogr.*, **248**, 49, 1982.

67. K. Jinno, Correlation between the retention of cardiac glycosides in reversed-phase high-performance liquid chromatography with diphenylsilyl stationary phase, their structure and biological activity, *J. Chromatogr.*, **264**, 485, 1983.

68. A. D. Nunn, Structure–distribution relationships of radiopharmaceuticals. Correlation between the reversed-phase capacity factors for Tc-99m phenylcarbamoyl-methyliminodiacetic acids and their renal elimination, *J. Chromatogr.*, **255**, 91, 1983.

69. A. D. Nunn, M. D. Loberg, and R. A. Conley, A structure–distribution-relationship approach leading to the development of Tc-99m mebrofenin: An improved cholescintigraphic agent, *J. Nucl. Med.*, **24**, 423, 1983.

70. K. Jinno, The use of capacity factors with micro-HPLC as a descriptors in quantitative-structure–activity relationships, *Anal. Lett.*, **1783**, 183, 1984.

71. K. Valkó, T. Friedmann, J. Báti, and A. Nagykáldi, Reversed-phase chromatographic system as a model for characterizing the offset rate of action of azidomorphines in guinea-pig ileum, *J. Liq. Chromatogr.*, **7**, 2073, 1984.

72. E. Noack, Investigation on structure–activity relationship in organic nitrates, *Meth. Find. Exptl. Clin. Pharmacol.*, **6**, 583, 1984.

73. W. A. Bruggeman, D. Weber-Fung, A. Opperhuizen, J. Van der Steen, A. Wijbenga, and O. Hutzinger, Absorption and retention of polydimethylsiloxanes (silicones) in fish: Preliminary experiments, *Toxicol. Environ. Chem.*, **7**, 287, 1984.

74. E. A. Coats, H.-P. Cordes, V. M. Kulkarni, M. Richter, K.-J. Schaper, M. Wiese, and J. K. Seydel, Multiple regression and principal component analysis of antibacterial activities of sulfones and sulfonamides in whole cell and cell-free systems of various DDS sensitive and resistant bacterial strains, *Quant. Struct.–Act. Relat.*, **4**, 99, 1985.

75. J. K. Baker, D. O. Rauls, and R. F. Borne, Correlation of biological activity and high-pressure liquid chromatographic retention index for a series of propranolol, barbiturate, and anthranilic acid analogues, *J. Med. Chem.*, **22**, 1302, 1979.

76. E. A. Coats, K. J. Shah, S. R. Milstein, C. S. Genther, D. M. Nene, J. Roesener, J. Schmidt, M. Pleiss, E. Wagner, and J. K. Baker, 4-Hydroxyquinoline-3-carboxylic acids as inhibitors of cell respiration. II. Quantitative structure–activity relationships of dehydrogenase enzyme and Ehrlich ascites tumor cell inhibitions, *J. Med. Chem.*, **25**, 57, 1982.

77. T. Sekiya, S. Yamada, S. Hata, and S.-I. Yamada, Pyrimidine derivatives. V. Structure–activity studies of antihypertensive quinazoline derivatives using the adaptive least-squares method, *Chem. Pharm. Bull.*, **31**, 2779, 1983.

78. D. A. Brent, J. J. Sabatka, D. J. Minick, and D. W. Henry, A simplified high-pressure liquid chromatography method for determining lipophilicity for structure–activity relationships, *J. Med. Chem.*, **26**, 1014, 1983.

79. T. Braumann, G. Weber, and L. H. Grimme, Quantitative structure–activity relationships for herbicides. Reversed-phase liquid chromatographic retention parameter, log k_w, versus liquid–liquid partition coefficient as a model of the hydrophobicity of phenylureas, s-triazines and phenoxycarbonic acid derivatives, *J. Chromatogr.*, **261**, 329, 1983.

80. R. Kaliszan, M. Pankowski, L. Szymula, H. Lamparczyk, A. Nasal, B. Tomaszewska, and J. Grzybowski, Structure–olfactory activity relationship in a group of substituted phenols, *Pharmazie*, **37**, 499, 1982.

81. P. Laffort and F. Patte, Solubility factors in gas–liquid chromatography. Comparison between two approaches and application to some biological studies, *J. Chromatogr.*, **126**, 625, 1976.

82. H. Lamparczyk, A. Radecki, and R. Kaliszan, Application of a geometric parameter defining molecular shape, for the quantitation of interaction of polycyclic aromatic hydrocarbons with enzyme systems, *Biochem. Pharmacol.*, **30**, 2337, 1981.

INDEX